MESO-SCALE SHEAR PHYSICS IN EARTHQUAKE AND LANDSLIDE MECHANICS

Meso-Scale Shear Physics in Earthquake and Landslide Mechanics

Editors

Yossef H. Hatzor

Department of Geological and Environmental Sciences,
Ben-Gurion University of the Negev, Beer-Sheva, Israel

Jean Sulem

Université Paris-Est, Ecole des Ponts ParisTech,
UR Navier, CERMES, Champs sur Marne, France

Ioannis Vardoulakis

Faculty of Applied Mathematics and Physics, Department of Mechanics,
National Technical University of Athens, Athens, Greece

CRC Press
Taylor & Francis Group
Boca Raton London New York

CRC Press is an imprint of the
Taylor & Francis Group, an **informa** business

Cover photo: A filled bedding plane in Masada rock slopes/Photo Yossef Hatzor

CRC Press
Taylor & Francis Group
6000 Broken Sound Parkway NW, Suite 300
Boca Raton, FL 33487-2742

First issued in paperback 2017

CRC Press/Balkema is an imprint of the Taylor & Francis Group, an informa business

© 2010 Taylor & Francis Group, London, UK

Typeset by Vikatan Publishing Solutions (P) Ltd., Chennai, India

No claim to original U.S. Government works

Published by: CRC Press/Balkema
 P.O. Box 447, 2300 AK Leiden, The Netherlands
 e-mail: Pub.NL@taylorandfrancis.com
 www.crcpress.com – www.taylorandfrancis.co.uk – www.balkema.nl

Library of Congress Cataloging-in-Publication Data

Meso-scale shear physics in earthquake and landslide mechanics / editors, Yossef H. Hatzor, Jean Sulem, Ioannis Vardoulakis.
 p. cm.
 Includes bibliographical references and index.
 ISBN 978-0-415-47558-7 (hardcover : alk. paper) -- ISBN 978-0-203-86290-2 (e-book)
1. Deformations (Mechanics) 2. Shear (Mechanics) 3. Faults (Geology) 4. Earthquakes.
5. Landslides. I. Hatzor, Yossef H. II. Sulem, J. (Jean) III. Vardoulakis, I. (Ioannis) IV. Title.

 QE604.M47 2010
 551.22--dc22

                                                                                    2009028817

ISBN: 978-0-415-47558-7 (hbk)
ISBN: 978-1-138-11267-4 (pbk)

Visit the Taylor & Francis Web site at
http://www.taylorandfrancis.com

and the CRC Press Web site at
http://www.crcpress.com

Table of contents

Preface

This book brings together a collection of invited articles on meso-scale mechanics, as a means to develop an understanding of the underlying physics dominating shear at material interfaces and within particulate systems during conditions of rapid shearing. This edited volume emanated from the "Batsheva Seminar on Shear Physics at the Meso-scale in Earthquake and Landslide Mechanics", sponsored by the Batsheva de Rothschild fund of the Israel Academy of Sciences and Humanities and co-sponsored jointly by the US Air Force Research Laboratory and the Ben-Gurion University of the Negev.

Identification of meso-scale phenomena occurring between microscopic and continuum length scales has been one of the most exciting developments in the last decades in understanding shear between material interfaces and in particulate systems, and is considered as the bridge between the two length scales for studying material response. At the meso-scale complexities arise due to the presence of structural elements like surface roughness, grains, internal boundaries between them, physical phenomena occurring at surfaces, formation of sub-grain elements during loading, the presence of fluids, interaction between different materials, and their combined effect on the response of the system. This research area has broad applications in Geosciences and Geoengineering. For example, the initiation of seismic slip along fault planes at great depths at rates nearing shock conditions and the initiation of deep seated landslides near the earth's surface. Additionally, the basic physics of thermo-poro-mechanical coupling can be elucidated through a meso-scale mechanics approach as a means of understanding the loss of shearing resistance when water and heat are trapped inside almost impervious shear layers under great pressures. In the case of seismic slip, tremendous amount of material slips at very high velocities (several meters per second) on ultra localized shear zones. Shear heating and fluid pressurization can be associated to phenomena such as phase transition and mineral decomposition and thus play a key role in the understanding of the energetics of earthquakes. In the case of deep seated landslides, thermo-poro-mechanical processes within localized shear zones control their triggering, their sliding velocity and consequently their runout.

The 21, peer-reviewed articles are grouped into five chapters that address theoretical, computational and experimental aspects of meso-scale mechanics of material interfaces and particulate systems as follows: 1) Dynamics of frictional slip, 2) Fault gauge mechanics, 3) Experimental fault zone mechanics, 4) Granular shear and liquefaction, and 5) Dynamics of landslides.

We wish to express our deep gratitude to Dr. Yossi Segal, Secretary of Natural Sciences, Israel Academy of Sciences; Dr. Major Wynn S. Sanders, Chief of Materials and Nanotechnology, European Office of Aerospace Research and Development—USAF, Professor Rivka Carmi (M.D.), President, Ben-Gurion University of the Negev, Professor Jimmy Weinblatt, Rector, BGU, and Professor Amir Sagi, Dean, Faculty of Natural Sciences, BGU, for their financial support. Finally, we also wish to thank Dr. Conrad Felice of Washington State University for co chairing the Batsheva seminar and for his resourceful assistance throughout the production of the seminar.

Yossef H. Hatzor
Jean Sulem
Ioannis Vardoulakis
Editors

I. *Dynamics of frictional slip*

Thermo- and hydro-mechanical processes along faults during rapid slip

James R. Rice
Department of Earth and Planetary Sciences and School of Engineering and Applied Sciences,
Harvard University - SEAS, Cambridge, MA, USA

Eric M. Dunham
Department of Geophysics, Stanford University, Stanford, CA, USA

Hiroyuki Noda
Division of Geological and Planetary Sciences,
California Institute of Technology, Pasadena, CA, USA

ABSTRACT: Field observations of maturely slipped faults show a generally broad zone of damage by cracking and granulation. Nevertheless, large shear deformation, and therefore heat generation, in individual earthquakes takes place with extreme localization to a zone <1–5 mm wide within a finely granulated fault core. Relevant fault weakening processes during large crustal events are therefore likely to be thermal. Further, given the porosity of the damage zones, it seems reasonable to assume groundwater presence. It is suggested that the two primary dynamic weakening mechanisms during seismic slip, both of which are expected to be active in at least the early phases of nearly all crustal events, are then as follows: (1) Flash heating at highly stressed frictional micro-contacts, and (2) Thermal pressurization of fault-zone pore fluid. Both have characteristics which promote extreme localization of shear. Macroscopic fault melting will occur only in cases for which those processes, or others which may sometimes become active at large enough slip (e.g., thermal decomposition, silica gelation), have not sufficiently reduced heat generation and thus limited temperature rise. Spontaneous dynamic rupture modeling, using procedures that embody mechanisms (1) and (2), shows how faults can be statically strong yet dynamically weak, and operate under low overall driving stress, in a manner that generates negligible heat and meets major seismic constraints on slip, stress drop, and self-healing rupture mode.

1 INTRODUCTION

There has been a surge of activity in recent years towards increased physical realism in description of the earthquake process. That includes insightful geological characterization of the fine structure of fault zones, new laboratory experiments that reveal response properties in rapid or large slip, and new theoretical concepts for modeling dynamic rupture. The purpose here is to review some of those new perspectives and their impact on how we think about earthquake rupture dynamics.

1.1 *Fault zone structure, friction and a quandary in seismology*

Field observations of maturely slipped faults show a generally broad zone of damage by cracking and granulation (Chester et al., 1993), but nevertheless suggest that shear in individual earthquakes takes place with extreme localization to a long-persistent slip zone, <1–5 mm wide, within or directly bordering a finely granulated, ultracataclastic fault core (Chester and Chester, 1998; Chester et al., 2003, 2004; Heermance et al., 2003; Wibberley and Shimamoto, 2003).

On the other hand, the shear strength along a fault may be represented by

$$\tau = f\bar{\sigma} \quad \text{where} \quad \bar{\sigma} = \sigma_n - p_f \tag{1}$$

Here f is the friction coefficient, $\bar{\sigma}$ is the effective normal stress, σ_n is the total normal stress clamping the fault shut, and p_f is the pore pressure along it. It is well known that lab estimates of f

for rocks (under sliding rates of, say, μm/s to mm/s) are usually high, $f \sim 0.60$–0.85 (e.g., Byerlee (1978)).

Given that fault slip zones seem to be so extremely thin, one must conclude that if those f prevail during seismic slip, with p_f that is much closer to hydrostatic than lithostatic, we should find the following: (a) measurable heat outflow near major faults, and (b) evidence of extensive melting along exhumed faults. However, neither effect (a) or (b) is generally found.

1.2 *Weak faults, vs. statically strong faults that dynamically weaken*

There are two general lines of explanation that have been explored to resolve this quandary. One line of explanation simply postulates that major faults are *weak*. That could be because fault core materials are simply different from most rocks and have very low f, e.g., like documented for some clays and talc. Alternatively, it could be because f is not necessarily low, but pore pressure p_f is high and nearly lithostatic over much of the fault, especially down-dip where σ_n is large.

It is not the purpose here to argue against such weak-fault lines of explanation, but rather to explore an alternative which we are led to by recent observations. That is that major faults are *statically strong* but *dynamically weaken* during seismic slip. Owing to the extreme thinness of slip zones, the relevant fault weakening processes during large crustal events are likely to be thermal and, given the damage zones and geologic evidence of water-rock interactions within them, it seems reasonable to assume fluid presence. Of the various dynamic weakening processes thus far identified, it has been argued that two should be singled out as being essentially universal, in that they are expected to be active and important from the start of seismic slip in crustal events (Rice, 2006; Rice and Cocco, 2007). These are as follows:

1. Flash heating and hence shear weakening of frictional micro-asperity contacts, a process which reduces f in rapid slip, and
2. Thermal pressurization of pore fluid, which reduces the effective stress; p_f increases, because the highly granulated fault gouge is of low permeability and the thermal expansion coefficient of water is much greater than that of the rock particles.

Other thermal weakening processes may set in at large enough slip or large enough rise in temperature T_f along the fault. These include the following:

3. Thermal decomposition at large rise in T_f in lithologies such as carbonates, thus liberating a fluid product phase at high pore pressure,
4. Formation of a gel-like layer at large slip in wet silica-rich fault zones, or some related process relying on the presence of silica and water, and
5. The ultimate thermal weakening mechanism, formation of a macroscopic melt layer along the fault at large enough slip and rise of T_f, *if* the above set has not limited the actual increase of T_f to levels lower that that for such bulk melting.

While we focus on processes (1) and (2), it is very important to understand (3), (4), and (5) and others not yet identified. Still, in a sense the latter are secondary, because one expects that some significant earthquake slip, and fault weakening, will already have occurred before they can become activated. Preliminary estimates (Rempel and Rice, 2006) of when (5) would set-in suggest that with hydrostatic p_f and representative material parameter ranges, (1) and (2) are sufficiently effective at shallow fault depths that slip in significant earthquakes could often be accommodated without an onset of macroscopic melting, but that deeper in a fault zone, where the initial $\sigma_n - p_f$ (which scales the rate of heat input), and the initial T_f, are higher, melt onset should occur during increasing slip of typical surface-breaking earthquakes. For quantification of the slips and parameter ranges involved, see Rempel and Rice (2006).

Flash heating, weakening process (1), is a mechanism that has been advanced to explain high speed frictional weakening in metals (Bowden and Thomas, 1954; Archard, 1958/59; Ettles, 1986; Lim and Ashby, 1987; Lim et al., 1989; Molinari et al., 1999). It is only relatively recently that it has been considered as a process active during earthquake slip (Rice, 1999, 2006; Beeler and Tullis, 2003; Beeler et al., 2008; Tullis and Goldsby, 2003a,b; Hirose and Shimamoto, 2005;

Noda et al., 2006; Noda, 2008). Because of the relatively low thermal conductivity of most rocks, and the relatively high shear stresses which they support at frictional micro-contacts, they are in fact susceptible to weakening by flash heating starting at sliding rates as low as 0.1 to 0.3 m/s, which is well less than the average slip rate of ~1 m/s (Heaton, 1991) inferred from seismic inversions for large earthquakes. Thermal pressurization, process (2), has independent roots in the literature on large landslides (Habib, 1967, 1975; Anderson, 1980; Voigt and Faust, 1982; Vardoulakis, 2002; Veveakis et al., 2007; Goren and Aharonov, 2009) and that on earthquakes (Sibson, 1973; Lachenbruch, 1980; Mase and Smith, 1985, 1987; Lee and Delaney, 1987; Andrews, 2002; Wibberley, 2002; Noda and Shimamoto, 2005; Sulem et al., 2005; Rice, 2006; Rempel and Rice, 2006; Ghabezloo and Sulem, 2008; Noda et al., 2009).

Process (3), thermal decomposition with generation of a high-pressure fluid phase (O'Hara et al., 2006; Han et al., 2007; Sulem and Famin, 2009) is, of course, a type of thermal pressurization. In considering process (2), the fluid phase is presumed to pre-exist in pore spaces within the fault gouge so that the pressurization begins as soon as slip and consequent frictional heating begin, whereas in (3) the fluid phase comes into existence only once enough slip, frictional heating, and temperature rise have accumulated to initiate the decomposition. Process (4) is based on findings from experiments at large but sub-seismic (in results reported thus far) slip that, in presence of water, frictional weakening at large slip is greatest for rocks of greatest silica content (Goldsby and Tullis, 2002; Di Toro et al., 2004; Roig Silva et al., 2004). The weakening is argued to be due to formation of an initially weak silica-gel layer through reaction of water with fine silica particles from fresh comminution along the shear zone. There are many studies of process (5), macroscopic melting in fault zones, of which the long-lived signature is noncrystalline pseudotachylyte veins along the fault surface and in side-wall injections. Recent contributions include Spray (1995), Tsutsumi and Shimamoto (1997), Fialko and Khazan (2004), Hirose and Shimamoto (2005), Sirono et al. (2006), and Nielsen et al. (2008).

2 DYNAMIC RUPTURE FORMULATION

2.1 *Elastodynamic methodology*

Noda et al. (2006, 2009) and Dunham et al. (2008) have begun to integrate weakening by flash heating and thermal pressurization into elastodynamic numerical methodology for spontaneous rupture development. The problems thus far addressed are of rupture along a planar fault zone within an effectively unbounded and homogeneous solid. For those, the implementation of an elastodynamic boundary integral equation (BIE), with a spectral basis set for the slip and stress distributions (Perrin et al., 1995; Geubelle and Rice, 1995) is extremely efficient and accurate. Results have been obtained for 2D anti-plane or in-plane strain. In the formulation, with the x axis passing along the fault plane, the shear stress $\tau(x, t)$ along the fault and the slip $\delta(x, t)$ are related by

$$\tau(x, t) = \tau_0(x, t) - (\mu/2c_s)V(x, t) + \phi(x, t) \tag{2}$$

where $V(x, t) \equiv \partial\delta(x, t)/\partial t$ is slip rate, μ is the shear modulus, c_s is the shear wave speed, and the functional $\phi(x, t)$ is given as a linear space-time convolution of an elastodynamic kernel, dependent on $x - x'$ and $t - t'$, with the slip $\delta(x', t')$ for all x', t' within the wave cone with vertex at x, t. Here $\tau_0(x, t)$ is some specified loading stress on the fault; it is the stress that would have been induced by the applied loadings if the fault had been constrained against any slip. We prescribe $\tau_0(x, t)$ as a uniform background stress τ^b for all time, plus some localized overstress applied at $t = 0$ to nucleate rupture. $\delta(x, t)$ and $\phi(x, t)$ are expanded in a Fourier basis set, so the convolution is expressed by

$$\begin{Bmatrix} \delta(x, t) \\ \phi(x, t) \end{Bmatrix} = \sum_{n=-N/2}^{N/2} \begin{Bmatrix} D_n(t) \\ \Phi_n(t) \end{Bmatrix} \exp(in\hat{k}x) \quad \text{with } \Phi_n(t) = \int_0^t C_n(t - t')D_n(t')dt' \tag{3}$$

Here N is a large even integer, the $C_n(t)$ are known real functions (Perrin et al., 1995; Geubelle and Rice, 1995), $\hat{k} = 2\pi/X$ where the periodic repeat length X of the truncated Fourier series is chosen large enough that waves from the periodic replications of the rupture event do not arrive to neighboring replications in the time of interest, and the $D_{-n}(t)$ are complex conjugates of the $D_n(t)$ with D_0 and $D_{N/2}$ being real. Eq. (3) is equivalent to a real Fourier cosine and sine series truncated at N terms (with no sine term when $n = N/2$). Through FFT procedures, the histories $D_n(t)$ are determined by $\delta(x, t)$ at N sample points, equally spaced by $\Delta x = X/N$, i.e., by the histories $\delta(j\Delta x - X/2, t)$ for $j = 0, 1, 2, \ldots, N-1$, and conversely. That greatly speeds calculations. The $\Phi_n(t)$ and $\phi(j\Delta x - X/2, t)$ are similarly related.

2.2   *Friction law with weakening by flash heating and thermal pressurization*

The simple flash heating model reviewed here (Rice, 1999, 2006; Beeler and Tullis, 2003; Beeler et al., 2008) was intended to approximately determine the expression for f of eq. (1) for conditions of sustained sliding at some speed V. The f so derived must be regarded as a *steady state* value, written here as f_{ss} and is regarded as a function of slip rate, $f_{ss} = f_{ss}(V)$, although it is also a function of the spatially averaged (over patches of fault area large enough to include many contacts) temperature T_f of the fault plane, which evolves with ongoing slip and time. In the model (Rice, 2006) it is assumed that contact temperature $T_c = T_f$ when a contact pair first forms, but then as the contact slides during its brief lifetime (which is D/V for a contact asperity of diameter D; see Fig. 1), T_c rises substantially above T_f. The rise is due to the intense localized heat generation at the contact, at rate $\tau_c V$, where τ_c is the contact shear strength (typically of order 0.1 times shear modulus μ at low T_c; see discussion in Rice (2006)). τ_c is assumed to have negligible variation as T_c increases, but then to abruptly decrease to a much lower weakened value τ_w when T_c reaches a "weakening" temperature T_w. Within the model, based on 1D heat conduction at the sliding contact like in Archard (1958/59), the slip rate such that an asperity of diameter D would begin to weaken only just as it is slid out of existence is then (Rice, 1999, 2006)

$$V_w = (\pi \alpha_{th}/D)(\rho c(T_w - T_f)/\tau_c)^2 \tag{4}$$

Here ρc is volumetric specific heat and $\alpha_{th} = K/\rho c$ is thermal diffusivity (K is thermal conductivity). Estimates based on rock properties and assumed D of order 10 μm, as well as comparison to experimental results of Tullis and Goldsby (2003a, b) and Beeler et al. (2008) (by rotary shear of a rock annulus in an Instron frame), suggest that V_w is of order 0.1 to 0.3 m/s for rocks such as quartzite, feldspar, granite and gabbro when T_f = room temperature. V_w is expected to be less at higher T_f. Thus the model takes contacts to be strong for all their lifetime at low slip rates, $V < V_w$, but to be strong for only a fraction of their lifetimes at high slip rates, $V > V_w$, that fraction being V_w/V.

Neglecting the actual statistical distribution of contact diameters, and taking D as a representative value, these concepts lead to the steady-state friction coefficient

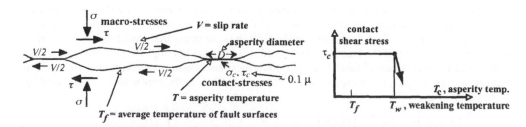

Figure 1.   Simple representation of asperity contacts and their strength loss for flash heating model.

$$f_{ss}(V) = \begin{cases} f_0 & V \leq V_w \\ f_w + (f_0 - f_w)V_w/V & V \geq V_w \end{cases} \tag{5}$$

where f_0 is the low-speed friction coefficient and $f_w = f_0 \tau_w/\tau_c$ is the value to which it would reduce if all contacts were in the weakened state. Note that despite the notation $f_{ss}(V)$, the dependence mentioned on ambient fault temperature enters from the dependence of V_w on T_f, eq. (4). Rough estimates from the Tullis and Goldsby (2003a, b) experiments, which covered a limited velocity range, up to slightly less than $V = 0.4$ m/s, are that $f_0 \approx 0.64$ and $f_w \approx 0.12$ for quartzite, $f_0 \approx 0.82$ and $f_w \approx 0.13$ for granite, and $f_0 \approx 0.88$ and $f_w \approx 0.15$ for gabbro. However, the f_w involve significant extrapolation and, in experiments of Yuan and Prakash (2008) on quartzite from the same source and annular configuration, but in a Kolsky bar dynamic torsion apparatus, f was only very slightly below 0.20 at slip rates as high as 2–4 m/s. Eq. (5) and the small $V_w \sim 0.1$ m/s for quartzite, suggest that f should nearly coincide with f_w at such rates, thus that $f_w \approx 0.18$–0.20.

In fact, we cannot simply assume f to be a decreasing function of V in eq. (1) because that makes the problem of sliding between elastic continua ill-posed (there is a short wavelength divergence in response to small initial perturbations from steady sliding). The problem is remedied (Rice et al., 2001) mathematically when we look to experiments and embed the description of variations of f in rate and state friction concepts. Thus, with a "slip" version of state evolution, f is assumed to be given by the form (Rice, 1983)

$$\frac{df}{dt} = \frac{a}{V}\frac{dV}{dt} - \frac{V}{L}(f - f_{ss}(V)) \tag{6}$$

where a is the direct effect coefficient in rate and state modeling ($a \approx 0.01$ at room T for quartzite and granite, and it scales in approximate proportion to absolute T), and L is a slip distance adequate to renew the asperity contact population, typically taken as 5–20 μm in our studies, as guided by observed state evolution slip distances in rate and state friction experiments. The smallness of L makes the simulations extremely challenging in terms of present-day computers and, as of the writing, the longest rupture lengths simulated are ~30 m (Dunham et al., 2008; Noda et al., 2009).

In the numerical simulations (Noda et al., 2009; Dunham et al., 2008) we replace the constant term f_0 in eq. (5) with $f_{LV}(V)$, which is the weak logarithmic function of V describing slow-rate

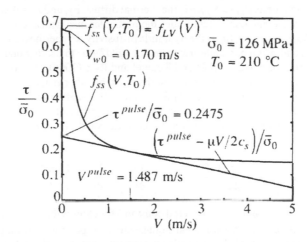

Figure 2.   From Noda, Dunham, and Rice (2009), their figure 1. Plot of f_{ss} based on Tullis and Goldsby (2003a, b) parameters for granite, adjusted to conditions at mid-seismogenic depth, ~7 km, for crustal strike slip earthquakes. Plot is based on assumed ambient effective normal stress $\bar{\sigma}_0 = 126$ MPa and temperature $T_0 = 210°$C; as slip develops, heating increases T_f over T_0, and pore fluid pressurization will reduce $\bar{\sigma}$ from $\bar{\sigma}_0$. The Zheng and Rice (1998) stress level τ^{pulse}, important to understanding whether rupture takes the self-healing pulse or classical shear crack mode, is also shown.

friction at steady-state. That is, $f_{LV}(V) = f_0 - (b - a)\ln(V/V_0)$ in the standard notation, where $b - a \approx 0.002$–0.004 for rate-weakening frictional surfaces in granite. Here V_0 can be chosen arbitrarily in that low-speed regime (say, as 1 μm/s) with corresponding adjustment of f_0 to correspond to the correct f_{ss} at that rate. Fig. 2, taken from Noda et al. (2009), plots the resulting f_{ss} using parameters which are thought to be representative for granite, at conditions corresponding to a mid-seismogenic zone depth for crustal strike-slip earthquakes. Also, while irrelevant in our simulations, for some purposes the a/V term in eq. (6) should be regularized near $V = 0$ in a manner consistent with the thermal activation basis for the direct effect (Rice et al., 2001).

2.3   *Inclusion of thermal pressurization of pore fluid in friction formulation*

In order to close our system of governing equations which, so far, consists of eqs. (1), (2), (3) and (6), it is necessary to relate the histories of p_f and T_f to that of slip and stressing along the fault. That is because p_f directly enters eq. (1) and T_f is a parameter on which $f_{ss}(V)$ depends (see eqs. (5) and (4)). The new ingredients which close the system are the equations of conservation of energy, with inclusion of conductive heat transfer, and of conservation of fluid mass with inclusion of Darcy fluid seepage and poro-thermo-elastic considerations. These amount to a pair of coupled PDEs for the fields of pore pressure p and temperature T near and on the fault. The values of p and T thus determined at the fault surface itself are the respective p_f and T_f that we seek.

In writing the conservation laws we neglect certain apparently minor terms (e.g., advective heat transfer by moving fluid), and recognize that the gradients of pore pressure p and temperature T very near the fault are generally very much larger in the direction perpendicular to the fault (the z direction) than in the x or y directions which are parallel to it (such may sometimes not be a valid assumption immediately at the moving rupture tip). Thus (e.g., Rice (2006)) we have

$$\rho c \frac{\partial T}{\partial t} = \frac{\partial}{\partial z}\left(\rho c \alpha_{th} \frac{\partial T}{\partial z}\right) + \tau \frac{\partial \gamma^{pl}}{\partial t} \quad \text{and} \quad \beta\left(\frac{\partial p}{\partial t} - \Lambda \frac{\partial T}{\partial t}\right) + \frac{\partial n^{pl}}{\partial t} = \frac{\partial}{\partial z}\left(\beta \alpha_{hy} \frac{\partial p}{\partial z}\right) \quad (7)$$

Here $\dot{\gamma}^{pl}$ is the inelastic fault-parallel shear strain rate and $\dot{n}^{pl}$ is the inelastic dilatancy rate (n itself is the volume of pore space per unit aggregate volume of porous material, that aggregate volume being measure in some reference state before the deformation episode considered). Also, ρc is the specific heat per unit volume, $\alpha_{th} = K/\rho c$ is the thermal diffusivity, and K is thermal conductivity; β is a porous medium storage coefficient under the particular mechanical constraints near a fault zone (see Rice (2006)), $\alpha_{hy} = k/\beta \eta_f$ is the hydraulic diffusivity, k is permeability, and η_f is viscosity of the pore fluid; and Λ is a parameter representing dp/dT due to heating under undrained, elastically reversible conditions. Rice (2006) and Rempel and Rice (2006) compile estimates of these various parameters at mid-seismogenic depths in the crust, based principally on data of Wibberley (2002) and Wibberley and Shimamoto (2003) for gouge of the Median Tectonic Line Fault (Japan) under a range of confining stresses, and on tabulated thermophysical data for water and minerals.

Rempel and Rice (2006) also compare fully non-linear solutions of eqs. (7) to the linearized versions

$$\frac{\partial T}{\partial t} = \alpha_{th} \frac{\partial^2 T}{\partial z^2} + \frac{\tau}{\rho c} \frac{\partial \gamma^{pl}}{\partial t} \quad \text{and} \quad \frac{\partial p}{\partial t} - \Lambda \frac{\partial T}{\partial t} + \frac{1}{\beta} \frac{\partial n^{pl}}{\partial t} = \alpha_{hy} \frac{\partial^2 p}{\partial z^2} \quad (8)$$

(with coefficients ρc, α_{th}, β and α_{hy} considered constant) for cases with *a priori* specified histories of $\partial \gamma^{pl}/\partial t$ and $\partial n^{pl}/\partial t$, e.g., representing earthquake slip at a specified constant rate in time, with the aim of estimating (using eq. (1) but assuming a constant low f as motivated by flash heating) the relation between stress τ and slip δ during seismic rupture. They find that the procedure adopted by Rice (2006), of iteratively choosing the constant values of coefficients in the linearized PDEs of eqs. (8), as certain path averages in p, T space of those same coefficients, when regarded as known functions of p, T, along the p, T path predicted by eqs. (8), gives a tolerable match to results for τ *versus* δ based the full non-linear solutions to eqs. (7). In our spontaneous elastodynamic analyses presented here we have used eqs. (8), without that iterative choice of coefficients, but rather with coefficients based on the ambient p, T at the mid-seismogenic depth (~ 7 km) considered.

The conceptually (but not computationally) simplest case for eqs. (7) or (8) is that of slip on a mathematical plane. In that case $\partial\gamma^{pl}/\partial t = V(x,t)\delta_{\mathrm{Dir}}(z)$, where V is the local slip rate and $\delta_{\mathrm{Dir}}(z)$ is the Dirac function. However, actual shear zones, when highly localized to a prominent slip surface, can nevertheless distribute deformation over regions that may extend over a few 10s to a few 100s of μm. Because the difference between such small but finite thicknesses and zero thickness is sometimes not negligible on the seismic time scale in our modeling, a Gaussian distribution of shear heating $\tau\,\partial\gamma^{pl}/\partial t$, like in Andrews (2002), over a zone of nominal thickness $2w$ is assumed. We write

$$\tau\frac{\partial\gamma^{pl}}{\partial t} = \frac{\tau(x,t)V(x,t)}{\sqrt{2\pi}\,w}\exp\left(-\frac{z^2}{2w^2}\right) \tag{9}$$

which reduces to $\tau\,\partial\gamma^{pl}/\partial t = \tau(x,t)V(x,t)\delta_{\mathrm{Dir}}(z)$ as $w \to 0$.

In the numerical studies (Noda et al., 2008; Dunham et al., 2008) of spontaneous rupture based on this formulation (which have thus far taken $\dot{n}^{pl} = 0$), explicit finite difference (in z and t) versions of eqs. (8) are solved at each elastodynamic gridpoint location $x_j = -X/2 + jX/N$ along the rupture, ultimately to give, with help of the other governing equations, $T_f(x_j,t)$ and $p_f(x_j,t)$. For what we think to be appropriate ranges $\alpha_{th} \sim 1$ mm^2/s and $\alpha_{hy} \sim 1$–10 mm^2/s (Rice, 2006; Rempel & Rice, 2006), it turns out that when the elastodynamic time steps Δt are already short enough to resolve the state evolution of eq. (6) with $L \sim 5$–$20\,\mu$m, explicit finite difference solution of eqs. (8) requires significantly shorter time steps, to meet the requirement is that diffusion grid spacing be sufficiently small that the error associated with the spatial discretization of the diffusion equations is comparable the error in the elastodynamic system. Thus Noda et al. (2008) devised a procedure based on a quadratic interpolant of $\phi(x,t)$ within an elastodynamic time step Δt, to achieve second-order accuracy. The interpolant is constructed from $\phi(x,t-\Delta t)$, $\phi(x,t)$, and $\phi(x,t+\Delta t)$ for use between t and $t+\Delta t$, so that the condition of eq. (2), $\tau = \tau_0 - (\mu/2c_s)V + \phi$, along with eq. (1), eq. (6) with eqs. (4) and (5) (with f_0 replaced by $f_{LV}(V)$), and eqs. (8) are satisfied, to numerical accuracy, within each of the smaller time steps for diffusion. This is highly accurate but very demanding computationally because of the small but presumably realistic $L \sim 5$–$20\,\mu$m and realistic diffusivities, ~ 1–10 mm^2/s, that we use.

2.4   *Theoretical background on strong rate-weakening and self-healing slip pulses*

The adopted friction description involves strong rate-weakening (see Fig. 2). To provide background for understanding when strong rate-weakening will lead to rupture in the mode of a self-healing slip pulse, versus a classical enlarging shear crack, we digress here to review results from Zheng

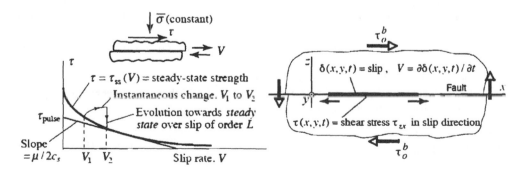

Figure 3.   Friction with strong rate dependence. It is assumed here, to simplify, that the steady-state friction $f_{ss}(V)$ is a function of slip rate V only, and that the effective normal stress $\overline{\sigma}$ is constant; then the steady-state shear stress $\tau_{ss}(V) = f_{ss}(V)\overline{\sigma}$. (Modified from Rice (2001).)

and Rice (1998), in part following a recapitulation in Rice (2001), discussed with reference to Fig. 3. Some representative results from Dunham et al. (2008) and Noda et al. (2009), based on the formulation in the earlier parts of this Section 2, are presented in Section 3 to follow.

Studies by Cochard and Madariaga (1994, 1996), Perrin et al. (1995), and Beeler and Tullis (1996) had shown that strong rate-weakening could lead to rupture in the self-healing mode. The objective in Zheng and Rice (1998) was then to establish guidelines for when this type of rupture mode would in fact result. To simplify enough that sharp results could be established, it was assumed in that development that the steady state shear strength $\tau_{ss}(V)$ was a function *only* of slip rate V, which could, e.g., result if $\overline{\sigma}$ is constant and $f_{ss}(V)$ is a function only of V. Then, $\tau_{ss}(V) = f_{ss}(V)\overline{\sigma}$. However $\tau_{ss}(V)$ does not give the expression for τ for variable V, because $\tau = f\overline{\sigma}$ and f evolves according to eq. (6). Thus the response to a sudden change in V is as depicted on the left in Fig. 3, a response which also assures well-posedness in problems of frictional sliding between elastic continua (Rice et al., 2001).

Now consider a fault surface which we treat as the boundary $z = 0$ between two identical half spaces (Fig. 3, at right). An initial shear stress $\tau_0(x, y) = \tau_0^b$, a constant level too small to cause failure, acts everywhere on S_∞ (the entire x, y plane) except in small nucleation region S_{nucl} which will be overstressed to start the rupture. The stress level τ_{pulse} (the same as what Noda et al. (2009) denoted τ^{pulse} in their diagram shown here in Fig. 2) is marked in Fig. 3. τ_{pulse} is defined as the largest value that τ_0^b could have and still satisfy $\tau_0^b - \mu V/2c_s \leq \tau_{ss}(V)$ for all $V > 0$.

Suppose that $\tau_0^b < \tau_{pulse}$. As will now be seen, that effectively precludes the possibility that rupture could occur on S_∞ in the form of an indefinitely expanding shear crack. Note that

$$\tau_0^b < \tau_{pulse} \text{ implies that } \tau_{ss}(V) - (\tau_0^b - \mu V/2c_s) > 0 \quad \text{for all } V > 0. \tag{10}$$

Use is made of an elastodynamic conservation theorem (Zheng and Rice, 1998)

$$\int\int_{S_\infty} [\tau(x, y, t) - \tau_0(x, y) + \mu V(x, y, t)/2c_s] \, dx\, dy = 0 \tag{11}$$

which holds throughout the rupture; $\tau_0(x, y)$ is the stress field at $t = 0$ when rupture is nucleated.

Assume that, with $\tau_0^b < \tau_{pulse}$, rupture has been locally nucleated and grows on S_∞ in the form of an *indefinitely expanding shear crack*. Such an assumption can be shown, as follows, to lead to a definite *contradiction* in the case of mode III (anti-plane) slip, and to seemingly implausible result in general, meaning that we must reject the assumption that an indefinitely expanding crack-like rupture is possible when $\tau_0^b < \tau_{pulse}$. To see why, note that the integrand in eq. (11) everywhere along the rupturing surface $S_{rupt}(t)$ where $\tau \approx \tau_{ss}(V)$, except for S_{nucl} and for small regions at the rupture front affected by the rate/state regularization (so that τ may depart significantly from $\tau_{ss}(V)$ in those regions; they are small because L is in the range of a few to a few tens of microns; see discussions in Zheng and Rice (1998) and Noda et al. (2009) for further quantification), is equal to

$$\tau_{ss}(V) - \tau_0^b + \mu V/2c_s = \tau_{ss}(V) - (\tau_0^b - \mu V/2c_s) > 0 \tag{12}$$

where the inequality follows from eq. (10). Thus, denoting by $S_{out}(t)$ $(=S_\infty - S_{rupt}(t))$ the region of S_∞ lying outside the rupture at time t, and noting that $V = 0$ there, we must by eq. (11) then have, for any sufficiently large rupture,

$$\Delta F_{out}(t) \equiv \int\int_{S_{out}(t)} [\tau(x, y, t) - \tau_0^b] \, dx\, dy < 0 \quad (\text{and } \Delta F_{out}(t) \to -\infty \text{ as } S_{rupt}(t) \to \infty) \tag{13}$$

where $\Delta F_{out}(t)$ is the change in total shear force (positive in the direction of initial shear stressing) supported *outside* the ruptured zone. Inequality (13), however seems implausible: We expect ruptures to result in an increase $\Delta F_{out}(t)$ in the net force carried outside themselves, or at least to not decrease the force, and in the 2D case of mode III rupture it is provably the case that $\Delta F_{out}(t)$ can never correspond to a decrease.

We therefore conclude that an indefinitely expanding rupture in crack-like mode cannot occur if $\tau_0^b < \tau_{pulse}$, and that any ruptures which do then occur would have to be of another type for which not all of the ruptured surface was slipping for all time. i.e., they would have to be of the self-healing type. That is consistent with a range of calculations for different stress levels and forms of $\tau_{ss}(V)$; crack-like ruptures are not found, only self-healing slip pulses, when $\tau_0^b < \tau_{pulse}$.

The guidelines just outlined on when self-healing versus crack-like ruptures occur are also consistent with laboratory studies in which both types of ruptures were generated in different conditions (Lykotrafitis et al., 2006; Lu et al., 2007).

Thus, imagine a fault on which tectonic stress is being increased slowly, and on which, at various times, ruptures nucleate from localized patches of the fault where the ratio $\tau/(\sigma_n - p_f)$ is high (consistent with the fault material being "statically strong"). Such nucleating regions could occur because τ is locally high from a stress concentration at a boundary between creeping and locked lithologies, or because p_f is locally high, or because σ_n is low due, e.g., to a dilational twist in the fault trace or to intersection with a secondary fault that slipped in the past and locally reduced σ_n. In general, when the average stress on the fault is low, we expect that none of these events will lead to a rupture that spreads over the entire fault. But ultimately, as average stress on the fault is gradually increased, one may do so. The first type of rupture encountered which can propagate without limit, once nucleated, will inevitably be of the self-healing type. Thus the presently reviewed results argue strongly for an association of strong rate-weakening on major tectonic faults with the occurrence of large earthquakes on those faults in the self-healing rupture mode (Lapusta and Rice, 2003), a mode which is generally thought consistent with seismic observations. A relevant quote from Zheng and Rice (1998) is as follows:

"The Gutenberg-Richter frequency versus size statistics of earthquakes tells us that for every 1 earthquake that achieves, say, magnitude 3 size but does not arrest and, rather, grows to magnitude 4, there are approximately 10 that do arrest at that smaller size, and so on for other magnitudes, at least within the range for which there is power law scaling with $b \approx 1$. An interpretation of such results is to say that faults are chronically understressed, so that most ruptures simply fail to become large. Such understressing is likely to be very heterogeneous and not like the uniform τ_0^b considered here. Nevertheless, if natural faults are indeed velocity weakening at seismic slip rates, so that our present analysis applies, then it is plausible to make the association that these faults are lightly stressed and perhaps understressed in the precise meaning of the term here. In such case, then, we could understand that the self-healing mode of rupture would be a pervasive one, because the stresses are too low on average to allow the cracklike mode and can do so only at places of local stress concentration where rupture nucleates."

They also add the caution that strong rate-weakening is not the only feasible explanation of self-healing ruptures:

"We emphasize that this study has been on understanding the rupture mode in the presence of velocity-weakening friction on a fault of spatially uniform properties between identical linear elastic solids. As explained in the Introduction, other mechanisms of self-healing pulse generation exist and involve, for example, strong spatial nonuniformity of frictional weakening properties within the fault zone or dissimilarity of elastic properties across the fault."

3   SUMMARY OF SOME REPRESENTATIVE RESULTS AND THEIR IMPLICATIONS

3.1   *Rupture modes, phase boundaries, and (low) background stress levels to sustain rupture*

Figures 4 and 5 show results from Noda et al. (2009), for the same model based on granite properties at high slip rates as shown here in Fig. 2. The parameter r in Figs. 4 and 5, and in Fig. 6 next, measures how much we assume that damage induced by the high off-fault and fault-parallel stresses and strains very near the rupture tip (beyond the elastic range) alter the near-fault poromechanical properties. Parameter $r = 0$ corresponds to poromechanical properties of intact fault gouge at

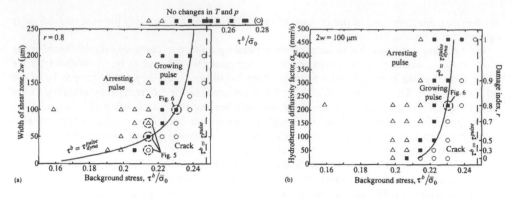

Figure 4.   From Noda, Dunham and Rice (2009), their figure 4; the figure numbers printed on the diagrams correspond to figures which follow. (a) Rupture mode identified as a function of background stress and width of the shear zone, for a given set of hydro-thermal properties. (b) Rupture mode identified as a function of background stress and hydro-thermal properties (parameterized by r; see text) for a given width of the shear zone. The horizontal bar at the top of (a) corresponds to a case for which T and p variations are neglected so that the theory of Section 2.4 applies with τ^{pulse} shown as the dashed vertical line. The solid lines in (a) and (b) are a plots of a parameter devised by Noda et al. (2009) that approximately generalizes the τ^{pulse} concept to the realistic cases with variable T and p.

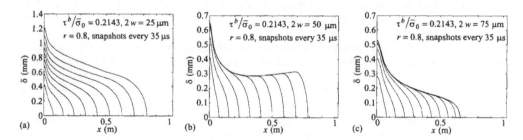

Figure 5.   From Noda, Dunham and Rice (2009), their figure 2. Illustration of the three rupture modes, (a) enlarging shear crack, (b) growing self-healing pulse, and (c) arresting self-healing pulse. Initial development of slip (*vs.* distance along fault at successive times, half of fault shown) during dynamic rupture, nucleated by localized overstressing, for different cases from Fig. 4, all at the same background stress level but with different widths of the shear zone.

temperature and effective stress conditions corresponding to 7 km depth (mid-seismogenic depth), as estimated from various sources by Rice (2006). Parameter $r = 1$ corresponds to the "highly damaged" fault walls case of Rice (2006), for which it was estimated, rather arbitrarily, that damage at the passing rupture front causes the near-fault materials to have a ten-fold increase in permeability and a doubling of a poroelastic compressibility measure.

Figure 6 based on Noda et al. (2009) shows results for the longest rupture it has thus far (as of April 2009) been possible to simulate. A companion case of rupture of the same length, showing crack-like rupture, has also been done; it shows much greater slip at a given rupture length.

3.2   Some implications

The results are preliminary, but due to current day computational limitations it will likely be much time before substantially longer ruptures can be treated routinely. At least that is so in studies following our aim to strictly adhere to parameters (including friction state transition length L, shear zone thickness $2w$, and those characterizing extreme rate-weakening at rapid slip rates) which we

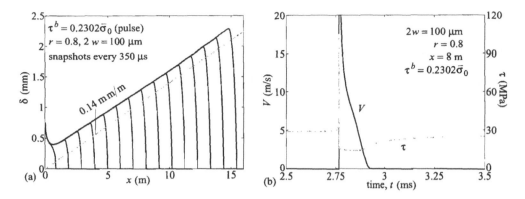

Figure 6. From Noda, Dunham and Rice (2009), portions of their figures 7 and 9. At left: Slip vs. distance at successive times for a self-healing rupture, shown when achieving an overall length of ∼30 m (at which calculation was ended). Note that an approximate self-similarity seems to be achieved. At right: Shear stress and slip velocity vs. time at a particular point on the rupture, 8 m from the center. The ratio $\tau/\bar{\sigma}_0$ is ∼0.23 for the background stress field (which nevertheless allows the rupture to grow dynamically, once nucleated), but a ratio $\tau/\bar{\sigma}_0$ of ∼0.85 is achieved locally at the rupture tip, in consistency with Fig. 2 and the effects of the a term in eq. (6). The stress drops quickly (in a manner that is fully resolved numerically) as slip develops and then, after self-healing at this location, climbs back towards the initial stress, so that the drop from initial to final static stress is considerably less than the drop from the initial stress to a representative stress level during rapid slip.

think represent reality as shown in the laboratory and in geological characterizations of natural faults.

Nevertheless, our results show definitively that faults modeled in that way can operate at low overall shear stress levels compared to Byerlee friction thresholds, yet produce runaway ruptures. For those ruptures, Byerlee-like conditions are achieved fleetingly at the moving rupture front, before dynamic weakening takes command.

Our results are in general consistency with the low overall fault stress levels inferred in tectonic modeling of strike slip faults in California (e.g., Bird and Kong (1994)). We typically find that faults can begin to sustain runaway ruptures at background stress levels corresponding to $\tau/\bar{\sigma}_0 \approx$ 0.20–0.25. That compares quite favorably to the estimate of Hickman and Zoback (2004), by direct stress measurements in the SAFOD pilot borehole near the San Andreas Fault at Parkfield, that at the deepest location measured, 2.1 km depth, a ratio $\tau/\bar{\sigma}_0 \approx$ 0.2–0.3 is inferred for the nearby San Andreas Fault.

As Fig. 6 shows, the results out to 30 km overall rupture length suggest an approximate self-similarity (like in a simpler self-healing pulse model by Nielsen and Madariaga (2003), and as conjectured from natural observations by Manighetti et al. (2005)). If we optimistically assume that such scaling would extend approximately to the multi-km dimensions of large surface-breaking ruptures, then we would predict a ratio of slip to rupture length of order 14 cm/km, which is quite reasonable in terms of earthquake phenomenology. Also, while the shear stress on the fault in Fig. 6 is low as slip develops (high dynamic stress drop), the final static stress drop is ∼3 MPa, which is quite in the range of such stress drops for crustal earthquakes. The stress on the fault can be seen in fig. 6 to gradually decrease as slip accumulates. That is essentially all due to thermal pressurization (which would seem to break strict self-similarity), but it means that for much longer ruptures, with correspondingly greater slip, that the average ratio $\tau/\bar{\sigma}_0$ during sliding in large events is expected to be in a range of ∼0.1 or less, which will not produce a perceptible heat-flow anomaly. And of course, our prediction, for reasons elaborated in Section 2.4, is that large ruptures will normally be in a self-healing mode, given the strong rate-weakening of friction.

While only results from laboratory rock observations and geological characterizations of fault zones were used to construct the model, the resulting dynamic rupture predictions made by it have no obvious inconsistencies with what can be learned from seismology.

ACKNOWLEDGMENTS

The study was supported at Harvard University by NSF-EAR Award 0510193 and by the Southern California Earthquake Center as funded by Cooperative Agreements NSF EAR-0106924 and USGS 02HQAG0008 (the SCEC contribution number for this paper is 1276).

REFERENCES

Anderson, D.L. (1980), An earthquake induced heat mechanism to explain the loss of strength of large rock and earth slides. *Int. Conf. on Engineering for Protection from Natural Disasters*, Bangkok, January 7–9, 1980, edited by P. Karasudhi, A.S. Balasubramaniam, and W. Kanok-Nukulchai, pp. 569–580, John Wiley, Hoboken, N.J.

Andrews, D.J. (2002), A fault constitutive relation accounting for thermal pressurization of pore fluid, *J. Geophys. Res.*, 107(B12), 2363, doi:10.1029/2002JB001942.

Archard, J.F. (1958/1959), The temperature of rubbing surfaces, *Wear*, 2, 438–455.

Beeler, N.M., and T.E. Tullis (1996), Self-healing pulse in dynamic rupture models due to velocity-dependent strength, *Bull. Seismol. Soc. Am.*, 86, 1130–1148.

Beeler, N.M., T.E. Tullis, M.L. Blanpied and J.D. Weeks (1996p), Frictional behavior of large displacement experimental faults, *J. Geophys. Res.*, 101, No. B4, 8697–8715.

Beeler, N.M., and T.E. Tullis (2003), Constitutive relationships for fault strength due to flash-heating, in *SCEC Annual Meeting Proceedings and Abstracts*, vol. XIII, p. 66, Southern California Earthquake Center.

Beeler, N.M., T.E. Tullis, and D.L. Goldsby (2008), Constitutive relationships and physical basis of fault strength due to flash heating, *J. Geophys. Res.*, 113, B01401, doi:10.1029/2007JB004988.

Bird, P., and X. Kong (1994), Computer simulations of California tectonics confirm very low strength of major faults, *Geol. Soc. Am. Bull.*, 106(2), 159–174.

Bowden, F.P., and P.H. Thomas (1954), The surface temperature of sliding solids, *Proc. Roy. Soc. Lond., Ser. A*, 223, 29–40.

Byerlee, J. (1978), Friction of rocks, *Pure Appl. Geophys.*, 116, 615–626.

Chester, F.M., and J.S. Chester (1998), Ultracataclasite structure and friction processes of the Punchbowl fault, San Andreas system, California, *Tectophys.*, 295, 199–221.

Chester, F.M., J.S. Chester, D.L. Kirschner, S.E. Schulz, and J.P. Evans (2004), Structure of large-displacement, strike-slip fault zones in the brittle continental crust, in *Rheology and Deformation in the Lithosphere at Continental Margins*, edited by G.D. Karner, B. Taylor, N.W. Driscoll, and D.L. Kohlstedt, pp. 223–260, Columbia Univ. Press, New York.

Chester, J.S., and D.L. Goldsby (2003), Microscale characterization of natural and experimental slip surfaces relevant to earthquake mechanics, SCEC Ann. Prog. Rep., Southern California Earthquake Center.

Cochard, A., and R. Madariaga (1994), Dynamic faulting under rate-dependent friction, *Pure Appl. Geophys.*, 142, 419–445.

Cochard, A., and R. Madariaga (1996), Complexity of seismicity due to highly rate-dependent friction, *J. Geophys. Res.*, 101, 25, 321–25, 336.

Di Toro, G., D.L. Goldsby, and T.E. Tullis (2004), Friction falls toward zero in quartz rock as slip velocity approaches seismic rates, *Nature*, 427, 436–439, doi:10.1038/nature02249.

Dunham, E.M., H. Noda, and J.R. Rice (2008), Earthquake ruptures with thermal weakening and the operation of major faults at low overall stress levels, *Eos Trans. AGU, 89*(53), *Fall Meet. Suppl.*, Abstract T21D-07.

Ettles, C.M. (1986), The thermal control of friction at high sliding speeds, *J. Tribology, Trans. ASME*, 108, 98–104.

Fialko, Y., and Y. Khazan (2004), Fusion by earthquake fault friction: Stick or slip?, *J. Geophys. Res.*, 110, B12407, doi:10.1029/2005JB003869.

Ghabezloo, S., and J. Sulem (2008), Stress dependent thermal pressurization of a fluid-saturated rock, *Rock Mech. Rock Engng.*, doi:10.1007/s00603-008-0165-z.

Geubelle, P.H., and J.R. Rice (1995), A spectral method for three-dimensional elastodynamic fracture problems, *J. Mech. Phys. Solids*, 43, 1791–1824.

Goldsby, D.L., and T.E. Tullis (2002), Low frictional strength of quartz rocks at subseismic slip rates, *Geophys. Res. Lett.*, 29(17), 1844, doi:10.1029/2002GL015240.

Goren, L., and E. Aharonov (2009), On the stability of landslides: A thermo-poro-elastic approach, *Earth Planet. Sci. Lett.*, 277, 365–372, doi:10.1016/j.epsl.2008.11.00.

Habib, P. (1967), Sur un mode de glissement des massifs rocheaux. *Comptes Rendus Hebd. Seanc. Acad. Sci.*, Paris, 264, 151–153.

Habib, P. (1975) Production of gaseous pore pressure during rock slides. *Rock Mech.*, 7, 193–197.

Han, R., T. Shimamoto, T. Hirose, J.-H. Ree, and J. Ando (2007), Ultralow friction of carbonate faults caused by thermal decomposition, *Science*, 316(5826), 878–881, doi:10.1126/science.1139763.

Heaton, T.H. (1990), Evidence for and implications of self-healing pulses of slip in earthquake rupture, *Phys. Earth Planet. In.*, 64, 1–20.

Heermance, R., Z.K. Shipton, and J.P. Evans (2003), Fault structure control on fault slip and ground motion during the 1999 rupture of the Chelungpu Fault, Taiwan, *Bull. Seismol. Soc. Am.*, 93(3), 1034–1050.

Hickman, S., and M. Zoback (2004), Stress orientations and magnitudes in the SAFOD pilot hole, *Geophys. Res. Lett.*, 31, L15S12, doi:10.1029/2004GL020043.

Hirose, T., and T. Shimamoto (2005), Growth of a molten zone as a mechanism of slip weakening of simulated faults in gabbro during frictional melting, *J. Geophys. Res.*, 110, B05202, doi:10.1029/2004JB003207.

Lachenbruch, A.H. (1980), Frictional heating, fluid pressure, and the resistance to fault motion, *J. Geophys. Res.*, 85, 6097–6122.

Lapusta, N., and J.R. Rice (2003), Low-heat and low-stress fault operation in earthquake models of statically strong but dynamically weak faults, *Eos Trans. AGU*, 84(46), *Fall Meet. Suppl.*, Abstract S51B-02.

Lee, T.C., and P.T. Delaney (1987), Frictional heating and pore pressure rise due to a fault slip, *Geophys. J. Roy. Astr. Soc.*, 88(3), 569–591.

Lim, S.C., and M.F. Ashby (1987), Wear mechanism maps, *Acta Metallurgica*, 35, 1–24.

Lim, S.C., M.F. Ashby and J.F. Brunton (1989), The effect of sliding conditions on the dry friction of metals, *Acta Metallurgica*, 37, 767–772.

Lu, X., N. Lapusta and A.J. Rosakis (2007), Pulse-like and crack-like ruptures in experiments mimicking crustal earthquakes, *Proc. Nat'l. Acad. Sci. USA*, 104, 48, 18931–18936, doi:10.1073/pnas.0704268104.

Lykotrafitis, G., A.J. Rosakis, and G. Ravichandran (2006), Self-healing pulse-like shear ruptures in the laboratory, *Science*, 313(5794), 1765–1768, doi:10.1126/science.1128359.

Manighetti, I., M. Campillo, C. Sammis, P.M. Mai, and G. King (2005), Evidence for self-similar, triangular slip distributions on earthquakes: Implications for earthquake and fault mechanics, *J. Geophys. Res.*, 110, B05302, doi:10.1029/2004JB003174.

Mase, C.W., and L. Smith (1985), Pore-fluid pressures and frictional heating on a fault surface, *Pure Appl. Geophys.*, 122, 583–607.

Mase, C.W., and L. Smith (1987), Effects of frictional heating on the thermal, hydrologic, and mechanical response of a fault, *J. Geophys. Res.*, 92, 6249–6272.

Molinari, A., Y. Estrin and S. Mercier (1999), Dependence of the coefficient of friction on sliding conditions in the high velocity range, *J. Tribology, Trans. ASME*, 121, 35–41.

Nielsen, S., and R. Madariaga (2003), On the self-healing fracture mode, *Bull. Seismol. Soc. Am.*, 93(6), 2375–2388, doi:10.1785/0120020090.

Nielsen, S.B., G.D. Toro, T. Hirose, and T. Shimamoto (2008), Frictional melt and seismic slip, *J. Geophys. Res.*, 113, B01308, doi:10.1029/2007JB005122.

Noda, H. (2004), Numerical simulation of rupture propagation with thermal pressurization based on measured hydraulic properties: Importance of deformation zone width, *Eos Trans. AGU*, 85(47), *Fall Meet. Suppl.*, Abstract T22A-08.

Noda, H. (2008), Frictional constitutive law at intermediate slip rates accounting for flash heating and thermally activated slip process, *J. Geophys. Res.*, 113, B09302, doi:10.1029/2007JB005406.

Noda, H., and T. Shimamoto (2005), Thermal pressurization and slip-weakening distance of a fault: An example of the Hanore fault, southwest Japan, *Bull. Seismol. Soc. Am.*, 95(4), 1224–1233, doi:10.1785/0120040089.

Noda, H., E.M. Dunham, and J.R. Rice (2006), Self-healing vs. crack-like rupture propagation in presence of thermal weakening processes based on realistic physical properties, *Eos Trans. AGU*, 87(52), *Fall Meet. Suppl.*, Abstract S42A-06.

Noda, H., E.M. Dunham, and J.R. Rice (2009), Earthquake ruptures with thermal weakening and the operation of major faults at low overall stress levels, *J. Geophys. Res.*, 114, B07302, doi:10.1029/2008JB006143.

O'Hara, K., K. Mizoguchi, T. Shimamoto, and J.C. Hower (2006), Experimental frictional heating of coal gouge at seismic slip rates: Evidence for devolatilization and thermal pressurization of gouge fluids, *Tectonophys.*, 424, 109–118.

Perrin, G., J.R. Rice, and G. Zheng (1995), Self-healing slip pulse on a frictional interface, *J. Mech. Phys. Solids*, 43, 1461–1495.

Rempel, A.W., and J.R. Rice (2006), Thermal pressurization and onset of melting in fault zones, *J. Geophys. Res.*, 111, B09314, doi:10.1029/2006JB004314.

Rice, J.R. (1983), Constitutive relations for fault slip and earthquake instabilities, *Pure Appl. Geophys.*, 121(3), 443–475, doi:10.1007/BF02590151.

Rice, J.R. (1999), Flash heating at asperity contacts and rate-dependent friction, *Eos Trans. AGU*, 80(46), *Fall Meet. Suppl.*, F6811.

Rice, J.R. (2001), New perspectives in crack and fault dynamics, in *Mechanics for a New Millennium* (Proceedings of the 20th International Congress of Theoretical and Applied Mechanics, 27 Aug–2 Sept 2000, Chicago), eds. H. Aref and J.W. Phillips, Kluwer Academic Publishers, pp. 1–23.

Rice, J.R. (2006), Heating and weakening of faults during earthquake slip, *J. Geophys. Res.*, 111(B5), B05311, doi:10.1029/2005JB004006.

Rice, J.R., N. Lapusta, and K. Ranjith (2001), Rate and state dependent friction and the stability of sliding between elastically deformable solids, *J. Mech. Phys. Solids*, 49, 1865–1898.

Rice, J.R., and M. Cocco, Seismic fault rheology and earthquake dynamics, in *Tectonic Faults: Agents of Change on a Dynamic Earth*, eds. M. R. Handy, G. Hirth and N. Hovius (Dahlem Workshop 95, Berlin, January 2005, on *The Dynamics of Fault Zones*), Chp. 5, pp. 99–137, The MIT Press, Cambridge, MA, USA.

Roig Silva, C., D.L. Goldsby, G. Di Toro and T.E. Tullis (2004), The role of silica content in dynamic fault weakening due to gel lubrication. *Eos Trans. AGU*, 85(47), *Fall Meet. Suppl.*, Abstract T21D-06.

Sibson, R. H. (1973), Interaction between temperature and pore-fluid pressure during earthquake faulting: A mechanism for partial or total stress relief, *Nature*, 243, 66–68.

Sirono, S., K. Satomi, and S. Watanabe (2006), Numerical simulations of frictional melting: Small dependence of shear stress drop on viscosity parameters, *J. Geophys. Res.*, 111, B06309, doi:10.1029/2005JB003858.

Spray, J. (1995), Pseudotachylyte controversy; fact or friction?, *Geol.*, 23, 1119–1122.

Sulem, J., and V. Famin (2009), Thermal decomposition of carbonates in fault zones: Slip-weakening and temperature limiting effects, *J. Geophys. Res.*, 114, B03309, doi:10.1029/2008JB006004.

Sulem, J., I. Vardoulakis, H. Ouffroukh, and V. Perdikatsis (2005), Thermo-poro-mechanical properties of the Aigion Fault clayey gouge—application to the analysis of shear heating and fluid pressurization, *Soils Found.*, 45(2), 97–108.

Tsutsumi, A., and T. Shimamoto (1997), High velocity frictional properties of gabbro, *Geophys. Res. Lett.*, 24, 699–702.

Tullis, T.E., and D.L. Goldsby (2003a), Flash melting of crustal rocks at almost seismic slip rates, *Eos Trans. AGU*, 84(46), *Fall Meet. Suppl.*, Abstract S51B-05.

Tullis, T.E., and D.L. Goldsby (2003b), Laboratory experiments on fault shear resistance relevant to coseismic earthquake slip, SCEC Ann. Prog. Rep., Southern California Earthquake Center.

Vardoulakis, I. (2002), Dynamic thermo-poro-mechanical analysis of catastrophic landslides, *Geotechnique*, 52(3), 157–171.

Veveakis, E., I. Vardoulakis and G. Di Toro (2007), Thermo-poromechanics of creeping landslides: The 1963 Vaiont slide, northern Italy, *J. Geophys. Res.*, 112, F03026, doi:10.1029/2006JF000702.

Voight, B. and C. Faust (1982), Frictional heat and strength loss in some rapid landslides. *Geotechnique*, 32, 43–54.

Wibberley, C.A.J. (2002), Hydraulic diffusivity of fault gouge zones and implications for thermal pressurization during seismic slip, *Earth Plant. Space*, 54(11), 1153–1171.

Wibberley, C.A.J., and T. Shimamoto (2003), Internal structure and permeability of major strike-slip fault zones: the Median Tectonic Line in Mie Prefecture, southwest Japan, *J. Struct. Geol.*, 25(1), 59–78, doi:10.1016/S0191-8141(02)00014-7.

Yuan, F., and V. Prakash (2008), Use of a modified torsional Kolsky bar to study frictional slip resistance in rock-analog materials at coseismic slip rates, *Int. J. Solids Structures*, 45, 4247–4263.

Zheng, G., and J.R. Rice (1998), Conditions under which velocity-weakening friction allows a self-healing versus a cracklike mode of rupture, *Bull. Seismol. Soc. Am.*, 88, 1466–1483.

Slip sequences in laboratory experiments as analogues to earthquakes associated with a fault edge

Shmuel M. Rubinstein, Gil Cohen & Jay Fineberg
The Racah Institute of Physics, The Hebrew University of Jerusalem, Givat Ram, Jerusalem, Israel

Zéev Reches
School of Geology and Geophysics, University of Oklahoma, Norman, OK, USA

ABSTRACT: Natural faults are intrinsically heterogeneous where jogs, edges and steps are common. We experimentally explore how fault edges may affect earthquake and slip dynamics by applying shear to the edge of one of two flat blocks in frictional contact. We show that slip occurs via a sequence of rapid rupture events that arrest after a finite distance. Successive events extend the slip size, transfer the applied shear across the block, and cause progressively larger changes of the contact area along the contact surface. Each sequence of events dynamically forms an asperity near the edge and largely reduces the contact area beyond. These sequences of rapid events all culminate in slow slip events that lead to major, unarrested slip along the entire contact surface. These results show that a simple deviation from uniform shear loading configuration can significantly and qualitatively affect both earthquake nucleation processes and the evolution of fault complexity.

1 INTRODUCTION

Faults are often modelled as planar interfaces separating two elastic half-spaces that are driven by a spatially uniform shear that is imposed by the motion of tectonic plates [Das, 2003; Lapusta and Rice, 2003; Rice and Ben-Zion, 1996] On the other hand, the loading of a natural fault is the superposition of the uniform shear due to remote loading (e.g. plate motion), and nonuniform loading generated by local structural features. Heterogeneities (e.g. steps, jogs, asperities and edges) [Ben-Zion and Sammis, 2003; Harris and Day, 1993; Shaw and Dieterich, 2007; Wesnousky, 2006], abound within the seismogenic zone. This complexity may govern some of the dominant properties of earthquakes.

In this paper we experimentally explore the influence of one type of nonuniformity: loading on the edge of a fault which, initially, is smooth and uniform. Our laboratory "fault" is formed by two elastic blocks separated by a roughened, but optically flat, frictional interface. Shear force is applied to one edge of the slider block (Figure 1a) while a uniform normal stress is remotely applied. This loading configuration is a simplified model for the inhomogeneous loading that is likely to occur at an edge or asperity along an otherwise planar fault. In order to highlight the unique contributions of the nonuniform application of shear to the resulting fault dynamics, this model focuses on the effects of the nonuniform component and ignores the uniform component of the applied shear.

We believe that similar effects such as those described in the paper occur in-situ, due the common presence of structural perturbations within natural faults. Configurations in which edge-loading may play an important role are common along active faults in the earth's crust, and include: Slip along a segment within a long fault that loads the neighbouring segments at the edge of the slipped region (e.g. north Anatolian fault [Stein et al., 1997], the physical edges formed between abutting segments (e.g. the intersection of the Susitna Glacier and Denalli faults [Aagaard and Heaton, 2004], and by asperities and steps along faults [Harris and Day, 1993; Johnson et al., 1994; Lay et al., 1982; Sagy et al., 2007; Shaw and Dieterich, 2007; Wesnousky, 2006]. As these examples show, the loading of large crustal faults is frequently modelled by a combination of "basal loading"

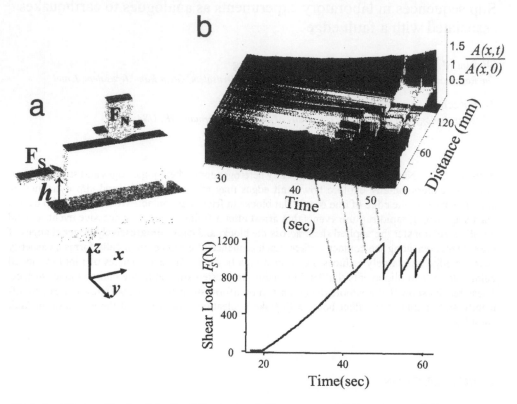

Figure 1.   The transition to stick-slip sliding is preceded by a sequence of discrete arrested slip events. **a,** A schematic illustration of the base and slider blocks and load application **b,** (bottom) Applied shear force, F_S, as a function of time with the (top) corresponding spatio-temporal evolution of $A(x,t)$. Each small discrete drop in F_S corresponds to a rapid slip event that arrests within the interface. Each arrested slip event generates significant changes in $A(x,t)$. Arrested events, together with the corresponding drop in shear load, are denoted by the arrows. The large stress drops mark stick-slip motion. Interface length, $L = 140$ mm, $F_N = 3.3$ kN.

on the crust base, and "edge loading" at the fault edge [Lachenbruch and Sass, 1980; Matsuura and Sato, 1997; Reches et al., 1994].

2   EXPERIMENTAL METHODS

The experimental setup used is described in detail in [Rubinstein et al., 2004; Rubinstein et al., 2006]. We performed real-time measurements of the *true* area of contact, $A(x,y,t)$, along the entire interface separating two polymethyl-methacrylate blocks whose $(x{:}y{:}z)$ dimensions were 300:30:27 mm for the static ("base") block and either 140:6:75 mm or 200:6:75 mm for sliding ("slider") block. x, y, and z are, respectively, the sliding, sample width, and normal loading directions. The optically flat base-slider interface was roughened to 1 μm rms. For the range $(1 < F_N < 4$ kN$)$ of normal load, (F_N), applied, $A(x,y,t)$ varied from 0.35–1.35% of the interface's nominal contact area [Dieterich and Kilgore, 1994]. $A(x,y,t)$ was measured by illuminating the contact area by a laser sheet whose incident angle was well beyond the angle for total internal reflection from the interface. Thus light is transmitted only at points of contact, with an intensity at each point (x,y) proportional to $A(x,y,t)$. $A(x,y,t)$ was imaged at rates up to 100,000 frames/sec. The data acquisition was designed to capture both slow processes at the quasi-static time scales governed by the loading rate and rapid, crack-like, processes whose entire duration takes place in the sub-msec range. As the onset dynamics are governed by one-dimensional rupture fronts [Rubinstein et al., 2004], $A(x,y,t)$ was averaged in y, yielding $A(x,t)$ to 1280 pixel resolution. Thus, $A(x,t)$ yields a local measurement of the contact area, where each pixel measures the integrated contact area a 0.1 mm × 6 mm region (with the higher resolution in the direction of motion). At the initiation of each experiment, before the application of shear the slider was oriented relative to the base to form an initial contact area that was, statistically, spatially uniform (utilizing the $A(x,y)$ measurements to guide the positioning). Upon completion of this initial positioning, F_N was applied. The corresponding value of $A(x,t = 0)$ was then used to normalize subsequent measurements of $A(x,t)$ to allow us to measure the changes in $A(x,t)$ resulting from the dynamics at each point x.

At $t = 0$, the shear force, F_S, was applied in the x direction to one edge (the "trailing" edge at $x = 0$) of the slider at a height $z = h$ (2 < h < 18 mm) above the interface (see Figure 1a). F_S was increased from zero at constant rates (range 1–20 µm/s) until, at $F_S = \mu_S F_N$, stick-slip sliding initiated. The precise means by which F_S was applied was unimportant (e.g. via rigid blocks of various dimensions), as long as h was defined as the mean height of the applied shear force. In this loading system [Rubinstein et al., 2006], any slip of the trailing edge immediately results in a sharp drop of F_S. While the leading edge (at $x = L$) is stationary, drops of F_S mirror the stress release across the interface.

3 RESULTS

3.1 *Slip sequences and stress history*

Concurrent measurements of contact area $A(x,t)$, and shear load $F_S(t)$ for a typical experiment (Figure 1b) reveal that large stick-slip events are the culmination of a complex history of precursory slip events. The $F_S(t)$ curve in Figure 1b reveals a discrete sequence of small sharp stress drops that occur at stress levels well below the peak values of $F_S(t)$. These small stress drops (of ~ 0.01–$0.02 \cdot F_S$) result from the propagation of a sequence of rapid, crack-like arrested slip events. The measurements of $A(x,t)$ at short times (Figure 2a-top) show that the initial slip events begin at the trailing edge, and propagate at "sub-Rayleigh" speeds, typically between 60–80% of the Rayleigh wave speed (V_R), before abruptly arresting (Figure 2b). These initial events (Figure 2a-top) are associated with slipping segments of length l that are relatively small compared to the entire fault size, L. We find that l obeys a linear scaling relation [Rubinstein et al., 2007], $l \propto F_S L / F_N$ (Figure 3a). Once l approaches 0.4–0.5L, this scaling breaks down and the initial dynamics undergo a qualitative change that marks a transition [Rubinstein et al., 2004; Rubinstein et al., 2006] into a new stage. The slip events in this stage also initiate at the trailing edge as rapid sub-Rayleigh slip events, but do not simply arrest. Instead, these larger events trigger a "slow" front that propagates at speeds over an order of magnitude slower (~ 50 m/sec in Figure 2b) than the sub-Rayleigh velocities of the triggering events. These slow fronts can propagate stably for some time, and either traverse the remainder of the interface or transition back to sub-Rayleigh fronts, as shown in Figure 2a (bottom). Significantly, overall motion (sliding) between the blocks initiates only after either a slow or subsequently triggered sub-Rayleigh front has reached the leading edge.

The discrete sequence of such arrested slip events, described by Figure 1, is *only* observed when shear is imposed at the sample's trailing edge. It is not observed when, for example, a uniform shear stress is imposed at a remote boundary parallel to the interface. With the trailing edge loading, each sequence initiates via a slip event of finite length, l_0, with l increasing by discrete increments, Δl, of constant length for each successive slip.

Figure 3b shows that the size of both l_0 and Δl is proportional to h, the height above the interface where F_S is applied at the edge. Note that h, however, has no effect on the overall scaling of l (Figure 3a). Since l scales linearly with F_S, the fixed value of Δl (for a given h) indicates that slip events occur at fixed intervals, ΔF_S, in F_S. Thus, h, which determines Δl, governs also the magnitude of the intervals, ΔF_S, between successive events. For a constant shear loading rate, (as in our experiments) the temporal periods between events are proportional to h.

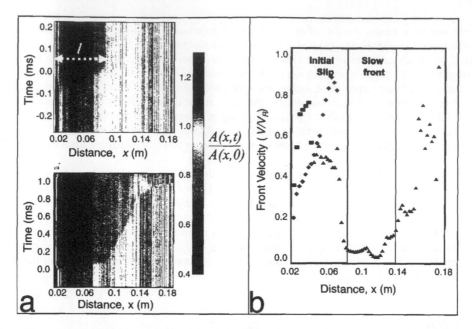

Figure 2.   **a**, (top) An arrested slip event of length, *l*, generated at stresses well below the onset of stick-slip. (bottom) The transition to stick-slip motion at the peak value of F_S. Here rapid slip arrests and triggers a slow front. At $x = 0.14$ m the slow front nucleates a rapid slip that traverses the remainder of the interface. Color bar indicates the change in $A(x, t)$, relative to the initial, uniform value $A(x, t = 0)$ when $F_S = 0$. $A(x, t)$, measured at 14 μsec intervals in two different events. **b**, The slip propagation velocities (as a function of *x*) of arrested events (diamonds and squares) and the transition (triangles) to stick-slip motion. Events depicted by diamonds and triangles correspond to (a).

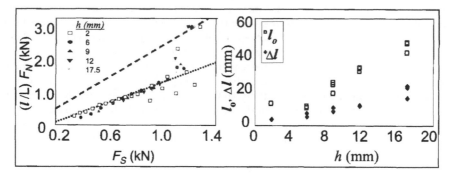

Figure 3.   Scaling behaviour of the lengths *l* of successive slip events. **a**, $l/L \cdot F_N$ as a function of F_S, where F_S was applied at different heights, *h*, above the interface. *h* does not influence the $l \propto L \cdot F_S/F_N$ scaling [Rubinstein et al., 2007] (dotted line). Here, $F_N = 3$ kN and $L = 140$ mm. This scaling breaks down at the transition to large events leading to stick-slip motion, described by the Amontons-Coulomb law (dashed line). **b**, The values of both the initial slip length, l_0 (squares), and the incremental extension of each slip event, Δl (diamonds), increase linearly with *h*. l_0 saturates at low *h* suggesting that a minimal length is needed for development of instability. Different points at the same *h* correspond to different F_N.

3.2   *Contact area and fault strength*

We now consider the evolution of the contact area $A(x, t)$. Prior to the first event, $A(x, t)$ is spatially uniform. The passage of each successive precursory slip event (Figure 4) significantly alters the contact area, and hence changes the local fault strength. With each successive event, the contact area increases in a region of width D, that is adjacent to the sample's trailing edge. This process

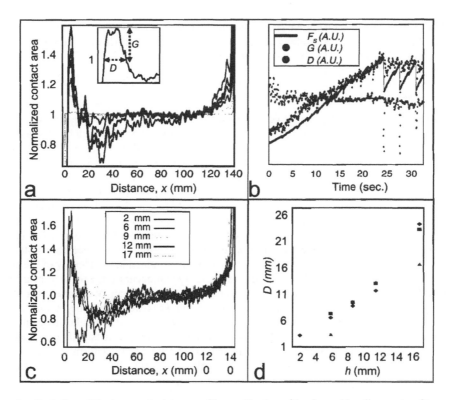

Figure 4. Evolution of the true contact area profile. **a**, $A(x, t)$ profiles formed by slip events of increasing length during a single experiment. Deformations of the initially uniform contact profile are both amplified and extended with each slip. (Inset) Close-up of $A(x, t)$ profile adjacent to the trailing edge depicting the width D and strength G of the asperity that is spontaneously formed there. **b**, F_S, D and G as a function of time for the experiment shown in (a) F_S, D and G are in arbitrary units to facilitate comparison. The asperity width D, stays nearly constant while G increases monotonically with F_S. **c**, D is determined h; $A(x, t)$ in experiments where h was varied. The profiles are qualitatively similar in appearance. **d**, D, increases nearly linearly with h. All profiles in (**c**) were obtained for F_S at 75% of the value needed for the onset of stick-slip. $A(x, t)$ in **a** and **c**, is normalized with respect to its spatially uniform value, $A(x, t = 0)$, at the start of the experiment.

dynamically forms an asperity (a localized area whose resistance to slip is much greater than its surroundings [Lay et al., 1982]). Although this asperity continuously strengthens with F_S (Figure 4b), its size, D, remains nearly constant throughout each experiment. After overcoming this asperity, each slip both extends the length of the reduced contact area region created by its predecessors, and further reduces $A(x, t)$ by a significant amount. This systematic decrease of $A(x, t)$ creates a highly weakened region in the asperity's wake. The contact profile, established as a result of this process, is highly non-uniform by the time large-scale overall motion (stick-slip) occurs. As shown in Figure 4c, changes in h do not qualitatively affect the general shape of the contact area profiles. Quantitative analysis reveals, however (Figure 4d), that the size of D is proportional to h.

One may suspect that the variations of the contact area is due to the torque imposed by the F_S, as F_S is applied at a finite height h rather than at $h = 0$. We found that this effect is negligible over the range of used h. For example, the torque resulting from $h = 2$ mm yields only a 3% variation of the normal stress over the interface length, whereas $A(x, t)$ varies by over 50% (Figure 4).

Surprisingly, once a contact profile is created, it is retained by the system, remaining nearly unchanged both after large-scale slip and in successive stick-slip events [Rubinstein et al., 2007]. Large internal stresses are, therefore, also retained by the system, even after major slip events occur. As the contact area mirrors the normal stress values, the existence of this non-uniform profile also indicates that the normal stresses along the interface (or fault) are highly non-uniform. In particular, the normal stresses along significant regions of the interface are considerably weaker than the remotely applied values would imply.

3.3   *Synthesis of experimental results*

These experiments suggest an intuitive picture for the sequence of events leading to frictional sliding. Before the onset of slip, the loading at the trailing edge imposes a high shear stress region near the edge, whose magnitude decays over a length proportional to h. When F_S is sufficiently large, this highly stressed region yields and an initial crack-like slip event is generated. The slip traverses this region and arrests at a distance, l_0, where the shear stress level is below the slipping threshold. This event results in: (1) slip within the region l_0, where built-up shear stresses are released, (2) elastic deformation of the slider to compensate for the slip-induced contraction in the x direction. The deformation results in an outwardly protruding region of size $D \sim 1/3l_0$ in which the contact area increases, thereby dynamically forming an asperity (cf. Figure 4b). The inwardly bowed region over the remainder of the region l_0, reduces the normal force (=decrease of $A(x,t)$) (3) establishment of a high residual shear stress concentration entrained in the vicinity of the point of arrest of the slip event (This is due to the stress singularity that occurs at the tip of a shear crack.). Upon further increase of F_S, the barrier imposed by the asperity is again overcome and a new slip event is generated. This slip event will easily traverse the weakened region beyond the asperity. Arriving at the tip of the previously arrested event, the new slip event will add sufficient energy to release the energy stored within the high-stress region imposed previously, thereby enabling it to extend itself by Δl. This extension is accompanied by further elastic deformation of the block, thereby increasing the non-uniformity of $A(x,t)$. In this way, each slip event transfers the shear stress imposed at the boundary further along the interface.

4   DISCUSSION

The influence of fault complexity on the dynamics of rupture propagation and seismicity has been the subject of intensive past and recent research [Aki, 1979; Das, 2003; Lay et al., 1982]. Our experiments suggest that the fingerprint of an 'edge' can be traced, not only to the dynamics of a given rupture, but throughout the entire seismic cycle of a given fault. The results suggest that a geometrical inhomogeneity introduces a scale that may govern the nucleation, size and repeatability of earthquakes along a given fault.

The experiment paints an interesting picture of earthquake dynamics along faults that are loaded asymmetrically (e.g. at an edge or step). They indicate that stress transfer along such faults may be mediated by a periodic sequence of precursory events. This sequence of slip events culminates by the triggering of slowly propagating front, which leads to system size events. The experiments suggest that the early slip events of the periodic sequence (within the scaling regime shown in Figure 3a) "feel" an effectively infinitely long (unsegmented) fault. In contrast, the accelerated growth of l that marks the break of scaling prior to the onset of large events indicates that the dynamics are affected by the fault size during the nucleation phase [Ohnaka and Shen, 1999] of a large event. This accelerated increase in l is strongly suggestive of the accelerated seismic release that precedes some large earthquakes [Bufe and Varnes, 1993]. The results also imply that precursory sequences of events that initiate from a fault edge strongly modify the fault contact plane prior to a large event.

An excellent example of such slip sequences along a fault edge in the crust is portrayed by the foreshock sequence of the 1998 Sendai Bay event, along the Nagamachi-Rifu fault, Japan [Umino et al., 2002]. In this field case, the main shock of M5.0 was preceded by 17 foreshocks ranging in magnitude from 1.7 to 3.8 with essentially identical seismic characteristics. The foreshocks sequence lasted about three days with the largest foreshock occurring six minutes prior to the main shock. Mechanical modelling of this sequence of events suggests that the fault was edge-loaded by non-seismic slip in the lower crust [Nakajima et al., 2006]. Accordingly, the foreshock hypocenters propagated upward along the locked part of the Nagamachi-Rifu fault. Finally, Umino et al noted that "A small ambiguous phase ... is observed in seismograms of both the M5.0 main shock and the M3.8 largest foreshock ..." [Umino et al., 2002]. This slow, low amplitude ambiguous phase is lacking in the other foreshocks and all aftershocks, and is likely the equivalent of the slow fronts observed in our experiments immediately *before* the main slip event (Fig. 2a). These slow fronts may be akin to the accelerated creep events that are anticipated to be part of the nucleation phase of major earthquakes [Dieterich and Kilgore, 1996; Ohnaka and Shen, 1999]. Thus, in spite of scale and complexity differences, we note the following similarities between the 1998 Sendai Bay events and our experiments: (1) qualitatively similar (edge) loading conditions, (2) a distinct sequence of precursory events; (3) initiation of precursory events from nearly the same location and (4) a slow ("ambiguous") phase that occurs only before the main event.

In conclusion, these experiments have shown that the fact that shear is applied non-uniformly to a sliding system leads to complex, *systematic* behaviour that appears analogous to natural phenomena, whose source is currently not well understood. We believe that the analogies between our experimental results and seismic observations stem from their similar edge-loading configurations. As elements of edge-loading are common in faults at many scales [Sagy et al., 2007; Stein et al., 1997], it is therefore anticipated that this loading will generate stress distributions that are similar to the laboratory model, and, consequently, may lead to similar dynamics.

ACKNOWLEDGMENTS

The authors acknowledge the support of the Israel Science Foundation Grant No. 57/07, grant no. 2006288 awarded by the U.S.–Israel Binational Science Foundation, as well as NSF Continental Dynamics grant No. 0409605 (NELSAM). We also thank E. Brodsky, E. Aharonov, A Sagy, and R. Madariaga for their helpful remarks.

REFERENCES

Aagaard, B.T., and T.H. Heaton, Near-source ground motions from simulations of sustained intersonic and supersonic fault ruptures, *Bulletin of the Seismological Society of America*, 94(6), 2064–2078, 2004.

Aki, K., Characterization of Barriers on an Earthquake Fault, *Journal of Geophysical Research*, 84(NB11), 6140–6148, 1979.

Ben-Zion, Y., and C.G. Sammis, Characterization of fault zones, *Pure and Applied Geophysics*, 160(3–4), 677–715, 2003.

Bufe, C.G., and D.J. Varnes, Predictive Modeling of the Seismic Cycle of the Greater San-Francisco Bay-Region, *Journal of Geophysical Research-Solid Earth*, 98(B6), 9871–9883, 1993.

Das, S., Spontaneous complex earthquake rupture propagation, *Pure and Applied Geophysics*, 160(3–4), 579–602, 2003.

Dieterich, J.H., and B. Kilgore, Implications of fault constitutive properties for earthquake prediction, *Proceedings of the National Academy of Sciences of the United States of America*, 93(9), 3787–3794, 1996.

Dieterich, J.H., and B.D. Kilgore, Direct Observation of Frictional Contacts - New Insights for State-Dependent Properties, *Pure and Applied Geophysics*, 143(1–3), 283–302, 1994.

Harris, R.A., and S.M. Day, Dynamics of Fault Interaction - Parallel Strike-Slip Faults, *Journal of Geophysical Research-Solid Earth*, 98(B3), 4461–4472, 1993.

Johnson, A.M., R.W. Fleming, and K.M. Cruikshank, Shear Zones Formed Along Long, Straight Traces of Fault Zones During the 28 June 1992 Landers, California, Earthquake, *Bulletin of the Seismological Society of America*, 84(3), 499–510, 1994.

Lachenbruch, A.H., and J.H. Sass, Heat-Flow and Energetics of the San-Andreas Fault Zone, *Journal of Geophysical Research*, 85(NB11), 6185–6222, 1980.

Lapusta, N., and J.R. Rice, Nucleation and early seismic propagation of small and large events in a crustal earthquake model, *Journal of Geophysical Research-Solid Earth*, 108(B4), 2205, 2003.

Lay, T., H. Kanamori, and L. Ruff, The Asperity Model and the Nature of Large Subduction Zone Earthquakes, *Earthquake Prediction Research*, 1(1), 3–71, 1982.

Matsuura, M., and T. Sato, Loading mechanism and scaling relations of large interplate earthquakes, *Tectonophysics*, 277(1–3), 189–198, 1997.

Nakajima, J., A. Hasegawa, S. Horiuchi, K. Yoshirnoto, T. Yoshide, and N. Umino, Crustal heterogeneity around the Nagamachi-Rifu fault, northeastern Japan, as inferred from travel-time tomography, *Earth Planets and Space*, 58(7), 843–853, 2006.

Ohnaka, M., and L.F. Shen, Scaling of the shear rupture process from nucleation to dynamic propagation: Implications of geometric irregularity of the rupturing surfaces, *Journal of Geophysical Research-Solid Earth*, 104(B1), 817–844, 1999.

Reches, Z., G. Schubert, and C. Anderson, Modeling of Periodic Great Earthquakes on the San-Andreas Fault – Effects of Nonlinear Crustal Theology, *Journal of Geophysical Research-Solid Earth*, 99(B11), 21983–22000, 1994.

Rice, J.R., and Y. Ben-Zion, Slip complexity in earthquake fault models, *Proceedings of the National Academy of Sciences of the United States of America*, 93(9), 3811–3818, 1996.

Rubinstein, S.M., G. Cohen, and J. Fineberg, Detachment fronts and the onset of dynamic friction, *Nature*, 430(7003), 1005–1009, 2004.

Rubinstein, S.M., G. Cohen, and J. Fineberg, Dynamics of precursors to frictional sliding, *Physical Review Letters*, 98(22), 2007.

Rubinstein, S.M., M. Shay, G. Cohen, and J. Fineberg, Crack like processes governing the onset of frictional slip, *Int. J. of Fracture*, 140, 201–212, 2006.

Sagy, A., E.E. Brodsky, and G.J. Axen, Evolution of fault-surface roughness with slip, *Geology*, 35(3), 283–286, 2007.

Shaw, B.E., and J.H. Dieterich, Probabilities for jumping fault segment stepovers, *Geophysical Research Letters*, 34(1), 2007.

Stein, R.S., A.A. Barka, and J.H. Dieterich, Progressive failure on the North Anatolian fault since 1939 by earthquake stress triggering, *Geophysical Journal International*, 128(3), 594–604, 1997.

Umino, N., T. Okada, and A. Hasegawa, Foreshock and aftershock sequence of the 1998 M 5.0 Sendai, northeastern Japan, earthquake and its implications for earthquake nucleation, *Bulletin of the Seismological Society of America*, 92(6), 2465–2477, 2002.

Wesnousky, S.G., Predicting the endpoints of earthquake ruptures, *Nature*, 444(7117), 358–360, 2006.

On the mechanism of junction growth in pre-sliding

A. Ovcharenko, G. Halperin & I. Etsion
Department of Mechanical Engineering, Technion, Haifa, Israel

ABSTRACT: The contact area evolution during pre-sliding (junction growth) of copper spheres loaded against a hard sapphire flat was recently investigated experimentally by Ovcharenko et al. (2008). Tests were performed with a specially developed test rig for real-time and in situ direct measurements. The results provide a new insight of the junction growth mechanism showing new points of the sphere surface that are coming into contact with the flat. It is found that junction growth at sliding inception can cause up to 45 percent increase in the initial contact area that is formed under normal preload alone. Good correlation is found between the present experimental results and a theoretical model for medium and high normal preloads.

1 INTRODUCTION

The real contact between two rough solids pressed together by a normal load, occurs at the summits of their micro-asperities. The elastic-plastic deformation of these contacting micro-asperities under combined normal and tangential loading plays a significant role in several problems of contact mechanics such as friction, adhesion, wear, electrical and thermal contact conductance to name a few. A key parameter in all theses problems is the real contact area. An accurate evaluation of the contact area and its evolution under varying normal and tangential loading is an essential step in solving different problems of tribology and contact mechanics.

In spite of the importance of this subject only a few experimental works were published so far on the contact area evolution due to tangential loading under a given normal preload during pre-sliding. Most of these works studied junction growth indirectly (e.g. McFarlane & Tabor (1950)), not in a real time (e.g. Parker & Hatch (1950)), and non *in situ* (e.g. Kayaba & Kato (1978)).

The term "junction growth" was introduced by Tabor (1959). The junction growth phenomenon was explained by Tabor as the need to maintain a constant von Mises stress at yielded contact points. Consequently a contact area, which already yielded plastically under a given normal load, must grow when subjected to additional tangential loading in order to reduce its mean contact pressure and be able to accommodate the additional shear stresses.

In spite of the accumulating experimental evidences reported in the literature on the junction growth, the actual mechanism of this phenomenon is not yet clearly understood. The main goal of this paper is therefore to describe the experimental study by Ovcharenko et al. (2008) of the junction growth phenomenon and its mechanism. This was accomplished by *in situ* and real-time direct and accurate optical measurement of the contact area evolution of a single spherical metallic elastic-plastic contact during pre-sliding, using the novel test rig described in Ovcharenko et al. (2006).

2 EXPERIMENTAL DETAILS

2.1 *Test apparatus description*

The test rig (Ovcharenko et al. (2006)) consists of four main modules that are shown schematically in Figure 1: An actuation module (I) that holds a transparent rigid flat and applies the tangential load to the spherical contact; a friction force measurement module (II), which holds a sphere and measures the tangential friction force applied to it by the flat; a normal force module (III), which applies and measures the normal loading of the contact; and an optical module (IV) for accurate measurement of the contact area evolution.

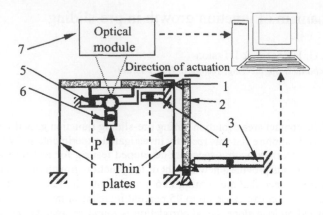

Figure 1.   Schematic representation of the test rig with 4 modules: I) Actuation module consisting of: parallelogram frame (1), mechanical lever (2), piezoelectric actuator (3), and proximity probe (4). II) Friction force measurement module (5). III) Normal force module (6). IV) Optical module (7).

Table 1.   Mechanical and geometrical properties of the copper spherical specimens, and the sapphire flat; (Y—yield strength, ν—Poisson's ratio, E—Young's modulus).

Symbol	Material	D [mm]	Y [MPa]	ν	E [GPa]	L_c [N]
◆	Copper	3	345	0.33*	139	0.14
?	Copper	5	345	0.33*	139	0.40
■	Copper	10	345	0.33*	139	1.60
●	Copper	15	345	0.33*	139	3.90
none	Sapphire	flat	2950*	0.27*	435*	–

* Values obtained from the literature.

2.2   *Specimens description*

The rigid hard flat of 6 mm thickness was made of sapphire (hardness $H = 19$ GPa, roughness $R_a = 3.5$ nm). The spherical specimens with diameter D ranging from 3 to 15 mm were machined from copper rod (UNS C10200, $H = 1.15$ GPa, $R_a = 15$ nm).

The mechanical properties of the spherical specimens and the sapphire flat, which were self measured or obtained from published literature, are summarized in Table 1. Also provided in the table are the diameters of the spheres along with their corresponding critical normal load, L_c, at plastic yield inception in full stick (for more details see Brizmer et al. (2006a)).

The four different diameters of the spheres provide a wide range of critical load values varying from 0.14 to 3.90 N (see Table 1). This allowed covering the entire elastic-plastic regime of deformation when the normal preload P varied from 10 to 190 N ($2.6 < P/L_c < 1360$). In the current experiments the maximum value of $P/L_c = P^*$ was limited to 500 which is already deep into the elastic-plastic regime of deformation (Brizmer et al. 2006b).

2.3   *Test procedure*

All the experiments were carried out at room temperature of 20–24°C and relative humidity of 33–43%. Each experiment was performed on a new spherical area and both the sphere and flat surfaces were cleaned by acetone prior to testing. The desired constant normal preload was applied for 30 sec, and then the tangential load was applied gradually at a constant rate of 2.25 N/s, until gross sliding incepted between the sphere and the flat. As was found in Ovcharenko et al. (2006) this low tangential load rate is sufficient to accurately capture the fine details of the contact area, friction force and relative displacement evolution during the pre-sliding and at the instant of sliding inception. Moreover, it was shown in Ovcharenko et al. (2006) that all these details are negligibly affected by the tangential load rate over a wide range between 0.03 to 37 N/s.

3   EXPERIMENTAL RESULTS

Figure 2 presents a special experiment that was performed in order to better understand the mechanism of the junction growth. Figure 2(a) shows a photograph of the surface of a 5 mm copper sphere, prior to any loading, with four lines of small micro indentation dots that were produced by micro hardness testing machine (Mitutoyo MVK-H1, Japan), using Vickers diamond pyramid and 0.1 N indentation load. The resulting typical dimensions of a single indentation were: lateral sides of 7 μm and depth of 0.4 μm. The average spacing between the centers of two adjacent indentations

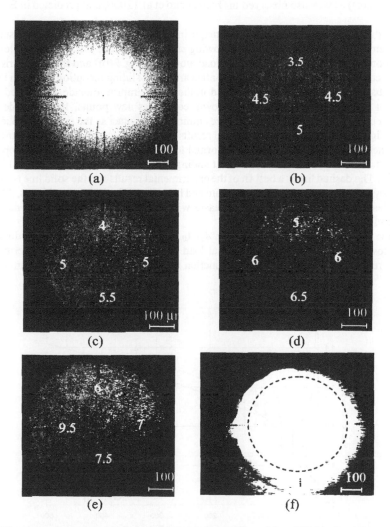

(a)   (b)

(c)   (d)

(e)   (f)

Figure 2.   The mechanism of junction growth with a 5 mm sphere at $P^* = 337$ (Ovcharenko et al. (2008)): Non in situ spherical surface image prior to loading showing micro indentation markers (a). In situ contact area images at tangential loads: $Q = 0$ (b), $0.5Q_{max}$ (c), $0.75Q_{max}$ (d), and Q_{max} (e), showing increasing number of indentations inside the contact area. Post test non in situ surface area imprint (f), with the dashed circle corresponding to the image in (b).

was about 10 μm. These micro indentation dots served as markers to allow tracing of points within or outside the evolving contact area. The number of indentations and their location was selected in such a way that just a few of them from each of the four lines will be included within the contact area under a selected normal load of $P^* = 337$. Figure 2(b) shows the *in situ* image of the contact area of the normally loaded sphere when $P^* = 337$ and tangential load $Q = 0$. Also shown are those indentation dots found within the contact area along with their counted number indicated next to each one of the four lines of indentations. Figures 2(c), 2(d) and 2(e) show the *in situ* images of the evolving contact area under increasing tangential loads of $Q = 0.5Q_{max}$, $Q = 0.75Q_{max}$ and $Q = Q_{max}$, respectively (Q_{max} being the maximum applied tangential load at sliding inception). As can be clearly seen from these sequential images the number of indentation dots within the growing contact area increases. Moreover, the changing numbers of indentation dots within the contact area in Figures 2(b)–2(d) show very little junction growth up to $Q = 0.5Q_{max}$ (Fig. 2(c)), and much higher growth rate thereafter with a maximum growth rate at the leading edge of the contact (Fig. 2(e)) as was also observed in Ovcharenko et al. (2006) and predicted in Brizmer et al. (2007).

Finally, the non *in situ* photographed image in Figure 2(f) shows the sphere surface after complete removal of the external loads following sliding inception. The dashed circle duplicates the initial contact area A_0 under the normal load alone (see Fig. 2(b)) and the entire bright imprint of the contact area corresponds to the final contact area A_s at sliding inception shown in Figure 2(e). From the observed indentation dots included in the bright imprint outside the dashed circle it is obvious that the mechanism of junction growth comprises new points, from outside the initial contact area under normal load alone, that are coming into contact as the tangential load gradually increases. Figure 3 presents the experimental results of the dimensionless junction growth, A_s/A_0, at sliding inception vs. the dimensionless normal load, P^* for the entire range of copper sphere diameters. A_s and A_0 are the contact areas at sliding inception, and under the normal load alone, respectively. The dashed line is a best fit of the experimental results and the solid line is a theoretical prediction by Brizmer et al. (2007). At medium and high dimensionless normal loads, $P^* > 50$, the dimensionless junction growth, A_s/A_0, increases with increasing P^* and the experimental results correlate well with the theoretical ones.

While the model predicts a minimum for A_s/A_0 at $P^* \approx 50$ and an increasing dimensionless junction growth as the dimensionless normal load continues to decrease, the experiment shows monotonic reduction of the dimensionless junction growth with almost negligible growth close to

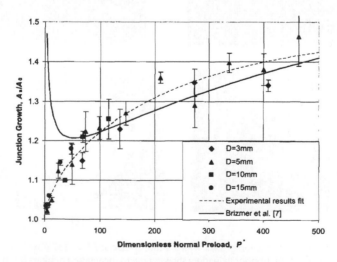

Figure 3.   Comparison of the experimental and theoretical results for the junction growth at sliding inception, A_s/A_0, as a function of the dimensionless normal preload P^* for different sphere diameters (Ovcharenko et al. (2008)).

the elastic regime of normal loads, $P^* \approx 1$. The difference between the theoretical and experimental results for $P^* < 50$ could be attributed to the surface roughness of the specimens, since the presence of surface asperities can drastically reduce the real contact area compared to that in the ideal model that assumes smooth sphere surface. Indeed, in a recent model by Cohen et al. (2009) where the sphere roughness was considered, a much better correlation with the complete experimental results by Ovcharenko et al. (2008) for $P^* < 50$ was obtained. It seems that $P^* = 50$ is a threshold load above which the asperities of the rough surface are flattened to the point where the smooth sphere surface assumption is more realistic.

4 CONCLUSION

A special marking technique was used to understand the mechanism of junction growth. No radial expansion of the initial contact area formed by the normal preload was observed but the contact area increase is due to new points of the sphere, outside the initial contact area, that are coming into contact with the flat.

Good correlation was found between the experimental results and the theoretical model of Brizmer et al. (2007) for the junction growth at sliding inception for normal preloads $P^* > 50$.

REFERENCES

Brizmer, V., Kligerman, Y. & Etsion, I. 2006a. The effect of contact conditions and material properties on the elasticity terminus of a spherical contact. *Int. J. of Solids and Structures* 43: 5736–5749.

Brizmer, V., Zait, Y., Kligerman, Y. & Etsion, I. 2006b. The effect of contact conditions and material properties on elastic-plastic spherical contact. *J. of Mechanics of Materials and Structures* 1: 865–879.

Brizmer, V., Kligerman, Y. & Etsion, I. 2007. A model for junction growth of a spherical contact. *J. Tribol. Trans. ASME* 129: 783–790.

Cohen, D., Kligerman, Y. & Etsion, I. 2009. The effect of surface roughness on static friction and junction growth of an elastic-plastic spherical contact. *J. Tribol. Trans. ASME* 131: 021404.

Kayaba, T. & Kato, K. 1978. Experimental analysis of junction growth with a junction model. *Wear* 51: 105–116.

McFarlane, J.S. & Tabor, D. 1950. Relation between friction and adhesion. *Proc. Roy. Soc. London A* 202: 244–253.

Ovcharenko, A., Halperin, G., Etsion, I. & Varenberg, M. 2006. A novel test rig for in situ and real time optical measurement of the contact area evolution during pre-sliding of a spherical contact. *Tribology Letters* 23: 55–63.

Ovcharenko, A., Halperin, G. & Etsion, I. 2008. In situ and real-time optical investigation of junction growth in spherical elastic-plastic contact. *Wear* 264: 1043–1050.

Parker, R.C. & Hatch, D. 1950. The static coefficient of friction and the area of contact. *Proc. Phys. Soc. London B* 63: 185–197.

Tabor, D. 1959. Junction growth in metallic friction: the role of combined stresses and surface contamination. *Proc. Roy. Soc. London A* 251: 378–393.

Nanoseismic measurement of the localized initiation of sliding friction

Gregory McLaskey & Steven D. Glaser

Department of Civil and Environmental Engineering, University of California, Berkeley, CA, USA

ABSTRACT: This paper presents theoretical and experimental results of nanosesimic waves generated from frictional sliding and recorded by an array of high-fidelity Glaser-NIST nanoseismic sensors. Absolute sensor calibration procedures are reported along with some preliminary experimental findings. Estimating seismic source characteristics given an array of recorded signals and the specimen's Green's function is an inverse problem formulation that has been successfully used for the study of earthquake source mechanisms, and to distinguish between earthquakes and underground explosions. When properly scaled, the study of nanoseismic sources can shed light on frictional behaviors at both very small and earthquake fault scales. The preliminary experiments, from a PMMA slider on a large PMMA plate, demonstrate that local force release on the friction surface can be detected on a number of different time and amplitude scales. For rapid load releases (on the order of 500 ns) the source location can be estimated to a few mm in space and μs in time for step-like source functions down to about 10 mN in amplitude. Load releases which occur less rapidly in time (on the order of 100 μs) or over a larger spatial area (larger than a few mm) are more difficult to identify and interpret. Absolute calibration on a PMMA plate shows that the sensors have a sensitivity of 15 mV/nm $\pm$4 dB over a frequency range of approximately 20 kHz to 1 MHz. The calibrated noise floor for the sensors is $\pm$15 pm.

1   INTRODUCTION

From the wear of our teeth, to earthquake hazards, to new nano-machines which require an understanding of friction at the atomic scale, the effects of friction are far reaching. At all scales, most of the mechanical energy sapped by friction is converted into thermal phonons at THz frequencies (e.g. Sevinçli 2008, Krim et al., 1991), but some small portion is converted to seismic waves which travel through the material and are manifested as the screech of brakes, the squeal of sliding rubber, earthquake ground motions, and nanoseismic vibrations. We have devised a laboratory-based nanoseismic measurement and interpretation technique which allows us to measure elastic waves radiated from load releases due to slippage on the micro- and possibly nano-scale, and the results may be scaled up to shed light on earthquake-scale sliding.

A number of different models of friction, describing a large range of length scales are illustrated in Figure 1. On one hand, a mechanistic study of friction will by necessity reduce to understanding inter-atomic forces and motions on the order of nm. On the other hand, the behavior of earthquake faults is measured over length scales of km, and at this scale (the macro-scale) all the way down to a human scale (the meso-scale), frictional force is defined by the so-called Coulomb (1779) friction model, the behavior of which has been directly observed since Da Vinci. As the micro-scale is approached, the friction model of Bowden & Tabor (1954) is more suitable. Their applications were lubricated shafts and cutters, gears and connecting arms. They posited quasi-static junctions, with an average area of contact, and a wear-free interface. The Bowden-Tabor model, however, cannot account for micro-mechanical behaviors and does not predict thousands of recent micro- and nano-scale measurements. For example, behaviors such as cold welding, plastic deformation (Muser 2001), and adsorbed sub-monolayers of so-called third species from the environment, e.g., oxidation, can control junction shear resistance (e.g. Krim, 1991). The junction behaviors scaling up from the nano (atomic bonds) and scaling down from the micro should at some point agree, yet few researchers even address this issue.

Apart from obvious differences in length scale, there is a phenomenologically self-similar behavior amongst many different scales. For example, a common atomistic model is the

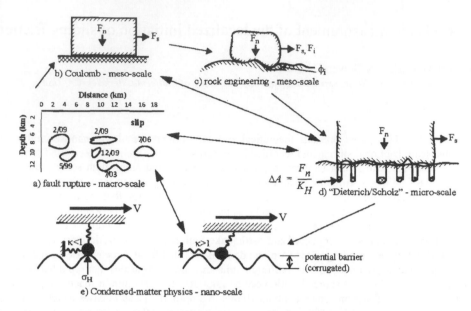

Figure 1.   Models of multi-scale sliding friction. Macro- and meso scale models are behavioral criteria that accurately predict behavior on that scale. The micro-scale model posits a physical mechanism.

Frenkel-Kontorova-Tomlinson model (Weiss & Elmer 1997), Fig 2a, where atoms in a body are considered as point bodies with spring-like inter-atomic bonds connecting to other atoms in all six Cartesian directions. On the other hand, a common seismological model is the Burridge-Knopoff (1967) model, Fig. 2b, where coherent bodily lengths are modeled as sliding blocks connected to other blocks by springs in all six Cartesian directions. The pinning of junctions on different scales have different apparent physical mechanisms but serve the same purpose—minimum energy attractors (Weiss & Elmer 1997). The physical scale at which the stick-slip transition appears to takes place gets smaller and smaller as instrumentation improves.

While a small specimen may appear to respond to sliding as a single block, slip on an earthquake fault cannot happen simultaneously along a km-scale length—stress information cannot move through the earth faster than the speed of sound. The fault must be broken down into multiple sections such as in the Burridge-Knopoff model. Very large sliding bodies such as faults are too large to be treated as a point mass. This limit is defined by the elastic coherence length, ξ. During slow-velocity sliding, only volume elements of dimensions about ξ will displace coherently (Persson 1998). The coherent length ξ becomes smaller as normal stress on the joint becomes larger or the materials become more compliant, and the multiple volume elements making up a sliding block can have individual stick-slip motions. This theory can explain much of earthquake fault mechanics, in particular questions brought up about self-healing slip and the foam rubber models (e.g. Hartzell & Archuleta 1979, Perrin et al., 1995, Heaton, 1990).

The local stress change at a breaking junction travels through the material as a stress wave. If all the atomic contacts making up a junction are slipping fast enough and with sufficient coherency, high frequency (well into the MHz range) stress waves will radiate away from the junction and through the material at the speed of sound. These nanoseismic events carry the time stress history of the junction rupture dynamics. Very careful recording and analysis of these signals can provide a glimpse of the in situ junction dynamics for a meaningful situation, providing an acoustic microscope of exquisite sensitivity—picometer displacements over a frequency bandwidth of 20 kHz to over 1 MHz. This paper reports on the initial sliding friction tests using our high-fidelity sensors and inversion techniques.

With the techniques outlined in this paper, we can measure changes in the local force field at the location of an individual breaking junction. Careful application of this toolset can provide new

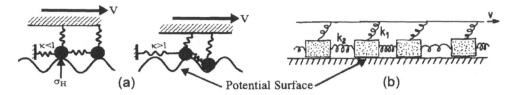

Figure 2. Condensed-matter physics and earthquake fault models of frictional interaction. A single contact junction is shown on the left, and multiple junction model is shown on the right -the Burridge-Knopoff model of 1967 (from Persson 1998).

insights into dynamic behaviors at the micro-scale and sub-micro "disconnect" scale which exists between micro- and nano-scales. These insights, in turn, may also pertain to the macro-scale since the effective length scale of experiments performed in the lab can be adjusted by scaling material properties, and normal loads. Brune (1973) modeled km of faults with cm of foam rubber.

2   BACKGROUND AND PROBLEM FORMULATION

The study of waves propagating through solids has its beginnings in the theoretical work dating back to the 19th century (e.g. Rayleigh 1887, Lamb 1904 , Love 1934), and more recent experimental work has confirmed theoretical findings (e.g. Aki & Richards 1980, White 1965). The elastodynamics of stresses and strains which govern elastic wave propagation in a solid body can be described by Navier's equation, which can be derived (Graff 1975) by substituting the appropriate definition for strain and the stress-strain relationship into the stress equations of motion:

$$\rho\frac{\partial^2 u_i}{\partial t^2} = f_i + (C_{ijkl}(u_{k,l} + u_{l,k})/2)_j \qquad (1)$$

where u_i is the displacement vector of a material point, ρ is the density of the material, C_{ijkl} is the fourth order elastic strain tensor, and f_i is a body force applied to the material. For this work, we seek solutions in the form of a Greens function $G_{in}(x, t, \xi, \tau)$, which is the displacement at point x at time t due to a unit impulse at location ξ in the direction n at time τ (Aki & Richards 1980). The Green's function is a symmetric second order tensor which must satisfy Equation 1, specifically:

$$\rho\frac{\partial^2 G_{in}}{\partial t^2} = \delta_{in}\delta(\mathbf{x} - \xi)\delta(t - \tau) + (C_{ijkl}G_{kn,l})_j \qquad (2)$$

The Green's function can be thought of as the impulse response function of an elastic material that maps the dynamic force field at location ξ to a mechanical disturbance u_i at the sensor site x. Once the Green's function is known, the material response to an arbitrary force time function can be calculated via convolution.

Under the definition of the Green's function, the displacement at the transducer location can be expressed as

$$u_i(\mathbf{x}, t) = \int\limits_{-\infty}^{\infty} \iiint\limits_V G_{in}(\mathbf{x}, t; \xi, \tau)f_n(\xi, \tau)dV dt \qquad (3)$$

where V is the source volume which contains non-zero portions of f_n. If the source volume is replaced by a point ξ^0, the Green's function can be expanded in a Taylor series about this point

(Stump & Johnson 1977). By taking only the first two terms of this series, we can write

$$u_i(\mathbf{x}, t) = G_{in}(\mathbf{x}, t; \xi^0, \tau) * F_n(\xi^0, \tau) + G_{in,j}(\mathbf{x}, t; \xi^0, \tau) * M_{nj}(\xi^0, \tau) \tag{4}$$

where $*$ represents convolution in time, the force vector

$$F_n(\xi^0, \tau) = \int_V f_n(\xi, \tau) dV \tag{5}$$

and the force moment tensor

$$M_{nj}(\xi^0, \tau) = \int_V x_j f_n(\xi, \tau) dV \tag{6}$$

Our objective is to represent the source using the vector-valued function $F_n(\tau)$ and the tensor valued function $M_{nj}(\tau)$ at the location ξ^w of a locked asperity, and to relate this to source characteristics and the mechanisms and dynamics of friction.

Following a transfer function calibration approach (fully described in Hsu & Breckenridge 1981), the transducer output $v(t)$ can be expressed as the linear convolution of the surface displacement and the transducer's instrument response function:

$$v(t) = u_i(\mathbf{x}, t) * i(t) \tag{7}$$

By inserting Equation 4 into Equation 7 and applying a Fourier transform to the result, we find

$$V(\omega) = I(\omega)[F_3(\xi^0, \varpi)G_{in}(\mathbf{x}, \omega; \xi^0, \varpi) + G_{in,j}(\mathbf{x}, \omega; \xi^0, \varpi)M_{nj}(\xi^0, \varpi)] \tag{8}$$

and the complex transfer function of the transducer can be found by inversion:

$$I(\omega) = V(\omega)u_i^{x_0}(\omega)^{-1} \tag{9}$$

Equation 9 can be reformulated into the form

$$\mathbf{V}(\omega) = \mathbf{I}(\omega)\mathbf{M}(\xi^0, \varpi)\mathbf{G}(\mathbf{x}, \omega; \xi^0, \varpi) \tag{10}$$

where $\mathbf{V}(\omega)$ is a vector which includes the real and imaginary components of the voltage output at a particular frequency ω and has dimensions $2n$ for the case of an array of n sensors, $\mathbf{I}(\omega)$ is a $2n$ by $2n$ diagonal matrix whose elements are the real and imaginary components of the instrument response function for sensor n at frequency ω, $\mathbf{M}(\xi^w, \omega)$ is a vector of length $2m$ which contains the real and imaginary components of the force vector and moment tensor elements which are to be determined, and $\mathbf{G}(x, \omega\xi^0, \varpi)$ is a $2n$ by $2m$ matrix composed of the real and imaginary components of the appropriate Green's function for sensor n and force/moment element m at frequency ω (see also Stump & Johnson 1977).

Under this formulation, the m elements of the force vector and moment tensor at each Fourier frequency can be found by inversion given the voltage outputs from an array of n sensors, the instrument response function for each sensor, and the appropriate Green's function for each combination of the force/moment elements and sensor locations.

A number of strategies may be employed to solve for the Green's functions either analytically (typically with the use of Fourier and Laplace transforms) or numerically. Closed form solutions currently exist for the whole space (e.g. White 1965, Aki & Richards 1980), half space (Pekeris 1955), and infinite plate (Johnson 1974). For this work, a solution scheme known as generalized ray theory has been used for an infinite plate geometry (Helmberger 1974, Ceranoglu & Pao 1981). This solution was checked against the theoretical solutions of Knopoff (1958) and Pekeris (1955).

3 SENSOR ARRAY

A number of different types of sensors have been employed for the detection of small-amplitude high-frequency surface displacements, such as those expected to result from frictional sources. These sensors include those developed for ultrasonic, microseismic, and acoustic emission applications. Following the solution strategy described in the previous section, absolute measurement of surface displacements at an array of sensor locations is required, and it is also desirable to use a sensor which has excellent sensitivity, a wide bandwidth, and near-flat frequency response. While much of the quantitative experimental work to date has made use of capacitive (e.g. Breckenridge et al., 1975, Kim & Sachse 1986, Hsu & Hardy 1978) or optical (Eisenhardt et al., 1999, Scruby & Drain 1990) transducers, the sensitivity requirements warrant the use of piezoelectric devices. Most piezoelectric sensors take advantage of some mechanical resonance to gain high sensitivity at the expense of loss of bandwidth and signal distortion, but a piezoelectric sensor with a large backing mass and conical piezoelectric element was designed in the late 1970s to provide a more faithful transduction of surface displacement (Proctor 1982, Greenspan 1987). This sensor is reported to have an extremely flat amplitude response between 100 kHz and 1 MHz. The Glaser-NIST transducers employed for this study are based on this original concept but were further developed in our laboratory. (For more details see Glaser et al., 1998, To & Glaser 2005, McLaskey et al., 2007.)

3.1 *Sensor calibration*

The sensor transfer function calibration approach used for this work follows the traditional experimental design carried out by acoustic emission researchers at the NBS in the late 1970s (Breckenridge et al., 1975). In a typical calibration of this type, a number of assumptions are made whereby it is assumed that the transducer response function $i(t)$ can be modeled as a linear time-invariant system which maps a mechanical disturbance $u_i(\mathbf{x}, t)$ to a transducer output $v(t)$ (Hsu & Breckenridge 1981). This system's transfer function can be estimated in the frequency domain from Equation 9 by directly dividing the complex-valued spectral estimates of the sensor output by those of the input (surface displacements at the location of the sensor).

To characterize the sensor in this manner, the exact mechanical disturbance (i.e. the displacement $u_i(\mathbf{x}, t)$ of the surface of the specimen which would exist in the absence of the sensor) must be known. This is found theoretically by using a "known" source and a test block for which the Green's function is known. The test block used in these calibration experiments was a massive, polymethylmethacrylate (PMMA) plate 50 mm thick and 940 mm square. The sensor location was chosen to be in the center of the plate directly beneath the source location, on the opposite side of the plate. Traveling at the compressional wave velocity in this material (measured to be 2.8 km/s), it takes approximately 330 μs for side reflections to return to the center of the plate where the source and sensor are located. In this time period, the plate can be treated as infinite and the Green's functions found from the infinite plate geometry can be applied to solve Equations 4 and 10.

The two "known" nanoseismic sources used for calibration purposes in this study are the fracture of a glass capillary tube and a ball impact. The capillary fracture is known to present a force time function into the test block which is very nearly equal to a step function with a rise time of less than 200 ns (Breckenridge et al., 1975). This source has been used by many researchers because the force at which the fracture occurs (and therefore the amplitude of the step) can be independently measured. The less-frequently used ball impact is known to introduce an impulse-like pulse into the material, and the change in momentum that the ball imparts to the transfer block can be independently measured. Starting with the work of Hertz (1882), a great deal of theoretical and experimental work concerning the collision of a sphere on a massive body has shown that the amplitude and frequency content of this pulse are functions of the ball size and impact velocity (Goldsmith 2001, Johnson 1985). Both of these sources are assumed to apply forces only in the direction normal to the surface of the test block.

The sensor output (experiment) and surface normal displacements (theory) due to a 0.4 mm diameter ruby ball dropped 0.325 m onto a 50 mm thick PMMA plate are shown in Figure 3(a). The sensor is located directly beneath the location of impact, on the opposite side of the plate. The theoretical displacements are found from the convolution of the Green's function (G_{33}, where direction 3 is the direction normal to the plate) and appropriate pulse found from Hertz theory (400 mN tall and 2.8 μs wide). The amplitude of the Fourier transform of these two time series is shown in Figure 3(b) compared to the amplitude of the Fourier transform of a pure noise signal.

The same comparison (sensor output and normal displacements) are shown in Figure 4 for the case of a 0.250 mm diameter glass capillary tube loaded on its side and fracturing under a load of 9 N. The sensor is again located directly beneath the location of impact, and the displacements are found from the convolution of the same Green's function as in Figure 3 with a step function of amplitude 9 N and rise time of 100 ns.

Following Equation 9, the instrument response function (or rather the complex transfer function of the sensor) can be found by dividing the Fourier transform of the sensor output (experiment) by the Fourier Transform of the normal displacements (theory). The amplitude (a) and phase (b) of the complex transfer function coefficients are plotted in Figure 5. This calibration shows that the sensor response is flat in amplitude to ±4 dB between 20 and 1000 kHz and nearly zero phase in the same frequency range. This transducer has sensitivity of about 15 mV/nm and the noise level is about ±15 pm. Both the ball drop and capillary fractures yield the same transfer function to within a few dB. This match validates both the step function capillary fracture model and the Hertzian impact model and illustrates the robustness of this calibration approach.

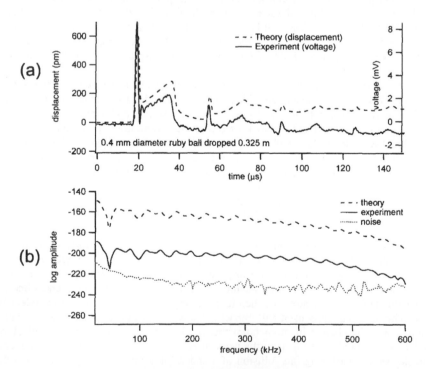

Figure 3.   The time series, (a), and amplitude of the Fourier transform, (b), of the sensor output (experiment) and surface normal displacements (theory) due to a 0.4 mm diameter ruby ball dropped 0.325 m onto a 50 mm thick PMMA plate.

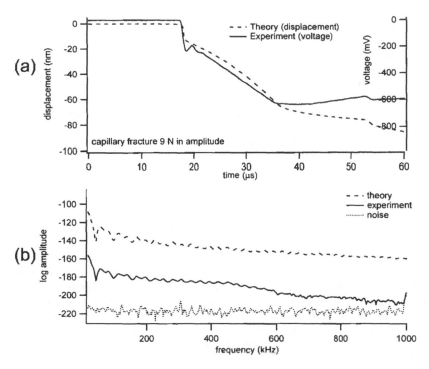

Figure 4.   The time series, (a), and amplitude of the Fourier transform, (b), of the sensor output (experiment) and surface normal displacements (theory) due to the fracture of a 0.25 mm diameter glass capillary tube on a 50 mm thick PMMA plate. The capillary tube was loaded on its side and fractured at a load of 9 N.

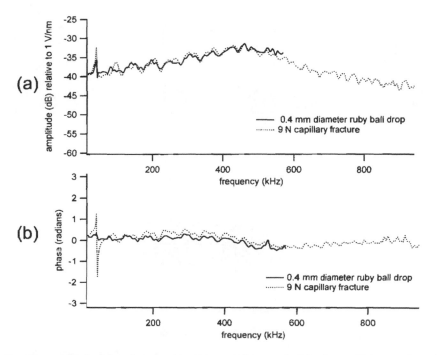

Figure 5.   The amplitude, (a), and phase, (b), of the complex transfer function coefficients estimated from both the ball impact and capillary fracture sources.

4   EXPERIMENTAL SETUP

A set of experiments were performed in order to validate the effectiveness of the solution strategy outlined in Section 2. While the results are entirely preliminary in nature, their existence demonstrates the feasibility of the method, and they highlight some of the strengths and limitations of the approach.

The experimental setup, depicted in Figure 6, includes the 50 mm thick PMMA base plate (a), a slider block (b) (dimensions 120 mm by 12 mm by 50 mm in the 1, 2, and 3 directions, respectively), an array of Glaser-NIST sensors (c) located on the bottom side of the base plate, and sensor (d) which measures the displacement of the back end of the slider block relative to the base plate. The slider block is loaded vertically with a normal force on the order of 35 N and then loaded horizontally at the back end of the block as shown by the arrow in Figure 6. When slip occurs between the slider and base, the local shear forces overcome the locked asperities on the friction surface. This force release, depicted at the location (e) of the previously locked asperity, causes elastic waves (f) to propagate through the base plate and slider block until they are detected by the array of sensors.

With a plate thickness of 50 mm, a 1 N step force will produce a P-wave amplitude of about 1.5 nm at the location of sensor 1, therefore with a 15 pm noise level the smallest detectable step force will be about 10 mN. The location e of the source of the detected waves can be found from triangulation based on the differences of arrival times of various wave phases arriving at the known locations of each sensor in the array (Salamon & Wiebols 1974, Baron & Ying, 1987, Stein & Wysession 2003). Once the source location is found, Green's functions can be calculated for each sensor-source pair, and the components of the force vector $F_n(\tau)$ and moment tensor $M_{nj}(\tau)$ can be solved using Equation 10.

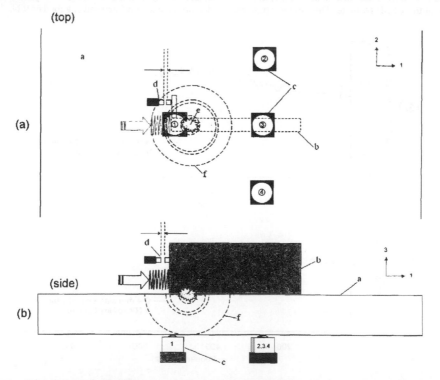

Figure 6.   Schematic of the experimental setup: a, base plate; b, slider block; c, Glaser-NIST sensors; d, displacement sensor; e, example friction source location; f, radiated transient elastic waves.

PMMA was chosen for both the slider and base materials because of its attenuative properties. Elastic waves above about 50 kHz propagating through this material will be highly attenuated after they have propagated more than a few hundreds of mm. This attenuation is great enough that all high frequency wave arrivals (greater than 50 kHz) felt by the sensors can be assumed to be direct arrivals from the frictional interface or nearby region. This keeps the wavefield free from high frequency reflections from the outside edges of the plate and allows the easy identification of high-frequency, short-wavelength elastic wave arrivals.

5   PRELIMINARY RESULTS

The results shown in Figure 7 were obtained following the experimental procedure described in the previous section. These sensor outputs were obtained at the onset of frictional sliding and are offset for clarity. The curve labeled 'displacement' and shown in grey is a plot of the output of the displacement sensor (object (d) from Fig. 6). The other four traces are from an array of Glaser-NIST sensors whose general locations are depicted in Figure 6. Two distinct wave arrivals can be seen from this plot, one at a approximately 2.5 ms and the other at about 4.88 ms. These two wave arrivals, plotted in Figures 6 and 7 at different time scales, are illustrative examples of two different and reoccurring types of phenomena observed during these preliminary tests. The first arrival to occur, depicted in Figure 8, is felt by all four sensors, but appears to be a low frequency arrival with a rise time on the order of 100 μs. For this event, the P- and S-wave arrivals (which should be

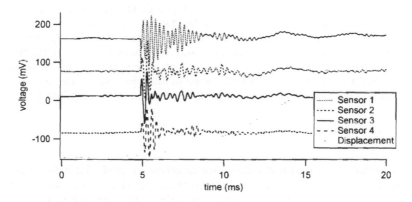

Figure 7.   Sensor outputs at the onset of frictional sliding (offset for clarity).

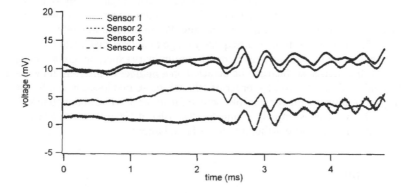

Figure 8.   Sensor outputs due to a low frequency wave arrival with a rise time on the order of 100 μs.

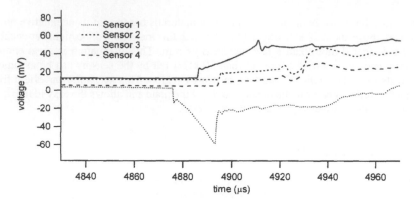

Figure 9.   Sensor outputs due to a more sudden, high frequency wave arrival with a rise time on the order of 500 ns.

separated by about 20–30 μs for the source-sensor distances of this test setup) are smeared together by a long temporal duration of the source function $F_n(\tau)$ and/or $M_{nj}(\tau)$.

The second, larger-amplitude wave arrival, depicted in Figure 9, is from a more sudden stress drop, with a rise time on the order of 500 ns. This wave arrival is of short enough duration that the P-wave and S-wave can be seen as distinct arrivals at about 4875 μs and 4892 μs, respectively, for Sensor 1. A head wave arrival can be seen on the traces from Sensors 2–4. Whereas the signals collected from the first, slower event shown in Figure 8 are difficult to interpret, the signals from the later, more rapid event contain a wealth of readily available information. For example, the location of the source of these waves was found to be of small spatial extend (less than a few mm) and located at the back end of the slider such as (e) of Figure 6. Additionally, the signals look very similar to the response to a step source, and based on the amplitude of the identifiable waves, this source is estimated to be approximately 200–400 mN in amplitude, which represents a small fraction (~3%) of the total applied shear load. Events of this type were detected on many different tests, even down to the smallest detectable amplitudes (about 10 mN in amplitude).

6   CONCLUSIONS

This work demonstrates that nanoseismic source inversion, coupled with high-fidelity sensing, can be used as a tool to study details of sliding friction, especially at the initiation of sliding. Observations were made as to phenomological similarities of frictional behaviors from the nano- to macro-scale. Solutions to the full elastodynamic equations of motion are presented in the form of a Green's function. Absolute calibration of the Glaser-NIST high-fidelity sensors used for the experiments was performed using the computed Green's functions and the known source functions from both a ball impact and a glass capillary fracture. When coupled to PMMA, the sensors have a sensitivity of 15 mV/nm ±4 dB over a frequency range of approximately 20 kHz to 1 MHz. The calibrated noise floor for the sensors is ±15 pm. The preliminary experiments demonstrate that local force release on the friction surface can be detected on a number of different time and amplitude scales. For rapid load releases (on the order of 500 ns) the source location can be estimated to a few mm in space and μs in time for step-like source functions down to about 10 mN in amplitude. Load releases which occur less rapidly in time (on the order of 100 μs) or over a larger spatial area (larger than a few mm) are more difficult to identify and interpret.

AKNOWLEDGEMENTS

This work was funded by NSF-GRF and NSF grant CMS-0624985.

REFERENCES

Aki, K., and Richards, P.G. 1980. *Quantitative Seismology: Theory and Methods*. Freeman: San Francisco.

Baron, J., and Ying, S. 1987. Acoustic Emission Source Location. In R. Miller, & P. McIntire (eds), *Nondestructive Testing Handbook Second Edition Vol. 5: Acoustic Emission Testing*, 136–154, American Society for Nondestructive Testing.

Bowden, F., and Tabor, D. 1950. *The Friction and Lubrication of Solids* Clarendon Press, Oxford.

Breckenridge, F., Tscheigg, C., and Greenspan, M. 1975. Acoustic emission: some applications of Lamb's Problem, *J. Acoustical Soc. Am.* 57(3), 626–631.

Brune, J.N. 1973. Earthquake modeling by stick-slip along precut surfaces in stressed foam rubber. *Bulliten of the Seismological Society of America*, 63, 2105–2119.

Burridge, R., and Knopoff, L. 1967. Model and theoretical seismicity. *Bulletin of the Seismological Society of America*, 57(3), 341–371.

Ceranoglu, A., and Pao, Y. 1981. Propagation of elastic pulses and acoustic emission in a plate. *Journal of Applied Mechanics: Transactions of the ASME* 48, 125–147.

Coulomb, C.A. 1779. *Theorie des Machines Simples. Librairie Scientifique et Technique*, Albert Blanchard: Paris.

Eisenhardt, C., Jacobs, L.J., and Qu, J. 1999. Application of laser ultrasonics to develop dispersion curves for elastic plates. *Journal of Applied Mechanics*. 66(4), 1043–1045.

Glaser, S., Weiss, G., and Johnson, L. 1998. Body Waves Recorded Inside an Elastic Half-space by an Embedded, Wideband Velocity Sensor. *Journal of Acoustic Society of America*, 104, 1404–1412.

Goldsmith, W. 2001. *Impact*, Dover Publications: New York.

Graff, K. 1975. *Wave Motion in Elastic Solids*. Oxford University Press: Mineola, NY.

Greenspan, M. 1987. The NBS conical transdcuer: analysis. *J. Acoust. Soc. Of America*. 81(1), 173–183.

Hartzell, S.H., and Archuleta, R.J. 1979. Rupture propagation and focusing of energy in a foam rubber model of a stick slip earthquake. *Journal of Geophysical Research*. 84(B7), 3623–3636.

Heaton, T. 1990. Evidence for and implications of self-healing pulses of slip in earthquake rupture. *Physics of the Earth and Planetary Interiors*. 64, 1–20.

Helmberger, D. 1974. Generalized ray theory for shear dislocations. *Bulletin of the Seismological Society of America*. 64(1), 45–64.

Hertz, H. 1882. Über die Berü hrung fester elastischer Körper. *J. Reine Angew. Mat.* 92, 156–171.

Hsu, N., and Breckenridge, F. Characterization of acoustic emission sensors. *Materials Evaluation*, 39, 60–68.

Hsu, N. & Hardy, S. 1978. Experiments in Acoustic Emission Waveform Analysis for Characterization of AE Sources, Sensors and Structures. *American Society of Mechanical Engineers, Applied Mechanics Division, AMD*, 85–106.

Johnson, K. 1985. *Contact Mechanics*. Cambridge University Press: Cambridge.

Johnson, L. 1974. Green's Function for Lamb's Problem. *Geophys. J. R. astro. Soc.* 37, 99–131.

Kim, K.Y., and Sachse, W. 1986. Characteristics of an acoustic emission source from a thermal crack in glass. *International Journal of Fracture*. 31, 211–231.

Knopoff, L. 1958. "Surface motions of a thick plate. *J. of Applied Physics*. 29(4), 661–670.

Krim, J., Solina, D., and Chiarello, R. 1991. Nanotribology of a Kr monolayer: a quartz crystal microbalance study of atomic-scale friction. *Physical Review Letters*, 66, 181–184.

Lamb, H. 1904. On the Propagation of Tremors over the Surface of an Elastic Solid. *Philosophical Transactions of the Royal Society of London A.* 203, 1–42.

Love, A. 1934. *A treatise on the mathematical theory of elasticity*. Cambridge Univeristy Press: London.

McLaskey, G., Glaser, S., and Grosse, C. 2007. Integrating broad-band high-fidelity acoustic emission sensors and array processing to study drying shrinkage cracking in concrete. *Proc. of SPIE* 6529, San Diego, Mar. 18–20.

Muser, M.H. 2001. Dry friction between flat surfaces: multistable elasticity vs. material transfer and plastic deformation. *Tribology Letters*. 10(1–2), 15–22.

Pekeris, C.L., 1955. The Seismic Surface Pulse. *Proceedings of the National Academy of Sciences* 41, 469–480.

Perrin, G., Rice, J.R., and Zheng, G. 1995. Self-healing slip pulse on a frictional surface. *Journal of the Mechanics and Physics of Solids*. 43, 1461–1495.

Persson, B. 1998. *Sliding Friction*. Springer: Berlin.

Proctor, T.M. 1982. An improved piezoelectric acoustic emission transducer. *Journal of Acoustic Society of America*. 71, 1163–1168.

Rayleigh, Lord, 1887. On waves propagating along the plane surface of an elastic solid. *Proc. London Mathematical Society*. 17, 4–11.

Salamon, M.D.G., and Wiebols, G.A. 1974. Digital location of seismic events by an underground network of seismometers using the arrival times of compressional waves. *Rock Mechanics*. 6, 141–166.

Scruby, C., and Drain, L. 1990. *Laser ultrasonics: techniques and applications*, CRC Press.

Sevinçli, H.,. Mukhopadhyay, S., Senger, R.T., and Ciraci, S. 2007. Dynamics of phononic dissipation at the atomic scale: dependence on internal degrees of freedom. *Physical Review B* 76(20), 205430.

Stein, S., and Wysession, M. 2003. *An introduction to seismology, earthquakes, and earth structure*, Blackwell: Massachusetts.

Stump, B., and Johnson, L. 1977. The determination of source properties by the linear inversion of seismograms. *Bulletin of the Seismological Society of America*. 67, 1489–1502.

To, A., and Glaser, S. 2005. Full waveform inversion of a 3-D source inside an artificial rock. *Journal of Sound and Vibration*. 285, 835–857.

Weiss, M., and Elmer, F.-J. 1997. Dry friction in the Frenkel–Kontorova–Tomlinson model: dynamical properties. *Zeitschrift Fur Physik B*. 104, 55–69.

White, J., 1965. Seismic Waves: Radiation, Transmission, and Attenuation, McGraw-Hill: New York.

Weakly nonlinear fracture mechanics: Experiments and theory

Eran Bouchbinder, Ariel Livne & Jay Fineberg
Racah Institute of Physics, Hebrew University of Jerusalem, Jerusalem, Israel

ABSTRACT: Material failure occurs at small scales in the immediate vicinity of the tip of a crack, in the so-called fracture process zone. Due to its generally microscopic size and high propagation velocity, the dynamical behavior of this elusive region has never before been directly observed. Here we present direct measurements of the deformation surrounding the tip of a dynamic mode I crack propagating in brittle elastomers. Both the detailed dynamics and fractography of these materials are identical to that of typical brittle amorphous materials such as soda-lime glass. We demonstrate that Linear Elastic Fracture Mechanics (LEFM) does not provide an accurate quantitative description of the measured deformation over fracture velocities ranging from 0.2 to $0.8c_s$. We therefore derive the leading nonlinear elastic corrections to the common LEFM asymptotic fields. The theory predicts $1/r$ displacement-gradients contributions, logarithmic corrections to the LEFM parabolic crack tip opening displacement and provides excellent quantitative agreement with the measured near-tip deformation. We further show that a dynamic lengthscale $\ell(v)$ arises naturally. The derived weakly nonlinear fracture mechanics may serve as a springboard for the development of a comprehensive theory of fracture dynamics.

1   INTRODUCTION

Understanding the dynamics of rapid cracks remains a major challenge in engineering, physics, material science and geophysics. For example, high velocity crack tip instabilities (Fineberg & Marder 1999, Livne et al. 2007) remain poorly understood from a fundamental point of view. Much of our understanding of how materials fail stems from Linear Elastic Fracture Mechanics (LEFM) (Freund 1998), which assumes that materials are *linearly* elastic outside of a small zone where all nonlinear and/or dissipative processes occur ("process zone"). A central facet of LEFM is that strains diverge as $r^{-1/2}$ at a crack's tip and that this singularity dominates all other strain contributions in this region. Linear elasticity should be expected to break down before dissipative processes occur. The small size and rapid propagation velocity of the near-tip region of brittle cracks have, however, rendered quantitative measurements of the near-tip fields elusive. As a result, under dynamic conditions the fundamental properties of the near tip region, where material separation is actually taking place, remain largely unknown. It is thus of prime importance to understand the basic physics needed to extend LEFM when the fracture process zone is approached.

Here we describe fracture measurements performed in polyacrylamide gels. These aqueous gels are amorphous, brittle and extremely compliant materials. Their high compliance is utilized to slow down the time scales of fracture processes by nearly 3 orders of magnitude. In achieving this, the study of the fracture dynamics of these gels provides a window into the complex processes near the fracture tip that are close to intractable in more "standard" experimental systems. We will first demonstrate that the Mode I fracture dynamics of these gels is identical to those of soda-lime glass and PMMA, materials that are commonly used in the study of brittle amorphous fracture. We then describe measurements of the structure of the near-tip region of a dynamic crack in gel fracture that would have been extremely difficult to observe in other materials. We demonstrate that LEFM does not provide an accurate quantitative description of the measured deformation fields in the near-tip region over fracture velocities ranging between 0.2 to $0.8c_s$.

To address these discrepancies, we derive the leading nonlinear elastic corrections to the common LEFM asymptotic fields by taking into account nonlinear corrections to linear elasticity, which *must* be relevant near the crack tip. This is achieved by perturbatively expanding the momentum balance equation for an elastic medium up to second order nonlinearities in the displacement gradients. The theory provides excellent quantitative agreement with the measured, near-tip, displacement and strain fields. We further show that no region of $r^{-1/2}$ dominance exists in this example, that "more-divergent" strain terms occur at a finite distance from the tip and that there exist logarithmic corrections to the parabolic crack tip opening displacement. In addition, a dynamical length-scale, associated with a nonlinear elastic zone, appears naturally. The theory may serve as a springboard for the development of a comprehensive theory of fracture dynamics.

2   THE FRACTURE OF POLYACRYLAMIDE GELS

2.1   *Experimental methods*

Detailed experimental study of fracture in the near-tip region is extremely difficult in "standard" brittle materials. In materials commonly used in experimental studies, fracture velocities are on the order of c_R, the Rayleigh wave speeds of the material (e.g. $c_R = 3440$ m/s in soda-lime

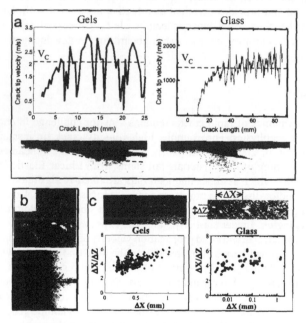

Figure 1.   Characteristic features of dynamic fracture are identical in both polyacrylamide gels and "standard" brittle materials, such as soda-lime glass (Livne et al. 2005). (a) The dynamics of two typical cracks in polyacrylamide gels (upper left) and soda-lime glass (upper right). In both materials cracks accelerate smoothly until the onset of the micro-branching instability at a velocity of $V_C = 0.4c_R$, where $c_R = 3340$ m/s in glass and 5.3 m/s in the gel used here. Beyond V_C, the crack velocity fluctuates as the single crack state undergoes a transition to a multi-crack state where the main crack coexists with microscopic branched cracks. Photographs of single micro-branches (in the XY plane) that extend beneath the fracture surface for gels (lower left) and glass (lower right). Micro-branches are identical in both materials with the same power-law form. (b) Photographs of the fracture surface of a gel (upper) and glass (lower) formed by cracks propagating from left to right at $\sim$0.5c_R. Chains of successive micro-branches are formed along the propagation direction (X) which are highly localized in the direction (Z) normal to propagation. Note also the existence of V-shaped markings on the fracture surfaces in both cases. These tracks were created by nonlinear perturbations propagating along the crack front, i.e. "front waves" (Livne et al. 2005). The perturbations are generated by the successive micro-branching events. (c) (top) "chains" of micro-branches (each "bubble" is a micro-branching site on the fracture surface) in gels (left) and glass (right). The spacing, Δx, between successive micro-branches scales with their width, Δz. The scaling is identical in both materials.

glass), while the process zone size is microscopic (~1 nm in glass). The coupling of these large velocities and extremely small scales makes the practical study of this region insurmountable. Thus, experimental study of this region has been limited to static or quasi-static cracks in extremely slow fracture (Bonamy et al. 2006). These problems can be surmounted by the use of polyacrylamide gels. In these soft materials the value of c_R is reduced by orders of magnitude, since the elastic constants in gels are 4–5 orders of magnitude lower than in standard brittle materials. By slowing down the fracture process to speeds amenable to study by means of fast cameras, the use of these gels enables us to study fracture dynamics with unprecedented spatial and temporal resolution.

The brittle gels used in these experiments are the transparent, neo-Hookean, incompressible elastomers that were used in (Livne et al. 2007). They are composed of 13.8% (weight/volume) total monomer and 2.6% bis-acrylamide cross-linker concentrations. The shear ($\mu = 35.2$ kPa) and Young's ($E = 3\,\mu$) moduli of these gels yield shear and dilatational wave speeds of $c_s = 5.90$ m/s and $c_d = 11.8$ m/s. The gels were cast between two flat glass plates. The dimensions of the samples used ranged from $200 \times 200 \times 2$ mm to $120 \times 120 \times 0.2$ mm in the X (propagation), Y (loading) and Z (thickness) directions. The face of one plate was randomly scratched with lapping powder. These scratches, of 16 μm mean depth, were imprinted on one of the gel faces. The resulting scratch pattern was used as a "tracer field" for visualization of the displacement field and did not affect the crack dynamics. The samples were fractured in Mode I, where samples were loaded under constant displacement (fixed grip) conditions. During crack propagation, the location of the crack tip was determined directly from the high-speed images acquired by a fast CMOS camera at frame rates up to 5000 frames/sec.

2.2 *Universality of brittle fracture in gels*

Although it is apparent that we can significantly reduce the Rayleigh wave speed c_R in poly-acrylamide gels, we must still address the question of whether the dynamics of fracture in these materials are at all similar to those observed in common brittle materials. The answer was provided in a series of recent experiments (Livne et al. 2005) in which the fracture dynamics of gels were quantitatively compared to those observed in both soda-lime glass and PMMA. Some of these results are summarized in Figure 1. These results conclusively showed that crack dynamics in gels are quantitatively identical to those observed in glass. As demonstrated in Figure 1, both of these very different materials undergo an instability (the "micro-branching instability") from a state of single-crack propagation to one in which the main crack is continuously generating short-lived microscopic side branches (Fineberg & Marder 1999).

3 MEASUREMENTS OF THE NEAR-TIP FIELDS OF RAPID CRACKS

Photographs of the crack tip opening profile, as presented in Figure 2, yield a direct measurement of the components u_x and u_y of the displacement field along the crack faces. We define a polar coordinate system that moves with the crack tip by $r = \sqrt{(x - vt)^2 + y^2}$ and $\theta = \tan^{-1}[y/(x - vt)]$, where (x, y) is the rest frame coordinate system. On the crack faces (i.e. $\theta = \pm\pi$), near the tip of the crack, linear elastic fracture mechanics relates $u_x(r, \theta = \pi)$, $u_y(r, \theta = \pm\pi)$ and the stress intensity factor K_I by $u_y^2(r, \theta = \pm\pi) \propto K_I^2 u_x(r, \theta = \pi)$. Thus, the curvature near the crack tip provides the value of K_I (or, alternatively, the fracture energy, Γ), up to known dynamical corrections. Figure 2b shows the values of the fracture energy as a function of the instantaneous crack velocity that was obtained in this way in a number of different experiments. The collapse of these derived values of $\Gamma(v)$ for all of the data indicates that this is an effective means to obtain precise values of the fracture energy.

Although this parabolic form is, generally, an excellent fit to the crack profile there are clear deviations from this expected crack-tip profile in the vicinity of the crack tip. The scale, δ, of these deviations is shown in the close-up of the crack tip in Figure 2a (right). While $\Gamma(v)$ (corresponding to the dissipation near the crack tip) increases relatively slowly with v, Figure 2c demonstrates that

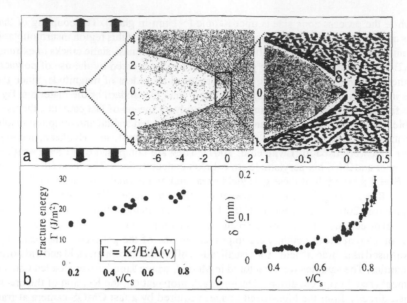

Figure 2.   (a) (left) A single crack traveling from left to right in a (to scale) uniaxially loaded sample. (center) Photograph of the profile at the tip of a typical dynamic crack where $v \simeq 0.7c_s$. A parabolic fit (dashed red curve) agrees well with the crack profile for a distance larger than 5 mm. (right) The x and y axes are labeled in mm, where the crack tip is located at (0,0). A closeup view reveals a deviation, δ, between the tip of the parabola and the crack tip. (b) The velocity-dependent fracture energy, $\Gamma(v)$, derived from K_I as determined by the parabolic form of the crack tip $u_y^2(r, \theta = \pm\pi) \propto K_I^2 u_x(r, \theta = \pi)$ predicted by LEFM. $A(v)$ is the universal kinematic factor derived in (Freund 1998). (c) δ as a function of v/c_s.

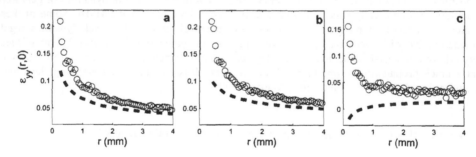

Figure 3.   The measured strain, $\varepsilon_{yy}(r, \theta = 0)$, (open circles) is compared to the theoretical (LEFM) prediction (dashed curve), see (Livne et al. 2008) for details. The discrepancy between the two increases with the crack velocity: (a) $v = 0.20c_s$, (b) $v = 0.53c_s$, (c) $v = 0.78c_s$. For the higher velocity (c), LEFM predicts a *negative* strain (compression) ahead of the crack tip.

the corresponding increase of $\delta(v)$ is much more rapid. This suggests that $\delta(v)$ is a measure of the size of the nonlinear elastic zone surrounding the crack tip, in which dissipative processes do not occur. Although the near tip region certainly includes this nonlinear zone, the precise structure of this important region is not known. To our knowledge, these are the first such measurements for a rapidly moving crack.

The discrepancies in the crack tip opening displacement, i.e. the deformation behind the crack tip (observed in Figure 2) lead us to consider the strain field component $\varepsilon_{yy} = \partial_y u_y$ along the line $\theta = 0$ which is immediately ahead of the crack tip. A comparison of LEFM predictions with the measured values is presented in Figure 3. The data show both clear discrepancies between LEFM predictions and measurements that become progressively larger as v increases and singular behavior of $\varepsilon_{yy}(r, \theta = 0)$ that appears stronger than the expected $r^{-1/2}$ dependence of LEFM. Below we will show that a quantitative description of the experimental data results from the incorporation of elastic nonlinearities in the material constitutive relation. These results are not limited to elastomers and are of general validity. Nonlinearity of the constitutive relation is as universal as linear elasticity and must be experienced by any material undergoing fracture. The theory, which generalizes LEFM to quadratic nonlinearities of the constitutive relation, resolves all of the discrepancies discussed above.

4 WEAKLY NONLINEAR THEORY OF DYNAMIC FRACTURE

The existence of a nonlinear elastic zone demonstrated above, as well as the discrepancies between the measured deformation and LEFM's predictions, motivated us to formulate a nonlinear elastic dynamic fracture problem under plane stress conditions. Consider the deformation field ϕ, which is assumed to be a continuous, differentiable and invertible mapping between a reference configuration x and a deformed configuration x' such that $x' = \phi(x) = x + u(x)$. The deformation gradient tensor F is defined as $F = \nabla \phi$ or explicitly $F_{ij} = \delta_{ij} + \partial_j u_i$. The first Piola-Kirchhoff stress tensor s, that is work-conjugate to the deformation gradient F, is given as $s = \partial_F U(F)$, where $U(F)$ is the strain energy in the deformed configuration per unit volume in the reference configuration (Holzapfel 2000). The momentum balance equation is

$$\nabla \cdot s = \rho \partial_{tt} \phi, \tag{1}$$

where ρ is the mass density. Under steady-state propagation conditions we expect all of the fields to depend on x and t through the combination $x - vt$ and therefore $\partial_t = -v \partial_x$. Recall, that the polar coordinate system that moves with the crack tip is related to the rest frame by $r = \sqrt{(x - vt)^2 + y^2}$ and $\theta = \tan^{-1}[y/(x - vt)]$. Thus, the traction-free boundary conditions on the crack faces are

$$s_{xy}(r, \theta = \pm\pi) = s_{yy}(r, \theta = \pm\pi) = 0 \tag{2}$$

To proceed, we note that in the measurement region discussed above the maximal strain levels are $0.2 - 0.35$ (see below) as the velocity of propagation varied from $0.20c_s$ to $0.78c_s$. These levels of strain motivate a perturbative approach where quadratic elastic nonlinearities must be taken into account. Higher order nonlinearities are neglected below, though they most probably become relevant as the crack velocity increases. We write the displacement field as

$$u(r, \theta) \simeq \epsilon u^{(1)}(r, \theta) + \epsilon^2 u^{(2)}(r, \theta) + \mathcal{O}(\epsilon^3), \tag{3}$$

where ϵ quantifies the (dimensionless) magnitude of the strain. For a general $U(F)$, s and ϕ can be expressed in terms of u of Equation (3). Substituting these in Equations (1)–(2) one can perform a controlled expansion in orders of ϵ.

To make the derivation concrete, we need an explicit $U(F)$ that corresponds to the experiments described above. The polymer gel used in these experiments is well-described by a plane stress incompressible Neo-Hookean constitutive law (Livne et al. 2005), defined by the energy functional (Knowles 1983)

$$U(F) = \frac{\mu}{2}[F_{ij}F_{ij} + \det(F)^{-2} - 3] \tag{4}$$

Using this explicit $U(F)$, we derive the first order problem in ϵ

$$\mu\nabla^2 u^{(1)} + 3\mu\nabla(\nabla \cdot u^{(1)}) = \rho\ddot{u}^{(1)}, \tag{5}$$

with the boundary conditions at $\theta = \pm\pi$

$$r^{-1}\partial_\theta u_x^{(1)} + \partial_r u_y^{(1)} = 0, \quad 4r^{-1}\partial_\theta u_y^{(1)} + 2\partial_r u_x^{(1)} = 0 \tag{6}$$

This is a standard LEFM problem (Freund 1998). The near crack-tip (asymptotic) expansion of the steady state solution for Mode I symmetry is (Freund 1998)

$$\epsilon u_x^{(1)}(r, \theta; v) = \frac{K_I \sqrt{r}}{4\mu\sqrt{2\pi}}\Omega_x(\theta; v) + \frac{Tr\cos\theta}{3\mu} + \mathcal{O}(r^{3/2}),$$

$$\epsilon u_y^{(1)}(r, \theta; v) = \frac{K_I \sqrt{r}}{4\mu\sqrt{2\pi}}\Omega_y(\theta; v) - \frac{Tr\sin\theta}{6\mu} + \mathcal{O}(r^{3/2}) \tag{7}$$

Here T is a constant known as the "T-stress". Note that K_I and T cannot determined by the asymptotic analysis as they depend on the *global* crack problem. $\Omega(\theta; v)$ is a known universal vector function whose components are given as

$$\Omega_x(\theta; v) = \frac{8}{D(v)}\left[(1 + \alpha_s^2)\sqrt{\gamma_d}\cos\left(\frac{\theta_d}{2}\right) - 2\alpha_d\alpha_s\sqrt{\gamma_s}\cos\left(\frac{\theta_s}{2}\right)\right],$$

$$\Omega_y(\theta; v) = -\frac{8\alpha_d}{D(v)}\left[(1 + \alpha_s^2)\sqrt{\gamma_d}\sin\left(\frac{\theta_d}{2}\right) - 2\sqrt{\gamma_s}\sin\left(\frac{\theta_s}{2}\right)\right] \tag{8}$$

Here

$$\gamma_{s,d} = \sqrt{1 - \frac{v^2\sin^2\theta}{c_{s,d}^2}}, \quad D(v) = 4\alpha_d\alpha_s - (1 + \alpha_s^2)^2$$

$$\tan\theta_{d,s} = \alpha_{d,s}\tan\theta, \quad \alpha_{d,s}^2 \equiv 1 - v^2/c_{d,s}^2 \tag{9}$$

ϵ in Equation (3) can be now defined explicitly as $\epsilon \equiv K_I/[4\mu\sqrt{2\pi\ell(v)}]$, where $\ell(v)$ is a velocity-dependent length-scale. $\ell(v)$ defines the scale where only the order ϵ and ϵ^2 problems are relevant. It is a *dynamic* length-scale that marks the onset of deviations from a linear elastic constitutive behavior.

The solution of the order ϵ equation, i.e. Equations (8), can be now used to derive the second order problem in ϵ. The form of the second order problem for an incompressible material is

$$\mu\nabla^2 u^{(2)} + 3\mu\nabla(\nabla \cdot u^{(2)}) + \frac{\mu\ell g(\theta; v)}{r^2} = \rho\ddot{u}^{(2)} \tag{10}$$

The boundary conditions at $\theta = \pm\pi$ become

$$r^{-1}\partial_\theta u_x^{(2)} + \partial_r u_y^{(2)} = 4r^{-1}\partial_\theta u_y^{(2)} + 2\partial_r u_x^{(2)} + \frac{\kappa(v)\ell}{r} = 0, \tag{11}$$

where contributions proportional to T were neglected. The function $\kappa(v)$ is given by

$$\kappa(v) = \frac{48v^2\alpha_d^2(\alpha_s^2 - 1)}{c_s^2 D^2(v)}, \tag{12}$$

where $c_d = 2c_s$ for an incompressible material under plane stress conditions should be used. The function $g(\theta; v)$ will be discussed below.

The problem posed by Equations (10)–(11) has the structure of an effective LEFM problem with a body force $\propto r^{-2}$ and a crack face force $\propto r^{-1}$. Note that Equations (10)–(11) are valid in the range $\sim \ell(v)$, where ϵ^2 is non-negligible with respect to ϵ, but higher order contributions are negligible. Since one cannot extrapolate the equations to smaller length-scales, no real divergent behavior in the $r \to 0$ limit is implied. We stress that the structure of this problem is universal. Only $g(\theta; v)$ and $\kappa(v)$ depend on the second order elastic constants resulting from expanding a *general* $U(F)$ to second order in ϵ. For example, the $\propto r^{-2}$ effective body-force in Equation (10) results from terms of the form $\partial(\partial u^{(1)} \partial u^{(1)})$, which are generic quadratic nonlinearities.

We now focus on solving Equation (10) with the boundary conditions of Equations (11) for the explicit $g(\theta; v)$ and $\kappa(v)$ derived from Equation (4). Our strategy is to look for a particular solution of the inhomogeneous Equation (10) *without* satisfying the boundary conditions of Equations (11) and then to add to it a solution of the corresponding homogeneous equation that makes the overall solution consistent with the boundary conditions. We find that the inhomogeneous solution, $\Upsilon(\theta; v)$, is r-independent. The homogeneous solution is obtained using a standard approach (Freund 1998) by noting that the second boundary condition of Equations (11) requires that its first spatial derivative scales as r^{-1}. Putting together the homogeneous and inhomogeneous solutions, the complete solution of the Mode I problem reads

$$\epsilon^2 u_x^{(2)}(r, \theta; v) = \left(\frac{K_I}{4\mu\sqrt{2\pi}}\right)^2 \left[A \log r + \frac{A}{2} \log\left(1 - \frac{v^2 \sin^2\theta}{c_d^2}\right)\right.$$
$$\left. + B\alpha_s \log r + \frac{B\alpha_s}{2} \log\left(1 - \frac{v^2 \sin^2\theta}{c_s^2}\right) + \Upsilon_x(\theta; v)\right],$$

$$\epsilon^2 u_y^{(2)}(r, \theta; v) = \left(\frac{K_I}{4\mu\sqrt{2\pi}}\right)^2 [-A\alpha_d\theta_d - B\theta_s + \Upsilon_y(\theta; v)], \tag{13}$$

where $A = [2\alpha_s B - 4\partial_\theta \Upsilon_y(\pi; v) - \kappa(v)]/(2 - 4\alpha_d^2)$ (cf. Eq. (11)) and the remaining parameter B can be determined following the detailed procedure described in (Bouchbinder et al. 2009). A striking feature of Equations (13) is that they lead to strain contributions that vary as r^{-1}, which are "more-singular" than the $r^{-1/2}$ strains predicted by LEFM.

The analytic form of $\Upsilon(\theta; v)$ depends mainly on $g(\theta; v)$. The latter can be represented as

$$g_x(\theta; v) \simeq \sum_{n=1}^{N(v)} a_n(v) \cos(n\theta), \quad g_y(\theta; v) \simeq \sum_{n=1}^{N(v)} b_n(v) \sin(n\theta) \tag{14}$$

For $v = 0$ we have $N(0) = 3$ and the representation is *exact*, while for higher velocities it provides analytic approximations with whatever accuracy needed. For $v \simeq 0.8c_s$ only seven terms provide a representation that can be regarded exact for any practical purpose, see below. $\Upsilon(\theta; v)$ is then obtained in the form

$$\Upsilon_x(\theta; v) \simeq \sum_{n=1}^{N(v)} c_n(v) \cos(n\theta), \quad \Upsilon_y(\theta; v) \simeq \sum_{n=1}^{N(v)} d_n(v) \sin(n\theta), \tag{15}$$

where the unknown coefficients are determined by solving a set of linear equations. The coefficients $a_n(v), b_n(v), c_n(v), d_n(v)$, for the three velocities discussed in this paper, are given in the following table:

Coefficient	$v = 0.20c_s$	$v = 0.53c_s$	$v = 0.78c_s$	coefficient	$v = 0.20c_s$	$v = 0.53c_s$	$v = 0.78c_s$
a_1	−7.7519	−14.5566	−56.7222	c_1	−2.3373	−4.282	−16.1418
a_2	18.4329	29.4669	97.2713	c_2	1.8484	2.997	9.9644
a_3	−2.2316	−1.3945	13.919	c_3	−0.1872	−0.1481	0.6158
a_4	−0.1839	−2.3996	−22.2912	c_4	−0.0082	−0.1023	−0.8665
a_5	0.0359	0.2165	−3.6793	c_5	0.00153	0.0112	−0.0537
a_6	0.0018	0.1951	5.1904	c_6	−	0.00465	0.1012
a_7	−	−0.0224	0.9612	c_7	−	−0.000734	0.00691
a_8	−	−	−1.2521	c_8	−	−	−0.01473
a_9	−	−	−0.2443	c_9	−	−	−0.00114
a_{10}	−	−	0.3038	c_{10}	−	−	0.002681
b_1	−7.2255	−13.4871	−48.3842	d_1	−1.4254	−2.9899	−12.5438
b_2	18.2259	26.8884	75.4972	d_2	1.8472	2.9861	9.9132
b_3	−2.1964	−1.1999	10.2823	d_3	−0.18996	−0.1769	0.45195
b_4	−0.1813	−2.1378	−16.0797	d_4	−0.00812	−0.1038	−0.90651
b_5	0.0352	0.1876	−2.5542	d_5	0.00152	0.0125	−0.03742
b_6	0.0018	0.1708	3.5718	d_6	−	0.00439	0.1112
b_7	−	−0.0193	0.64926	d_7	−	−0.000736	0.00467
b_8	−	−	−0.8409	d_8	−	−	−0.01654
b_9	−	−	−0.1626	d_9	−	−	−0.000567
b_{10}	−	−	0.2014	d_{10}	−	−	0.002314

5   COMPARISON TO EXPERIMENTS

We now show that the second order solution of Equations (13) entirely resolves the discrepancies raised by trying to interpret the experimental observations in the framework of LEFM. In principle, the complete second order asymptotic solution, Equations (3), (8) and (13), contains two parameters (K_I and T) that cannot be determined from the asymptotic solution and therefore must be extracted from the experimental data. The parameter B was shown to be determined by the condition that the $1/r$ displacement-gradients singular contribution does not generate a spurious unbalanced force on a line encircling the crack tip (Bouchbinder et al. 2009). Moreover, it was shown that the theoretically determined B is consistent with the value of B obtained from the experimental data when B is treated as a free parameter. Therefore, in the present context we treat B as a free fitting parameter, bearing in mind the results of (Bouchbinder et al. 2009) which clearly show that B is in fact a theoretically determined function of K_I, in accord with the important concept of the autonomy of the near-tip nonlinear zone. The latter concept suggests that *all* the properties of the nonlinear ("inner") zone surrounded by the asymptotic LEFM K-fields are uniquely determined by the stress intensity factor K_I, such that once K_I is known, there remain no undetermined parameters in the "inner" nonlinear problem.

The parameters K_I, T and B were chosen such that Equations (3), (8) and (13) properly describe the measured $u_x(r, 0)$. Examples for $v/c_s = 0.20$, 0.53 and 0.78 are provided in Figure 4 (top). With K_I, T and B at hand, we can now test the theory's predictions for $\varepsilon_{yy}(r, 0)$ with *no adjustable free parameters*. The corresponding results are compared with both the measured data and LEFM predictions in Figure 4 (bottom). In general, the agreement with the experimental data is excellent. We stress again that in (Bouchbinder et al. 2009) it was explicitly demonstrated (for $v = 0.20c_s$) that very similar results are obtained when B is calculated theoretically instead of being determined experimentally. These results demonstrate the importance of the predicted r^{-1} strain terms near the crack tip. ℓ is estimated as the scale where the largest strain component reaches values of 0.10–0.15. For the data presented in Figure 4a, $\varepsilon_{yy} > \varepsilon_{xx}$, where $\varepsilon_{xx} = \partial_x u_x$ is obtained by differentiating u_x. Thus, ℓ can be read off of the bottom panel to be $\sim 0.5 - 1$ mm. Similar estimates can be obtained for every v, though not always does $\varepsilon_{yy} > \varepsilon_{xx}$, e.g. Figure 4c.

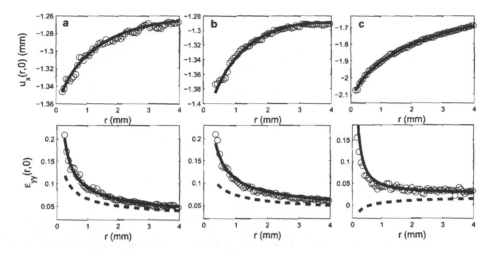

Figure 4.   Top: Measured $u_x(r,0)$ (circles) fitted to the x component of Equation (3) (solid line) for (a) $v = 0.20c_s$ with $K_I = 1070\text{Pa}\sqrt{m}$, $T = -3150\text{Pa}$ and $B = 18$. (b) $v = 0.53c_s$ with $K_I = 1250\text{Pa}\sqrt{m}$, $T = -6200\text{Pa}$ and $B = 7.3$ and (c) $v = 0.78c_s$ with $K_I = 980\text{Pa}\sqrt{m}$, $T = -6900\text{Pa}$ and $B = 26$. Bottom: corresponding measurements of $\varepsilon_{yy}(r,0) = \partial_y u_y(r,0)$ (circles) compared to the theoretical nonlinear solution (cf. Eq. (3)) *with no adjustable parameters* (solid lines); K_I, T and B are taken from the fit of $u_x(r,0)$. (dashed lines) LEFM predictions (analysis as in (Livne et al. 2008)) were added for comparison, cf. Figure 3.

For $v = 0.53c_s$ (Fig. 4b) the theory still agrees well with the measurements, although some deviations near the tip are observed. These deviations signal that higher order corrections may be needed, though second order nonlinearities still seem to provide the dominant correction to LEFM. For higher velocities, it is not clear, *a-priori*, that second order nonlinearities are sufficient to describe the data. In fact, the strain component $\varepsilon_{xx}(r,0)$ for $v = 0.78c_s$ reaches a value of ~0.35 in Figure 4c, suggesting that higher order nonlinearities may be important. Nevertheless, the second order theory avoids a fundamental failure of LEFM; at high velocities ($v > 0.73c_s$ for an incompressible material) LEFM predicts (dashed line in Fig. 4c) that the contribution proportional to K_I in $\varepsilon_{yy}(r,0)$ (derived from Eqs. (7)) becomes *negative*. This implies that $\varepsilon_{yy}(r,0)$ *decreases* as the crack tip is approached and becomes *compressive*. This is surprising, as material points straddling $y = 0$ must be separated from one another to precipitate fracture. Thus, the second order nonlinear solution (solid line), though applied beyond its range of validity, already induces a qualitative change in the character of the strain. This is a striking manifestation of the breakdown of LEFM, demonstrating that elastic nonlinearities are generally unavoidable, especially as high crack velocities are reached. The results of Figs. 4a–c both provide compelling evidence in favor of the developed theory and highlight inherent limitations of LEFM. We note that $\ell(v)$ increases with increasing v, reaching values in the mm-scale at very high v.

Our results indicate that the widely accepted assumption of "K-dominance" of LEFM, i.e. that there is always a region where the $r^{-1/2}$ strain term dominates all other contributions, is violated here. The results presented in Figure 4 explicitly demonstrate that quadratic nonlinearities become important in the same region where a non-negligible T-stress exists. As elastic nonlinearities intervene before the $r^{-1/2}$ term dominates the strain fields, the contributions of *both* of these terms must be taken into account as one approaches the crack tip. Since the values of the T-stress are system specific, this observation is valid for the specific experimental system under study. They do indicate that the assumption of "K-dominance" is not always valid.

Let us now consider the Crack Tip Opening Displacement (CTOD) when r is further reduced. Equations (13) predict the existence of log-terms in $\phi_x(r,\theta)$. These terms, which are negligible at $\theta = \pi$ on a scale $\ell(v)$, must become noticeable at smaller scales. Although this region is formally beyond the range of validity of the expansion of Equation (3), we would still expect the existence of a CTOD contribution proportional to $\log r$ to be observable. We test this prediction in Figure 5 by

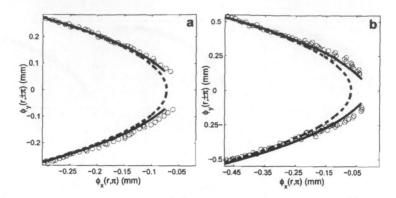

Figure 5.   Measured crack tip profiles ($\phi_y(r, \pm\pi)$ vs. $\phi_x(r, \pi)$) (circles). Shown are the parabolic LEFM best fit (dashed line) and the profiles predicted by the second order nonlinear corrections (solid line). (a) $v = 0.2c_s$ and (b) $v = 0.53c_s$. T and B are as in Figure 4. In contrast to the $\sim 20\%$ discrepancy in values of K_I obtained in (Livne et al. 2008), the respective values $K_I = 1170 \, \text{Pa}\sqrt{m}$ and $K_I = 1300 \, \text{Pa}\sqrt{m}$ correspond to within 9% and 4%, respectively, of K_I obtained from $u_x(r, 0)$ using the nonlinear theory, cf. Figure 4.

comparing the measured small-scale CTOD to both the parabolic LEFM form and the second order nonlinear solution *with no adjustable parameters*. We find that these log-terms, whose coefficients were determined at a scale $\ell(v)$, capture the initial deviation from the parabolic CTOD at $\theta = \pm\pi$ to a surprising degree of accuracy. This result lends further independent support to the validity of Equations (13).

6   CONCLUDING REMARKS

In this paper we have shown that the second order solution presented in Equations (13) resolves in a self-consistent way all of the discrepancies with LEFM highlighted by the direct measurements of the deformation near the tip of a rapid mode I crack. This solution is universal in the sense that its generic properties are independent of geometry, loading conditions and material parameters. We would entirely expect that *any* material subjected to the enormous deformations that surround the tip of a crack experiences *at least* quadratic elastic nonlinearities, prior to the onset of the irreversible deformation that leads to failure. Our results show that these deformations, which are the vehicle for transmitting breaking stresses to crack tips, must be significantly different from the LEFM description, especially at high v.

One may ask why we should not consider still higher order elastic nonlinearities. We surmise that quadratic elastic nonlinearities may be special, as they mark the emergence of a dynamic length-scale $\ell(v)$ that characterizes a region where material properties—like local wave speeds, local response times and anisotropy—become *deformation dependent*. This line of thought seems consistent with the observations of Refs. (Buehler et al. 2003, Buehler & Gao 2006). As supporting evidence for this view, we note that the geometry-independent wave-length of crack path oscillations discussed in (Livne et al. 2007, Bouchbinder & Procaccia 2007) seems to correlate with the mm-scale $\ell(v)$ at high v. Therefore, our results may have implications for understanding crack tip instabilities.

ACKNOWLEDGEMENTS

This research was supported by grant 57/07 of the Israel Science Foundation. E.B. acknowledges support from the Horowitz Center for Complexity Science and the Lady Davis Trust.

REFERENCES

Bonamy D. et al. 2006, Experimental investigation of damage and fracture in glassy materials at the nanometre scale, International Journal of Materials & Product Technology 26(3–4): 339–353.

Bouchbinder E. & Procaccia I. 2007, Oscillatory Instability in Two-Dimensional Dynamic Fracture, Phys. Rev. Lett. 98: 124302–4.

Bouchbinder E., Livne A. & Fineberg J. 2009, The $1/r$ singularity in weakly nonlinear fracture mechanics, arXiv:0902.2121. To appear in J. Mech. Phys. Solids.

Buehler M.J., Abraham F.F. & Gao H. 2003, Hyperelasticity governs dynamic fracture at a critical length scale, Nature 426: 141–146.

Buehler M.J. & Gao H. 2006, Dynamical fracture instabilities due to local hyperelasticity at crack tips, Nature 439: 307–310.

Fineberg J. & Marder M. 1999, Instability in dynamic fracture, Phys. Rep. 313: 1–108.

Freund L.B. 1998, Dynamic Fracture Mechanics, Cambridge: Cambridge University Press.

Holzapfel G.A. 2000, Nonlinear Solid Mechanics, Chichester: Wiley.

Knowles J.K. & Sternberg E. 1983, Large deformation near a tip of an interface-crack between 2 neo-Hookean sheets, J. Elasticity 13: 257–293.

Livne A., Cohen G. & Fineberg J. 2005, Universality and Hysteretic Dynamics in Rapid Fracture, Phys. Rev. Lett. 94: 224301–4.

Livne A., Ben-David O. & Fineberg J. 2007, Oscillations in Rapid Fracture, Phys. Rev. Lett. 98: 124301–4.

Livne A., Bouchbinder E. & Fineberg J. 2008, Breakdown of Linear Elastic Fracture Mechanics near the Tip of a Rapid Crack, Phys. Rev. Lett. 101: 264301–4.

II. *Fault gauge mechanics*

The effect of mineral decomposition as a mechanism of fault weakening during seismic slip

Jean Sulem
CERMES - UR Navier, Ecole des Ponts Paris Tech, Marne-la-Vallée, Cedex, France

Vincent Famin
Laboratoire Géosciences Réunion - IPGP,
Université de la Réunion, Saint Denis messag., Cedex, France

ABSTRACT: Recent studies have emphasized the role of thermal pore fluid pressurization as an important cause of slip-weakening. Those studies rely on the assumption that no fluid is produced nor consumed during seismic slip. There is, however, growing evidence that temperature-induced decomposition of minerals may be a significant source of fluids in fault rocks. In this paper, we model the mechanical effects of calcite thermal decomposition on the slip behavior of a fault zone during an earthquake. It is shown that the endothermic reaction of calcite decomposition limits the coseismic temperature increase to less than $\sim$800°C (corresponding to the initiation of the chemical reaction) inside the slip zone and that the rapid emission of CO_2 by decarbonation significantly increases the slip-weakening effect of thermal pressurization. The pore pressure reaches a maximum and then decreases due to the reduction of solid volume, causing a re-strengthening of the shear stress.

1 INTRODUCTION

During the rupture of a fault, an earthquake occurs because the frictional resistance to slip on the fault walls decreases with increasing slip, causing an acceleration of sliding. To quantify the energy dissipated by an earthquake and assess the hazard of future ruptures, it is critical to understand the mechanics of slip weakening, i.e. how and how much fault friction drops in due course of the rupture. As the frictional heat generated during an earthquake is the largest part of the total seismic energy budget, the estimation of the temperatures reached during an earthquake can provide important information about faulting mechanisms, such as frictional melting and thermal pressurization, and dynamic shear stress during the earthquake.

There is evidence of CO_2 release in several active crustal faults. In the Corinth rift (Greece) for example, chemical analyzes of water springs near the seismogenic Heliki and Aegion faults revealed an anomalously high content of dissolved CO_2 compared with the regional values (Pizzino et al. 2004). The surface trace of the San Andreas fault also displays a positive anomaly of CO_2 fluxes (Lewicki and Brantley, 2000), and this CO_2 comes from a shallow source, not from the mantle (Lewicki et al., 2003). Moreover, there is growing evidence that CO_2 release coincides with seismic slip in crustal faults, active and/or exhumed. In the vicinity of the Nojima fault (Japan), Sato and Takahashi (1997) reported that the HCO_3^- concentration of springs increased by 30 wt% immediately after the 1995 Kobe earthquake. This carbon discharge, together with other coseismic geochemical anomalies, decreased gradually to normal values in the following ten months. A micro infrared analysis of exhumed pseudotachylites (i.e. friction induced melts produced by seismic slip) from the Nojima fault revealed that shear melting destroyed the carbonates from the fault zone and released CO_2, thus providing an explanation to the coseismic CO_2 spikes in springs (Famin et al., 2008). The quantity of CO_2 released by friction melting during the 1995 Kobe earthquake was evaluated to 1.8 to $3.4\ 10^3$ tons. As for the San Andreas fault, the carbon isotopic signature of springs and fault rocks from Nojima is consistent with a decomposition of biogenic carbonates, not from a mantle origin (Ueda et al., 1999; Arai et al., 2001; Arai et al., 2003; Lin et al., 2003). In the Central Apennines, (Italiano et al., 2008) also reported enhanced fluxes of crustal CO_2 (i.e. not mantellic) during the 1997–1998 seismic crisis of major faults, and proposed that coseismic decarbonation was responsible for the CO_2 emission. Recent studies of the Chelungpu Fault (Taiwan) responsible for the 1999 ChiChi earthquake also showed that the fault core was depleted in carbon relative to the damage zone, and the depletion was attributed to a decarbonation induced by frictional heat (Hirono et al., 2006; Hirono et al., 2007). In addition, recent high velocity friction experiments on Carrara marble have shown that thermal decomposition of calcite due to frictional heating induces a pronounced fault weakening (Han et al., 2007). The production of co-seismic CO_2 is therefore attested by various field and experimental techniques, thus making the thermal decomposition of carbonates an important additional mechanism to be investigated among possible fault weakening processes (Sulem and Famin, 2009).

In this paper, we investigate the impact of heat-induced mineral destabilization on the frictional properties of a shear zone during seismic slip. Our study focuses on the kinetics of chemical decomposition of calcite (decarbonation) $CaCO_3 \rightarrow CaO + CO_2$ because carbonates are present in every fault zones from the ductile-brittle transition ($\sim$15 km) to the subsurface, and because positive CO_2 anomalies are observed in the vicinity of many active crustal faults. To take calcite destabilization and CO_2 degassing into account, we introduce the additional complexity of mineral volume loss and fluid production in the thermal pressurization model. The equations that govern the evolution of pore pressure and temperature inside the shear zone and the mass of emitted CO_2 are deduced from the mass and energy balance of the multi-phases saturated medium and from the kinetics of decarbonation. Our numerical simulations of seismic slip at depth of 7 km show that decarbonation has critical consequences on the slip-weakening behavior and heat production of a fault in carbonate rocks, especially for $M_w > 5$ earthquakes.

2   POSITION OF THE PROBLEM AND GOVERNING EQUATIONS

In this chapter, we introduce the chemical coupling of calcite volume loss and CO_2 production in the mechanical analysis of shear heating and fluid pressurization phenomenon (Rice, 2006, Veveakis et al., 2007, Sulem et al., 2007, Brantut et al., 2008).

Considering that the length scales in the direction parallel to the fault over which the thermo-poro-mechanical fields vary are much larger than in the direction normal to it, we analyse here a 1D problem. We consider a rapidly deforming and infinitely long shear band of thickness h consisting of fluid-saturated carbonate rock. This shear band begins to undergo slip δ at a time $t = 0$ with an imposed overall slip-rate $V = d\delta/dt$ in the x-direction, as shown in figure 1. For simplicity, it is assumed that the porosity of the fault zone is saturated with pure CO_2, even though other fluids such as H_2O may be dominant before an earthquake. This assumption does not yield significant

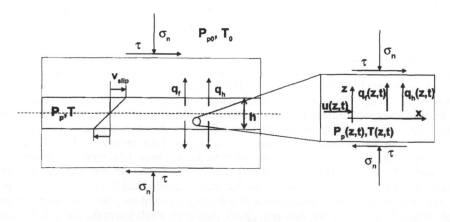

Figure 1.   Model of a deforming shear-band with heat and fluid fluxes.

error in the calculation because the compressibility factors and viscosities of H_2O and CO_2 are nearly equal within the pressure and temperature range considered here. For these conditions of pressure and temperature, carbon dioxide is in super-critical state.

Inside such a shear-band the pore pressure $P_p(t, z)$, the temperature $T(t, z)$, and the velocity $v(t, z)$ are assumed to be functions only of time t and of the position z in the direction normal to the band (figure 1).

2.1   Mass balance

Conservation of fluid mass is expressed by

$$\frac{\partial m_f}{\partial t} = \frac{\partial m_d}{\partial t} - \frac{\partial q_f}{\partial z} \tag{1}$$

where m_f is the total fluid mass per unit volume of porous medium (in the reference state), m_d is the mass of emitted CO_2 per unit volume and q_f is the flux of fluid. The total fluid mass per unit volume of porous medium is written as $m_f = \rho_f n$, where n is the pore volume fraction (Lagrangian porosity) and ρ_f is the density of the fluid (here supercritical CO_2). The left hand side of equation (1) is obtained by differentiating this product:

$$\frac{\partial m_f}{\partial t} = n \frac{\partial \rho_f}{\partial t} + \rho_f \frac{\partial n}{\partial t} \tag{2}$$

The derivatives of the right hand side of equation (2) are given by:

$$d\rho_f = \rho_f \beta_f dP_p - \rho_f \lambda_f dT \tag{3}$$

and

$$\frac{\partial n}{\partial t} = n\beta_n \frac{\partial P_p}{\partial t} + n\lambda_n \frac{\partial T}{\partial t} + \frac{\partial n_d}{\partial t} \tag{4}$$

where P_p is the pore pressure of the fluid, β_f and λ_f are the compressibility and the thermal expansion coefficient of the pore fluid respectively, β_n is the pore volume compressibility and λ_n is the thermal expansion coefficient of the pore volume, which is equal to the thermal expansion coefficient of the solid fraction for thermo-poro-elastic materials.

In equation (4), $\frac{\partial n_d}{\partial t}$ is the rate of porosity change due to the decomposition of the solid phase and is expressed as:

$$\frac{\partial n_d}{\partial t} = -\frac{1}{\rho_{CaCO_3}} \frac{\partial m_{CaCO_3}}{\partial t} - \frac{1}{\rho_{CaO}} \frac{\partial m_{CaO}}{\partial t} = \left(\frac{1}{\rho_{CaCO_3}} \frac{M_{CaCO_3}}{M_{CO_2}} - \frac{1}{\rho_{CaO}} \frac{M_{CaO}}{M_{CO_2}} \right) \frac{\partial m_d}{\partial t} \tag{5}$$

where ρ_{CaCO_3} (resp. ρ_{CaO}) and M_{CaCO_3} (resp. M_{CaO}) are the density and the molar mass of $CaCO_3$ (resp. CaO), and m_d is the mass of emitted CO_2 per unit volume.

The rate of emitted CO_2 is evaluated using the kinetics of the chemical reaction of calcite thermal decomposition

$$CaCO_3 \rightarrow CaO + CO_2 \tag{6}$$

The relationship between the rate of emitted CO_2 and the temperature is expressed by the Arrhenius equation (L'vov, 2002)

$$\frac{\partial m_d}{\partial t} = \chi \rho_{CaCO_3} (1 - n) A \exp \left(-\frac{E_a}{RT} \right) \tag{7}$$

where χ is the ratio between the molar mass M_{CO_2} of CO_2 (44 g/mol) and the molar mass M_{CaCO3} of calcite (100 g/mol) if we assume that the total amount of calcite can be decomposed, A is a constant (pre-exponential term of the Arrhenius law), E_a is the activation energy of the reaction, R is the gas constant (8.31447 $JK^{-1}mol^{-1}$). In the following we will take the values corresponding to $CaCO_3$ mixed with silica: $E_a = 319000$ $Jmol^{-1}$, $A = 2.95 \times 10^{15}s^{-1}$ (Dollimore et al., 1996).

Using equations (2–4), the first term of equation (1) is thus evaluated as

$$\frac{\partial m_f}{\partial t} = n\rho_f(\beta_n + \beta_f)\frac{\partial P_p}{\partial t} - \rho_f n(\lambda_f - \lambda_n)\frac{\partial T}{\partial t} + \rho_f\frac{\partial n_d}{\partial t} \qquad (8)$$

The relationship between the change of solid mass (per unit volume) and the rate of emitted CO_2 is simply given by

$$\frac{\partial m_s}{\partial t} = -\frac{\partial m_d}{\partial t} \qquad (9)$$

Note that following a remark of Noda (private communication, 2009) equations (5) and (9) slightly differ from those written in Sulem and Famin 2009 (see Sulem et al. 2009).

The flux term in equation (1) is evaluated assuming Darcy's law for fluid flow with viscosity through η_f a material with permeability k_f

$$q_f = -\frac{\rho_f}{\eta_f}k_f\frac{\partial P_p}{\partial z} \qquad (10)$$

Substituting, (5), (8) and (10) into (1) gives the fluid mass conservation equation

$$\frac{\partial P_p}{\partial t} = \Lambda\frac{\partial T}{\partial t} + \frac{1}{n\rho_f(\beta_n + \beta_f)}\frac{\partial}{\partial z}\left(\rho_f\frac{k_f}{\eta_f}\frac{\partial P_p}{\partial z}\right) + \frac{1 - \rho_f\zeta/\rho_{CaCO_3}}{n\rho_f(\beta_n + \beta_f)}\frac{\partial m_d}{\partial t} \qquad (11)$$

where $\Lambda = \frac{\lambda_f - \lambda_n}{\beta_f + \beta_n}$ is the thermo-elastic pressurization coefficient under undrained conditions (Rice, 2006), $\rho_s = \rho_{CaCO_3}$ and $\zeta = \frac{M_{CaCO_3}}{M_{CO_2}} - \frac{\rho_{CaCO_3}}{\rho_{CaO}}\frac{M_{CaO}}{M_{CO_2}}$. With $M_{CaCO3} = 100$ g/mol, $M_{CaO} = 56$ g/mol, $M_{CO_2} = 44$ g/mol, $\rho_{CaCO_3} = 2.71$ g/cm^3 and $\rho_{CaO} = 3.35$ g/cm^3, $\zeta = 1.24$. The thermo-elastic pressurization coefficient Λ is pressure and temperature dependent because the compressibility and the thermal expansion coefficients of the fluid vary with pressure and temperature, and also because the compressibility of the pore space of the rock can change with the effective stress (Ghabezloo and Sulem 2009).

2.2 Energy balance equation

The thermal decomposition of calcite is endothermic. Therefore, the equation of conservation of energy is expressed as:

$$\rho C\frac{\partial T}{\partial t} = -\frac{\partial q_h}{\partial z} + \Psi_p - \frac{\Delta_r H_T^0}{\chi M_{CaCO_3}}\frac{\partial m_d}{\partial t} \qquad (12)$$

where ρC is the specific heat per unit volume of the fault material in its reference state, q_h is the heat flux, Ψ_p is the rate of mechanical energy dissipation due to inelastic deformation, $\Delta_r H_T^0$ is the enthalpy change of the reaction (i.e. the energy consumed by the reaction), which for calcite decomposition is equal to the activation energy E_a in the isobaric mode (L'vov, 2002; L'vov and Ugolkov, 2004; L'vov, 2007). Note that in the lack of experimental data, we make here the basic assumption that all the plastic work is converted into heat.

The heat flux is related to the temperature gradient by Fourier's law

$$q_h = -k_T \frac{\partial T}{\partial z} \tag{13}$$

where k_T is the thermal conductivity of the saturated material. In equation (12) it is assumed that the heat flux is entirely due to heat conduction, neglecting heat convection by the moving hot fluid. This assumption is justified by the low pore volume fraction and the low permeability of fault gouges.

If we neglect all dissipation in the fluid the rate of mechanical energy dissipation is written as

$$\Psi_p = \tau \frac{\partial v}{\partial z} \tag{14}$$

where v is the local fault parallel velocity and τ is the shear stress. In equation (14) the work done by the normal stress σ_n is considered as negligible as compared to the one done by τ at the large shear considered.

Substituting (13) and (14) into (12) gives the energy conservation equation

$$\frac{\partial T}{\partial t} = \frac{1}{\rho C} \frac{\partial}{\partial z} \left(k_T \frac{\partial T}{\partial z} \right) + \frac{1}{\rho C} \tau \frac{\partial v}{\partial z} - \frac{1}{\rho C} \frac{\Delta_r H_T^0}{\chi M_{CaCO_3}} \frac{\partial m_d}{\partial t} \tag{15}$$

2.3   Momentum balance equation

The 1D-momentum balance equation reads as

$$\frac{\partial \tau}{\partial z} = \rho \frac{\partial v}{\partial t} \tag{16}$$

As discussed by Rice (2006), the effect of even large accelerations like several times the acceleration of the gravity g is insignificant over the small length scales in the z-direction normal to the fault where the heat and fluid diffusion process are taking place during rapid slip and very high values of pressure and temperature gradients. For example, assuming an acceleration of $10\,g$ and a specific mass of the material of 2500 kg/m^3 would result in a change of 2.5×10^{-2} MPa/m for τ. The relevant length scale in z-direction is only few cm and thus the variation for τ can be neglected and mechanical equilibrium can be assumed.

$$\frac{\partial \tau}{\partial z} = 0 \tag{17}$$

Consequently, as the shear stress is constant in space, the Coulomb friction law cannot be assumed to be met in all deforming regions unless the pore pressure is also constant in space as it is the case in the undrained adiabatic limit.

It is thus assumed that the frictional resistance is proportional to the mean effective stress inside the band:

$$\tau(t) = f \left(\sigma_n - \frac{1}{h} \int_{-h/2}^{h/2} P_p(\xi, t) d\xi \right) \tag{18}$$

where f is the friction coefficient of the gouge and h is the width of the shear band.

2.4   *Summary of the governing equations*

The above coupled production/diffusion equations can be summarized as:

$$\frac{\partial P_p}{\partial t} = \Lambda \frac{\partial T}{\partial t} + \frac{1}{n(\beta_n + \beta_f)} \frac{\partial}{\partial z} \left(\frac{k_f}{\eta_f} \frac{\partial P_p}{\partial z} \right) + \frac{1 - \rho_f \zeta / \rho_s}{n \rho_f (\beta_n + \beta_f)} \frac{\partial m_d}{\partial t}$$

$$\frac{\partial T}{\partial t} = \frac{1}{\rho C} \frac{\partial}{\partial z} \left(k_T \frac{\partial T}{\partial z} \right) + \frac{1}{\rho C} f \left(\sigma_n - \frac{1}{h} \int_0^h P_p(\xi, t) d\xi \right) \frac{\partial v}{\partial z} - \frac{(1 - n)}{\rho C} \frac{\Delta_r H_T^0}{M_{CaO_3}} \rho_s A \exp \left(-\frac{E_a}{RT} \right)$$

$$\tag{19}$$

2.5   *Permeability law*

The decomposition of carbonate can induce substantial change in the porosity of the rock which affects the permeability. It is known that there is no unique relationship between porosity and permeability applicable to all porous media and that the geological evolution process of the pore space influence the permeability-porosity relationship. The empirical power law $k_f \propto n^\alpha$ is commonly used for geomaterials. The exponent α characterizes the porosity sensitivity of permeability and can take values ranging from 1 to 25 (David et al., 1994). The high values of α correspond in general to rocks with a high porosity whereas for low porosity rocks law values of α are obtained. The commonly used cubic Carman Kozeny permeability law is assumed here to take into account the effect of porosity change due to mineral decomposition on the permeability of the rock:

$$k_f = k_{f0} \left(\frac{1 - n_0}{1 - n} \right)^2 \left(\frac{n}{n_0} \right)^3 \tag{20}$$

where k_{f0} is the reference permeability corresponding to the reference porosity n_0. We would like to emphasize the fact that the assumed permeability law has a strong effect on the numerical results and that there is a need for experimental data on permeability changes resulting from the particular process of carbonate decomposition.

2.6   *Thermodynamical properties of supercritical CO_2*

The above governing equations (19) involve the thermodynamical properties of the pore fluid. The viscosity, the density, the compressibility and the thermal expansion of CO_2 in supercritical state have been fitted using the data published by the National Institute of Standards and Technology (http://webbook.nist.gov/chemistry/) and from the state equation proposed by Saxena and Fei (1987) for very high pressures and temperatures. Fitted curves are shown on Figure 2.

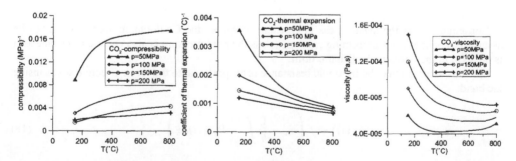

Figure 2.   Thermodynamical properties of supercritical CO_2.

3 NUMERICAL EXAMPLE

In the following, the system of equations (19) is solved numerically using an explicit 2nd order Runge-Kutta finite difference scheme. Among the parameters which appear in the considered problem, some have a major influence on the evolution of the system: the initial sate of stress and initial temperature, the thickness of the sheared zone, the initial permeability of the medium, the friction coefficient.

As pointed out by Rice (2006), recent field observations suggest that slip in individual events may then be extremely localized, and may occur primarily within a thin shear zone of a few millimeters thickness (e.g. Chester et al., 2004; Wibberley and Shimamoto, 2005). Recent high velocity friction experiments on natural clayey gouge have also shown that shear occurs in a very localized thin zone of few hundreds of microns (Boutareaud et al., 2008). In the following, we present a reference computation corresponding to some typical values of the parameters, and then the influence of some parameters is discussed and illustrated.

The reference computation is performed for a fault at 7 km depth in a carbonate rock. We consider a shear-band with a thickness $h = 5$ mm and we assume that the slip velocity is $V = 1\,m/s$. The initial temperature, initial pore pressure and total normal stress are assumed to be $T_0 = 210°C, P_{p0} = 70$ MPa, $\sigma_n = 180$ MPa. The thermal expansion coefficient is $\lambda_n = 2.4 \times 10^{-5}(°C)^{-1}$. For the pore volume compressibility we account for effect of porosity change due to the mineral decomposition and also for the effect of effective stress change due to pore pressure rise. The expression of the pore volume compressibility is obtained from poroelasticity theory and is given by (Rice 2006, Ghabezloo and Sulem, 2009)

$$\beta_n = \frac{1}{n}(\beta_d - (1+n)\beta_s) \tag{21}$$

where β_s is the compressibility of the solid phase ($1.25 \times 10^{-5}MPa^{-1}$ for calcite) and β_d is the drained compressibility of the porous rock. We consider here the empirical expression for the effective stress dependent drained compressibility as proposed by Wibberley and Shimamoto (2005) for the Median Tectonic line fault zone in Japan:

$$\beta_d(\text{in MPa}^{-1}) = 2.5 \times 10^{-4} \exp(-1.38 \times 10^{-2}(\sigma_n - P_p)), ((\sigma_n - P_p)\text{in MPa}) \tag{22}$$

The initial porosity of the rock is taken equal to 0.03. The density of the porous rock is taken equal to $\rho = 2.6$ g/cm^3, the density of the solid phase (calcite) is $\rho_s = 2.71$ g/cm^3, the density of lime is $\rho_{CaO} = 3.35$ g/cm^3 and the specific heat is $\rho C = 2.7$ MPa/(°C). The recent high velocity shear experiments on Carrara marble of Han et al. (2007) have shown that the friction coefficient decreases rapidly to values as low as 0.06 due to the thermal decomposition of calcite induced by frictional heating. Here, we take $f = 0.1$.

We assume that the initial permeability of the intact medium is $k_f = 10^{-19}m^2$.

The computed results are presented in figure 3. The evolution in time of the temperature and the pore pressure in the centre of the band is plotted in figure 3a. Considering the constant slip velocity of 1 m/s, this graph can also be seen as the evolution of temperature and pore pressure with accumulated slip. The corresponding shear stress is plotted in figure 3b. These results show the coupling effect of heat which induces first a pore pressure increase. When the decomposition of the carbonate rock begins at about 700°C, the temperature increase is drastically slowed due to the energy consumed in the endothermic chemical reaction. Two competing effects act on the evolution of the pore pressure: on one hand the production of CO_2 induces an additional fluid mass and thus a pressurization of the pore fluid; on the other hand the increase of porosity due to the solid decomposition induces an increase of the permeability of the medium which limits the pressurization. The pore pressure in the centre of the band exhibits a maximum of about 168 MPa which does not exceed the total normal stress acting (180 MPa) on the band which means that full liquefaction is not reached. As mentioned above, through the friction law the shear stress is related

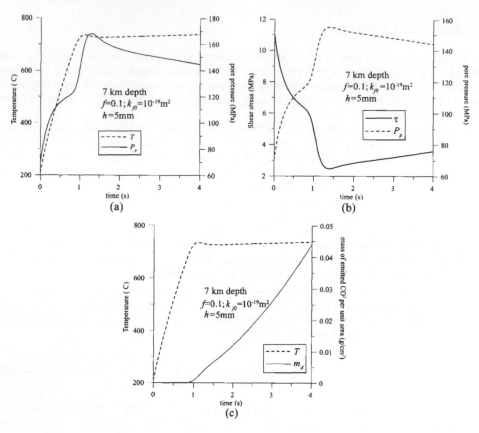

Figure 3.   Fault at 7 km depth: (a) Evolution of temperature and pore pressure in the centre of the shear band, (b) evolution of the shear stress, and (c) evolution of the mass of emitted CO_2.

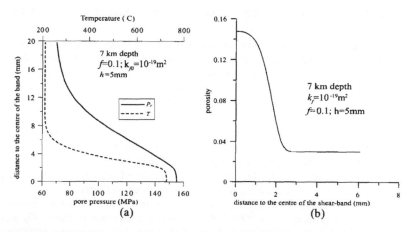

Figure 4.   Fault at 7 km depth: Pore pressure, temperature (a) and porosity (b) fields after 5 s.

to the mean effective stress inside the band (figure 3b). Consequently, the shear stress decreases rapidly during initial pressurization and then increases again. Thus, the mineral decomposition of the rock can be seen as a mechanism of fault weakening in a first stage then fault re-strengthening in a second stage.

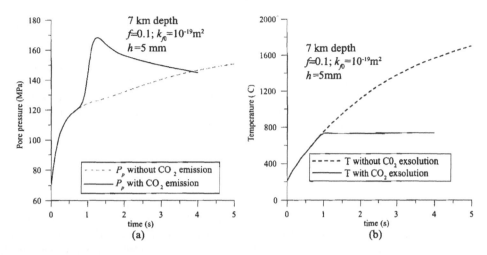

Figure 5. Fault at 7 km depth: Effect of CO_2 emission on the evolution of temperature (a) and pore pressure (b) in the centre of the shear band.

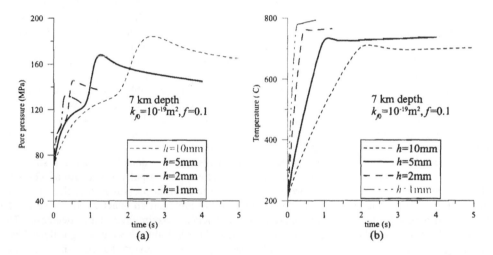

Figure 6. Effect of the shear band thickness: Evolution of temperature (a) and pore pressure (b) in the centre of the shear band.

The accumulated mass per unit area of the fault plane of emitted CO_2 is plotted on figure 3c. After a few seconds, the temperature reaches a quasi constant state and the production rate is almost constant. These results show that the mass of emitted CO_2 after 4 s is about 0.045 g/cm^2. For example if we assume that the area of the fault zone is 1 km^2, we obtain an amount of 450 t.

The pore pressure and temperature field around the centre of the band is shown in figure 4a for $t = 4\ s$ and the porosity is plotted in figure 4b. The plots in figure 4 show that the pore pressure and temperature increases are localized in the central zone of the band and that the porosity reaches about 15%. The corresponding permeability is about 10^{-17} m^2. The porosity is affected only in zones where the temperature exceeds 700°C which for the considered computation corresponds to a width of about 4 mm.

The effect of mineral decomposition is also shown in figure 5 where the results of the above computation are compared to the ones obtained without CO_2 degassing. In this latter case only thermal pressurization occurs. If no mineral decomposition occurs, the pore pressure increase is slower and for the considered parameters, full liquefaction is not attained and consequently the temperature increase is much stronger and melting of the rock can occur (the melting temperature is about 1600°C for calcite).

As discussed by Vardoulakis (2002), Rempel and Rice (2006) and Sulem et al. (2007) the actual thickness of the 'ultra-localized' zone of highly strained material is a key parameter in the analysis. The influence of the shear band thickness is shown in figure 6 for which the above computation is repeated assuming different values of h (1 mm, 2.5 mm, 10 mm). For thinner sheared zones, the pore pressure pulse is occurring earlier and is shorter and the corresponding temperature increase is stronger. However the maximum temperature is not so much affected ($\sim$750°C–800°C).

4   CONCLUSION

A first essential result of the above analysis is that the endothermic reaction of calcite decomposition limits the co-seismic temperature increase to less than $\sim$800°C (corresponding to the initiation of the chemical reaction) within a carbonate shear band under rapid slip. Decarbonation is only one of the possible thermal reactions of mineral decomposition. Phyllosilicates are common secondary minerals in faults and their reaction of thermal dehydration is also endothermic. Therefore, it is likely that the reaction-induced temperature limitation obtained in our model may apply qualitatively to most fault zones. In this case, mature faults with a long history of slip and a large cumulated displacement are likely more prone to reaction-induced temperature limitation than recent faults, because of their larger content in volatile-rich secondary minerals. This would provide another explanation to the notorious absence of positive heat flow anomaly on active crustal faults such as San Andreas (Lachenbruch and Sass 1980): a large part of the heat produced by friction would be consumed by endothermic reactions. Another consequence is that friction melting is hampered by endothermic calcite decomposition in carbonate fault rocks, and probably in other faults containing a sufficiently high proportion of volatile-rich secondary minerals. This is consistent with the relative scarcity of pseudotachylytes in mature faults such as the Punchbawl fault (Sibson 1975; Chester et al., 2004), and their occurrence in less evolved faults such as Nojima (Fujimoto et al., 1999; Ohtani et al., 2000; Tanaka et al., 2001; Otsuki et al., 2003).

The second essential implication of the above analysis is that decarbonation is a source of CO_2 that significantly increases the slip-weakening effect of thermal pressurization. The pore fluid pressure exhibits an initial phase of increase due to thermal pressurization, then a sudden acceleration of generated pore pressure when the solid decomposition is activated. However, the increase of permeability limits the pore pressure that reaches a maximum and then decreases. The numerical results reproduce this pore pressure pulse and the initial fault weakening followed by a re-strengthening of the shear stress. Our model may be particularly adapted to account for seismic slip in seismogenic normal faults of the Corinth rift (Greece) cutting across mesozoic carbonates at 4–6 km depth (Flotté et al., 2005). In particular the Heliki and the Aegion faults have produced several earthquakes with $M_w \geq 5$ in the past decades. The positive CO_2 anomaly in springs nearby these faults (Pizzino et al., 2004) might come from slip-induced decarbonation.

Thermal decomposition of rocks appears to be an important physical process in the phenomenon of thermal heating and pore fluid pressurization during seismic slip. The combined effects of frictional heating, temperature rise, endothermic mineral decomposition, pore pressure rise, porosity and permeability increase result in highly coupled and competing processes. Our model of thermal pressurization, taking into account these coupled processes, provides a more robust framework for estimating the dynamic friction of faults and the energy balance of earthquakes.

REFERENCES

Arai, T., Okusawa, T. and Tsukahara, H. 2001. Behavior of gases in the Nojima Fault Zone revealed from the chemical composition and carbon isotope ratio of gases extracted from DPRI 1800 m drill core, *The Island Arc*, 10 (3–4), 430–438.

Arai, T., Tsukahara, H. and Morikiyo, T. 2003. Sealing Process with Calcite in the Nojima Active Fault Zone Revealed from Isotope Analysis of Calcite, *Journal of Geography*, 112 (6), 915–925.

Boutareaud S., Calugaru D., Han R., Fabbri O., Mizoguchi K., Tsutsumi A., and Shimamoto T. 2008. Clay-clast aggregates: A new textural evidence for seismic fault sliding? Geophy. Res. Let., 35, L05302, doi: 10.1029/2007GL032554.

Brantut, N., Schubnel, A., Rouzaud, J.-N., Brunet, F. and Shimamoto, T. (2008). High-velocity frictional properties of a clay-bearing fault gouge and implications for earthquake mechanics, *J. Geophys. Res.*, 113, B10401, doi:10.1029/2007JB005551.

Chester, F.M., Chester, J.S., Kirschner, D.L., Schulz, S.E. and Evans, J.P. 2004. Structure of large-displacement, strike-slip fault zones in: Rheology and deformation in the lithosphere at continental margins eds. Karner, G.D., Taylor, B., Driscoll, N.W. & Kohlstedt, D.L., Columbia University Press, New York.

David, C., Wong, T-F., Zhu, W. and Zhang, J. 1994. Laboratory measurement of compaction-induced permeability change in porous rocks: Implications for the generation and maintenance of pore pressure excess in the crust. Pure and Applied Geophysics;143 (1–3), 425–456.

Dollimore, D., Tong, P. and Alexander, K.S. 1996. The kinetic interpretation of the decomposition of calcium carbonate by use of relationships other than the Arrhenius equation, Thermochimica Acta, 282/283, 13–27.

Famin, V., Nakashima, S., Boullier, A.-M., Fujimoto, K. and Hirono, T. 2008. Earthquake produce carbon dioxide in crustal faults, Earth and Planetary Science Letters, 265, 3–4 (30) 487–497.

Flotté, N., Sorel, D., Miller, C. and Tensi, J. 2005. Along strike changes in the structural evolution over a brittle detachment fault: Example of the Pleistocene Corinth-Patras rift Greece, Tectonophysics, 403 (1–4), 77–94.

Fujimoto, K., Tanaka, H., Tomida, N., Ohtani, T. and Ito, H. 1999. Characterization of fault gouge from GSJ Hirabayashi core samples and implications for the activity of the Nojima fault, in: The International workshop on the Nojima Fault core and borehole data analysis, edited by H. Ito, K. Fujimoto, H. Tanaka, and D. Lockner, pp. 103–109, Geological Survey of Japan, Tsukuba.

Ghabezloo, S. and Sulem, J. 2009. Stress dependent thermal pressurization of a fluid-saturated rock. Rock Mechanics and Rock Engineering, 42, 1–24.

Han, R., Shimamoto, T., Hirose, T., Ree, J.-H. and Ando, J. 2007. Ultralow friction of carbonate faults caused by thermal decomposition, Science, 316, 878–881.

Hirono, T., Ikehara, M., Otsuki, K., Mishima, T., Sakaguchi, M., Soh, W., Omori, M., Lin, W., Yeh, E., Tanikawa, W. and Wang, C.Y. 2006. Evidence of frictional melting from disk-shaped black material, discovered within the Taiwan Chelungpu fault system, Geophysical Research letters, 33 L19311, doi:10.1029/2006GL027329.

Hirono, T., Yokohama, T., Hamada, Y., Tanikawa, W., Mishima, T., Ikehara, M., Famin, V., Tanimizu, M., Lin, W., Soh, W. and Song, S.-R. 2007. A chemical kinetic approach to estimate dynamic shear stress during the 1999 Taiwan Chi-Chi earthquake, Geophysical Research Letters, 34 L19308, doi:10.1029/2007GL030743.

Italiano, F., Martinelli, G. and Plescia, P. 2008. CO_2 Degassing over seismic areas: The role of mechanochemical production at the study case of central Apennines, Pure Appl. Geophys. 165, 75–94.

Lachenbruch, A.H. and Sass, J.H. 1980. Heat Flow and Energetics of the San Andreas Fault Zone, J. Geophys. Res, 85 (11), 6185–6223.

Lewicki, J.L. and Brantley, S.L. 2000. CO_2 degassing along the San Andreas fault, Parkfield, California, Geophysical Research letters, 27 (1), 5–8.

Lewicki, J.L., Evans, W.C., Hilley, G.E., Sorey, M.L., Rogie, J.D. and Brantley, S.L. 2003. Shallow soil CO_2 flow along the San Andreas and Calaveras Faults, California, Journal Of Geophysical Research, Vol. 108, N°. B4, 2187.

Lin, A., Tanaka, N., Uda, S. and Satish-Kumar, M. 2003. Repeated coseismic infiltration of meteoric and seawater into deep fault zones: a case study of the Nojima fault zone, Japan, Chemical Geology, 202 (1–2), 139–153.

L'vov, B.V. 2002. Mechanism and kinetics of thermal decomposition of carbonates, Thermochimica Acta, 386, 1–16.

L'vov, B.V. 2007. Thermal Decomposition of Solids and Melts, doi:10.1007/978-1-4020-5672-7 pp., Springer Netherlands.

L'vov, B.V. and Ugolkov, V.L. 2004. Peculiarities of $CaCO_3$, $SrCO_3$ and $BaCO_3$ decomposition in CO_2 as a proof of their primary dissociative evaporation, Thermochimica Acta, 410, 47–55.

Ohtani, T., Fujimoto, K., Ito, H., Tanaka, H., Tomida, N. and Higuchi, T. 2000. Fault rocks and past to recent fluid characteristics from the borehole survey of the Nojima fault ruptured in the 1995 Kobe earthquake, soutwest Japan, J. Geophys. Res., 106 (B7), 16161–16171.

Otsuki, K., Monzawa, N. and Nagase, T. 2003. Fluidization and melting of fault gouge during seismic slip: Identification in the Nojima fault zone and implications for focal earthquake mechanisms, J. Geophys. Res., 108 (B4), doi:10.1029/2001JB001711.

Pizzino, L., Quattrochi, F., Cinti, D. and Galli, G. 2004. Fluid geochemistry along the Eliki and Aigion seismogenic segments Gulf of Corinth, Greece, C. R. Geoscience, 336, 367–374.

Rempel, A.W. and Rice, J.R. 2006. Thermal pressurization and onset of melting in fault zones, J. Geophys. Res, 111, B09314

Rice, J.R. 2006. Heating and weakening of faults during earthquake slip, J. Geophys. Res., 111, B05311

Sato, T. and Takahashi, M. 1997. Geochemical changes in anomalously discharged groundwater in Awaji Island -after the 1995 Kobe earthquake-. Chikyukagaku 31, 89–98.

Saxena, S.K. and Fei, Y. 1987. Fluids at crustal pressures and temperatures. 1. Pure species. Contribution to Mineralogy and Petrology 95, 370–375.

Sibson, R.H. 1975. Generation of pseudotachylyte by ancient seismic faulting, Geophysical Journal International, 43 (3), 775.

Sulem, J., Lazar, P. and Vardoulakis, I. 2007. Thermo-Poro-Mechanical Properties of Clayey Gouge and Application to Rapid Fault Shearing, Int. J. Num. Anal. Meth. Geomechanics, 31 (3), 523–540.

Sulem, J. and Famin, V. 2009 Thermal decomposition of carbonates in fault zones: slip-weakening and temperature-limiting effects, J. Geophys. Res., 114, B03309, doi:10.1029/2008JB006004.

Sulem, J., Famin, V. and Noda, H. 2009. A correction to "Thermal decomposition of carbonates in fault zones: slip-weakening and temperature limiting effects", Journal of Geophysical Research, 114, B06311, doi:10.1029/2009JB006576.

Tanaka, H., Fujimoto, K., Ohtani, T. and Ito, H. 2001. Structural and chemical characterization of shear zones in the freshly activated Nojima fault, Awaji Island, southwest Japan, J. Geophys. Res., 106 B5, 8789–8810.

Ueda, A., Kawabata, A., Fujimoto, K., Tanaka, H., Tomida, N., Ohtani, T. and Ito, H. 1999. Isotopic study of carbonates in Nojima fault cores, in The international workshop on the Nojima fault core and borehole data analysis, edited by H. Ito, K. Fujimoto, H. Tanaka, and D. Lockner, pp. 127–132, Geological Survey of Japan, Tsukuba.

Vardoulakis, I. 2002. Dynamic thermo-poro-mechanical analysis of catastrophic landslides, Géotechnique, Vol. 52, No. 3, 157–171.

Veveakis, E., Vardoulakis, I. and Di Toro, G. 2007. Thermoporomechanics of creeping landslides: The 1963 Vaiont slide, northern Italy, J. Geophys. Res., 112, F03026, doi:10.1029/2006JF000702.

Wibberley, C. and Shimamoto, T. 2005. Earthquake slip weakening and asperities explained by thermal pressurization, Nature 426 (4), 689–692.

Thermal mechanisms and friction laws determining the stability and localization during slip weakening of shallow faults

E. Veveakis, S. Alevizos & I. Vardoulakis

Department of Applied Mathematics and Physics, National Technical University of Athens, Athens, Greece

ABSTRACT: In this study we provide the mathematical formulation of the problem of simple shear of a biphasic soil material. The governing equations are derived from first principles of continuum mechanics and thermodynamics and several constitutive laws are studied with respect to their influence to the behaviour of the sheared material. It is shown that shear heating may lead to unstable accelerations and localization of the deformation and of dissipation. The theory is then applied to specific weakening mechanisms observed in faults, such as shear melting, thermal pressurization of the pore fluid and chemical decomposition of the minerals of the fault, and the characteristic times for each effect is extracted. A simple spring-block model is then allowed to move over a finite shear zone that exhibits the aforementioned behavior. The stability behavior of the different frictional regimes are analyzed and stick-slip instabilities appear even in velocity hardening regime.

1 INTRODUCTION

Field evidence from exhauming faults reveal that seismic events take place in even narrower zones within the shear-bands, formed from post-failure evolution (Chester & Chester, 1998). The mechanisms that lead to fault weakening and seismic slip in these zones are considered to be mainly thermal in origin, like thermal pressurization, flash heating and melting (Rice, 2006). A vast category of geomaterials exhibit rate-dependent behaviour during shear, thus influencing their mechanical properties and state with velocity. Indeed, it has been shown recently by Veveakis et al. (2007) that a fully saturated (biphasic) clay material may change, when sheared under constant load, its material properties after some temperature and velocity. In particular, the material may undergo a solid-liquid phase transition where the clay expels water due to heat and becomes liquefied, verifying that thermal pressurization could be one of the two main weakening mechanisms for earthquakes on mature faults.

Also, Vardoulakis (2002a, b), Goren and Aharonov (2007) and Veveakis et al. (2007) showed that the same concept of thermal pressurization could be the main weakening mechanism for creeping, deep seated landslides. Recently, Fialko and Khazan (2005) used a similar framework of shear heating and thermodynamics of phase transitions for the effect of melting, that may exhibit the same behavior with thermal pressurization, and provide the conditions at which melting may reduce the strength of a fault. In order to model such kind of behaviors it is common to account for rate and state dependency in the frictional behavior of the geomaterials (Dieterich, 1972). In fact, the need of accounting to rate dependent friction laws in order to model the stick-slip motions that characterize an earthquake was brought into light by Gu et al. (1986). Scholz (1998) added that the development of a full constitutive law for rock friction is crucial to understand earthquake phenomena such as seismogenesis and seismic coupling, pre- and postseismic phenomena, and the insensitivity of earthquakes to stress transients.

Chester and Chester (1998), based on field observations suggested that unstable seismic slip is an extremely localized effect, that occurs primarily in a thin shear zone (of the order of millimetre), which lies within a finely granulated (ultracataclastic) fault core of typically tens to hundreds millimetre (Figure 1a). As discussed thoroughly by Scholz (2002, pages 70–74) the structure of an ultrathin shear zone within the ultracaclastic core (and near the gouge-rock interface) is observed only in the case of a wet (i.e. saturated) gouge, under hot conditions. It is to be noted also that under these hot-wet conditions litle grain size reduction was observed, unlike the case of cool-wet

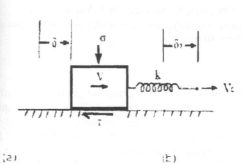

(a)   (b)

Figure 1.   (a) The Ultracataclastic core at the base of the sliding block with the velocity and temperature profiles. At the centre of the core we depict the primary slip surface, i.e. the zone at which the block effectively slides during the pressurization phase (Chester and Chester, 1998). (b) The spring-block model, sliding over the ultracataclastic core under constant load $\sigma = mg$.

or hot-dry conditions during the shearing of a quartz gouge ([21]). Under the hot-wet conditions a temperature weakening and velocity strengthening frictional behavior is observed, which, as argued by Chester (1995) is dominated by solution transfer processes, such as pressure solution.

Based on this observation, in this work we study the stability behavior of a frictional spring-block system (Figure 1b), by assuming rate and thermal dependence of friction. We consider the basal ultracataclastic zone to be filled by a fully saturated, viscous material. The solid phase of this material is assumed to have permanent contacts between the grains, forming thus a permanent solid skeleton. Since we allow the temperature to rise, little grain crushing is anticipated and this assumption remains valid throughout this study. When the ultracataclastic zone is active, the corresponding velocities and displacements inside the viscous material of the shear band provide a physical process known as visco-plastic flow. It is shown that shear heating of active faults is a post-failure localizing mechanism, in a narrow core zone within the initial fault core. In order to obtain a more realistic mathematical model of the various processes occuring in a slipping fault, and account for the observations of Chester (1995) for solution transfer processes, we emphasize on the mechanisms that are observed to dominate the later, unstable stage of the slip evolution (taking place in seismic velocity range, >1 m/s), in extremely localized zones. Such processes are the thermal pore fluid pressurization, shear melting, elasto-hydrodynamic pressurization or silica gel lubrication and finally the decomposition, of minerals in fault rocks. All these mechanisms are considered to be weakening mechanisms, reducing the shearing resistance of the fault plane when active. These mechanisms are considered to be trigerred either by the unstable evolution of the shear heating during quasi-static creep in smaller velocities (of the order of mm/day) or, independently, due to the unstable initial temperature profiles (example in the case of a locked fault).

2   THE SPRING-BLOCK MODEL

We consider the simplest and most commonly used model for discussion of the slip motion (Gu et al., 1986); the spring-block model (Figure 1). A rigid block is draged under a velocity, $V_0 = V_0(t)$, called drift velocity, sliding on a plane with velocity $V = V(t)$ under constant load $\sigma = mg$, where m is the block's mass. Attached to the block there is a spring of stiffness k. Instead of having a block-to-plane interface at the base of the block, in this framework we consider that the block slides on a thin basal shear of infinitesimally small—but finite—thickness, d. The stresses, as well as the temperature and velocity fields of the two structures are requested to be continuous in the common boundary of them, i.e. the upper boundary of the shear zone $z = d/2$. In addittion we assume a rate- and state- friction law of the form $\tau = \sigma\mu(V,\theta)$, where as the state variable θ we consider temperature. We will accept friction laws that are the form, $\mu(V,\theta) = f(V)g(\theta)$, i.e. that consists of two antagonistic mechanisms, that counterbalance each other (Vardoulakis, 2002b; Veveakis et al., 2007). The equations of motion of the problem become (all the quantities are dimensionless):

$$\dot{u} = l_1 F$$

$$\dot{F} = -u - l_2(f_u g V_0 \dot{u} + f g_\theta \theta_0 \dot{\theta}) \tag{1}$$

where we used the dimensionless quantities $V = V/V_0, t = t/\tau_r, F = F/kL, L = \tau_r V_0, \theta = \theta/\theta_0$, τ_r being the "relaxation time", i.e. the time at which the system shifts from one steady state to another, or the time that is needed for the system to recover a stress drop to a new steady state (Gu et al., 1986). Also, in the above expression $u = V - 1$, $l_1 = k\tau_r^2/m$ and $l_2 = \sigma/kL$, while f_u, g_θ are the partial derivatives of the friction coefficient with respect to velocity and basal temperature, respectively. In order to close the system of equations we need an evolution equation for temperature. However, the heat equation for the block will always yield an insensitive block to finite temperature variations as long as its heat capacity is large. Since it is realisticto restrict ourselves to bounded variations of friction with temperature, we may set in Eq. (1), $f g_\theta \theta_0 \dot{\theta} = \varepsilon \ll 1$, and perform an imperfect bifurcation analysis of the system (Veveakis et al., 2009).

If we focus only in the case $\dot{\theta} > 0$, the assumption of two antagonistic mechanisms that govern the frictional behavior of the basal material forces us to examine three different cases for the sign of f_u and ε:

1. Velocity and Thermal Hardening ($f_u, \varepsilon > 0$): In this case the system has two eigenvalues that have negative real parts, thus the system is stable.
2. Velocity and Thermal Softening ($f_u, \varepsilon < 0$): In this case both eigenvalues have positive real parts, thus the system is unconditionally unstable.
3. Velocity Hardening and Thermal Softening ($f_u > 0, \varepsilon < 0$): In this case, since $f_{u0} \sim O(1)$, when $kL/\sigma \geq \varepsilon$ the system is stable, while when $kL/\sigma < \varepsilon$ the system is unstable. In addition, when $f_{u0} \leq 2\frac{|\varepsilon| + \sqrt{|\varepsilon|^2 + l_1\, g_0^2}}{l_1 l_2\, g_0^2}$ the system's eigenvalues are complex, pertaining to oscillatory stability or instability.
4. Velocity Softening and Thermal Hardening ($f_u < 0, \varepsilon > 0$): This case is exactly the same like the previous case, when $kL/\sigma \geq \varepsilon$ the system is stable, while when $kL/\sigma < \varepsilon$ the system is unstable. It is to be noted that the stability condition $kL/\sigma < \varepsilon$ is actually the main difference from the previous studies (Gu et al., 1986), since it introduces a new instability area during Velocity hardening. Indeed, when $\varepsilon = 0$ this area is meaningless, and the system provides the same results with the model of Gu et al. (1986) only with rate dependency.

It is evident that through an imperfect bifurcation analysis we managed to derive stability regimes for the spring block with respect to a basal parameter, the basal temperature. Thus, in order to understand the importance and the physical meaning of the restriction $kL/\sigma < \varepsilon$ we must examine the behaviour of the basal material.

3 FRICTIONAL BEHAVIOR OF THE BASAL MATERIAL

3.1 *Friction laws*

Based on first principles we may derive the heat diffusion equation for the basal materal, if we assume it to deformed under constant volume (Veveakis et al., 2007):

$$\frac{\partial\theta}{\partial t} = \kappa_m \frac{\partial^2\theta}{\partial z^2} + \frac{\tau V}{j(\rho C)_m} \tag{2}$$

where $(\rho C)_m$ is the specific heat capacity of the biphasic mixture. We may assume different friction laws to determine the product τV. The most common are the Dieterich-Ruina law, the Arrhenius law and an exponential law. Assuming thus that Equation (2) may be written in the dimensionless form

$$\frac{\partial \theta}{\partial t} = \frac{\partial^2 \theta}{\partial z^2} + Gr \cdot \varphi(\theta) \tag{3}$$

where $Gr \propto \tau_d V_0 d / jk_m$ is the Gruntfest number (Veveakis et al, 2009) and $\varphi(\theta)$ is a function of temperature that depends on the friction law (In this case $\varphi(\theta) = e^\theta$ for exponential law, $\varphi(\theta) \propto e^{1/\theta}$ for Arrhenius law and $\varphi(\theta) = \theta^m$ for Dieterich-Ruina law).

For the 3 laws mentioned earlier we may perform a numerical bifurcation analysis to obtain the results depicted in Figure 2 (Veveakis et al, 2009). The behaviour depicted in Fig. 2((a) and (b)) represents the celebrated phenomenon of thermal runaway. For various values of the Gruntfest number the problem has two ($Gr < Grc$), one ($Gr = Grc$) or no ($Gr > Grc$) steady state solutions. For $Gr < Grc$ the lower branch is stable, while the upper part is unstable. It is to be noted that unlike the exponential and logarithmic laws, the Arrhenius law does not provide the property of thermal

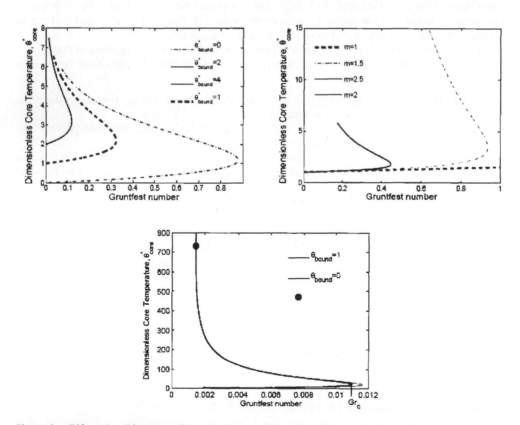

Figure 2.   Bifurcation Diagrams of (a) the Bratu problem. Results are plotted for various values of the boundary temperature. All the profiles present a turning point, for $Gr = Grc$. The lower branch is stable (since eig $(J) < 0$, J being the Jacobian of the dynamical system), while the upper branch is unstable (eig $(J) > 0$); (b) the Dieterich-Ruina law for various values of the power coefficient, m. It is to be noted that for m > 1 the model presents a turning point, exactly like the Bratu problem, while for m < 1 the solution is stable; (c) The Arrhenius law model, for various values of the boundary temperature, presents two turning points. The lower branch admit the same behavior with the two previous models, while the upper -stable- branch bounds the infinite increase of temperature (the black dot depicts the upper turning point).

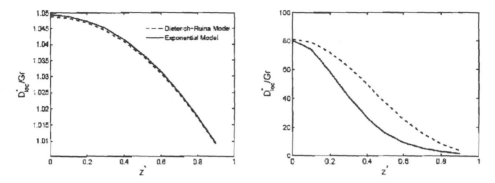

Figure 3. Plot of the Dissipation functions (1) $D_{loc}/Gr = e^\theta$ (2) $D_{loc}/Gr = e^{25}$ for an arbitrary (different) value of the Guntfest coefficient, before. Since $Gr > 0$ always, the figure depicts qualititavely the shape of each dissipation function (a) before and (b) after the bifurcation point. We notice that localization of the dissipation occurs after the bifurcation point.

runaway although it can be seen that for lower temperatures the behavior of this law is identical to the two previous; a lo wer, stable, branch is followed by an intermediate, unstable, branch. The difference however is that at elevated temperatures the Arrhenius law present a third, stable, branch which does not allow for infinite temperature growth and completes the S-shape response curve of Fig. 2(c). This thermal runaway is localized in a thin band across the centre of the shear zone. To depict that, we have plotted the profiles of the dissipation function inside the shear zone (Fig. 3). It is obvious that at the unstable branch the process localizes in to what is observed in field as the Principal Slip Surface (Chester & Chester, 1998). Since thermal runaway instability forces temperature to unbounded increase, something physically unacceptable, we must consider that some reaction is triggered at higher temperatures, forcing the shear strength to drop further, and temperature to stabilize. To model this, we must take under account the reaction kinetics of these proceses.

3.2 Reaction kinetics

Chemical reactions typically involve the absorption or release of heat. In geomechanical applications and especially in fault mechanics, mechanisms like the thermal pressurization, the phase transition (decomposition) of the solid phase (e.g. the calcite decomposition in some faults) or melting are processes that require considerations from reaction kinetics and thermodynamics in order to be modelled. An arbitrary reaction has a reaction rate of the Arrhenius form:

$$r \propto \exp\left(-\frac{E}{R\theta}\right) \qquad (4)$$

where E is the activation energy of the reaction and R is the universal gas constant. In the presence of a chemical reaction, the heat diffusion equation is written

$$\frac{\partial \theta}{\partial t} = \kappa_m \frac{\partial^2 \theta}{\partial z^2} + \frac{\tau V}{j(\rho C)_m} - r\frac{\Delta H}{j(\rho C)_m} \qquad (5)$$

where ΔH is the rate that the reaction absorbs heat (i.e. the specific enthalpy of the reaction). The mass balance equation for the reactant phase is written as:

$$\frac{\partial c}{\partial t} = D\frac{\partial^2 c}{\partial z^2} - r \qquad (6)$$

where D is the reactant's diffusion coefficient. If we assume the presence of a first order reaction, $r = Ace^{-\frac{E}{R\theta}}$, and that friction obeys Arrhenius law ($\mu = \mu_0 (V/V_0)^N e^{\frac{E_d}{R\theta}}$) with activation energy $E_d = xE$, we may write Equation (5) in the dimensionless form:

$$\frac{\partial \theta}{\partial t} = \frac{\partial^2 \theta}{\partial z^2} + Gr \left[\exp \left(\frac{A}{1 + \delta\theta} \right) \pm Da \right] \exp \left(Ar \frac{\delta\theta}{1 + \delta\theta} \right) \tag{7}$$

where the sign $(+)$ denotes exothermic and $(-)$ endothermic reaction, $Ar = e^{-\frac{E}{R\theta_0}}$ is the Arrhenius number of the system, $A = \left(1 - \frac{x}{N} \right) Ar$, and

$$Gr = m \frac{\beta_T \tau_d \dot{\gamma}_0}{jk_m} \left(\frac{d}{2} \right)^2 e^{-\frac{E}{R\theta_0}}$$

$$Da = \frac{|\Delta H| k_0 c_0}{\beta_T \tau_d \dot{\gamma}_0} \tag{8}$$

By neglecting the effect of reactant's diffusion, we may perform a numerical bifurcation analysis of Equation (7) we observe that indeed chemical reaction stabilizes the unbounded temperature increase (Fig. 4) and the unstable localization of the dissipation (Fig. 5).

A general characteristic time that governs the chemical process may be calculated to be (Veveakis et al, 2009):

$$t_{ch} = \frac{j (\rho C)_m}{|\Delta H| k_0 c_0} \frac{E}{R} \tag{9}$$

This characteristic time is similar to the time that was calculated to govern thermal pressurization by Wibberely & Shimamoto (2005) and Veveakis et al. (2007):

$$t_{ch} = \frac{j (\rho C)_m}{\mu_0 \lambda_m} \frac{d}{V_0} \tag{10}$$

where λ_m is the pressurization coefficient.

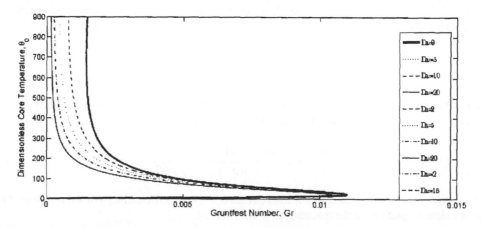

Figure 4.   Response diagram for various Da values, both for endothermic $Da < 0$ or exothermic $Da > 0$ reactions. We may see that the upper branches are stable, stabilizing thus the unbound temperature increase of the middle, unstable branch.

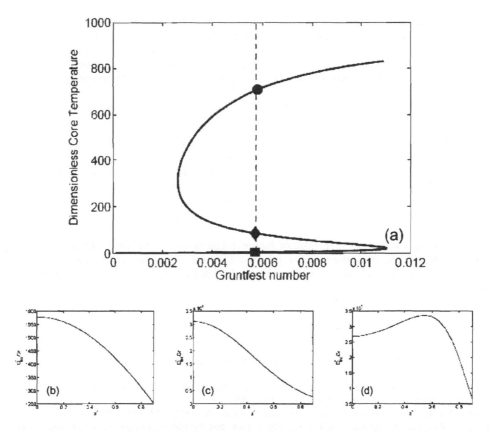

Figure 5. (a) Response diagram of Eq. (7) for $Da = -2$. Below this we present the diagram of the normalized Dissipation function for a specific Gruntfest number, at (b) the lower stable branch (square point of (a)), (c) the unstable middle branch (the rhomb point) and (d) the upper stable branch (the circle point).

4 CONCLUSIONS

The behavior of a frictional spring-block model with a rate and thermal dependence of friction is studied in this paper. The system presents a critical point of the temperature rate $\dot{\theta} > kL/\sigma$, above which it becomes unstable. By examining the frictional behavior of the basal material we observe that it also depicts unstable behavior for a specific range of the problem parameters, and of temperature (Fig. 4). Pass this point, temperature increases abruptly and a chemical reaction that weakens the fault is triggered. All this process takes place in a continuously localizing effective shear zone and the characteristic time that governs the process is the one shown in Equation (9).

The competition of rate- and state dependency of friction and the assumption in simple shear of a visco-plastic geomaterial leads to unstable plastic localization, when the strengthening mechanism fails to counterbalance the effect of the weakening one. This is a typical run-away instability of finite time. Motivated by the physics and mechanics of faults we could show here that this unstable behavior is capped, at elevated temperatures by the activation of endothermic chemical reactions. This was done here by means of a numerical bifurcation analysis of the steady problem. The analysis has shown clearly that at lower temperatures indeed friction dominates, while at elevated temperatures chemical reactions take the upper hand. The analysis has revealed the importance of set of dimensionless groups that determine fully the thermo-chemo-mechanical behavior of the fault under shear. Needles to say, that determining the actual values of these parameters for real case scenarios is but trivial. Finally we should make a comment concerning transient analyses, that were purposely omitted in this study. To that end we recall that the initial phase of shearing takes place at low temperatures and is friction-dominated. During this phase any of the considered here friction laws would provide qualitatively the same results. In order to highlight the importance of these dimensionless groups for the transient behavior we consider the case of the exponential friction law, that, in the absence of convective terms leads to the Frank-Kamenetskii equation,

$$\frac{\partial \theta}{\partial t} = \frac{\partial^2 \theta}{\partial z^2} + Gr \cdot e^{\theta} \tag{11}$$

In a recent paper (Veveakis et al., 2007) the authors showed that this equation has an asymptotic solution and that it describes a process which turns from stable to unstable at a dimensionless time . Past that point, slip localizes into an ultrathin core zone, temperature rises dramatically and following the present bifurcation analysis must triggers a chemical reaction at a later time, which in turn depends on the reactions activation temperature. Using the adiabatic approximation of the system equations (5) and (6) we may calculate the characteristic time scales of the various physical phases.

Formulating and treating, in the light of the present analysis, a numerical integration and asymptotic matching of the governing evolution pdes is still a work in progress, and the useful conclusions derived from this study could assist in understanding of the dominant physical processes of the problem.

REFERENCES

Chester F.M. A geologic model for wet crust applied to strike-slip faults. *J. Geophys. Res.—Solid Earth* 1995; 100, 13033–13044.

Chester F.M. and Chester J.S. Ultracataclasite structure and friction processes of the Punchbowl fault, San Andreas system, California. *Tectonophysics* 1998; 199–221.

Dieterich J.H. Time-dependent friction in rocks. *J. Geophys. Res* 1972; 377, 3690–3697.

Gu J.C., Rice J.R., Ruina A.L. and Tse S.T. Slip motion and stability of a single degree of freedom elastic system with rate and state dependent friction *J. Mech. Phys. Solids* 1986; 32 (3), 167–196.

Fialko Y. and Khazan Y. Fusion by earthquake fault friction: Stick or slip? *J. Geophys. Res.* 2005; 110, B12407, doi:10.1029/2005JB003869.

Goren L. and Aharonov, E. Long runout landslides: The role of frictional heating and hydraulic diffusivity. *Geophys. Res. Lett.* 2007; 34, L07301, doi:10.1029/2006GL028895.

Rice J.R. Heating and weakening of faults during earthquake slip. *J. Geophys. Res.* 2006; 111, B05311. doi:10.1029/2005JB004006.

Scholz C.H. Earthquakes and friction laws Nature 1998; 391, 37–42.

Scholz C.H. The Mechanics of Earthquake and Faulting. Gambridg University Press, 2002 (2nd Edition).

Veveakis E., Vardoulakis I. and Di Toro, G. Thermoporomechanics of creeping landslides: The 1963 Vaiont slide, northern Italy. *J. Geophys. Res.* 2007; 112, F03026, doi:10.1029/2006JF000702.

Vardoulakis I. Dynamic thermo-poro-mechanical analysis of catastrophic landslides. *Geotechnique* 2002a; 52, 157–171.

Vardoulakis I. Steady shear and thermal run-away in clayey gouges. *Int. J. Solids Struct.* 2002b; 39, 3831–3844.

Veveakis E., Alevizos S., Vardoulakis I. (2009) Chemical Reaction Capping of Thermal Runaway During Shear of Frictional Faults, Int. J. Mech Phys Solids, Submitted.

Cataclastic and ultra-cataclastic shear using breakage mechanics

Itai Einav & Giang D. Nguyen
School of Civil Engineering, The University of Sydney, Sydney, NSW, Australia

ABSTRACT: Grain crushing is an important phenomenon that occurs in cataclastic granular shear. As crushing progresses the grain size distribution (gsd) evolves. The effect of the gsd on the phenomenology of the material is well established. A theory of breakage mechanics was recently proposed based on first principles and the framework of thermodynamics, which can explain the phenomenology. Here, we clarify some conceptual aspects in relation to the property of dissipation. We demonstrate the theory's potential for modelling granular shear, with or without the incorporation of non-local integration terms. The latter consideration enables simulations to present the formation of ultracalaclasite in continuously sheared layers.

1 INTRODUCTION

Comminution is an important phenomenon that describes the evolution of the grain size distribution (gsd) due to grain crushing, and occurs in cataclastic granular shear. At the microscopic level grain crushing triggers fascinating physics such as self-organisation, fractal scaling and non-local redistribution effects. The effect of the gsd on the phenomenology of the material is well established. For more than a century, theories of comminution have been studied in terms of the energy consumed during the operation of mills. Subsequent models of comminution have refined the predictive capabilities of these early works for specific circumstances. Each of these models can be calibrated to fit predictions of the change in gsd during a particular operation. However, their fit is extremely sensitive to changes in geometrical constraints of an operation. A continuum mechanics theory of comminution is therefore required, since it may embrace through a single constitutive model the required ingredients for predicting how the gsd evolves in response to stress-strain variations, at any point in the problem domain, without having to recalibrate the model to changes in the boundary conditions.

Recently, such a theory was proposed that accounts for these requirements (Einav, 2007a). First principles of thermodynamics and statistical physical arguments were proposed, which enabled to incorporate an internal variable, called "breakage", that scales the distance of the current grain size distribution (gsd) from a reference initial gsd and an ultimate gsd. In analogy with the term "damage mechanics" in solid mechanics the current theory was named "breakage mechanics".

Models based on damage mechanics have been used extensively (e.g. Lyakhovsky et al., 1997; Ricard and Bercovici, 2003; Hamiel et al., 2004) to study the mechanical behaviour of rocks and formation of shear bands at appropriate scales (e.g. Fig. 1). In these models, the use of a scalar damage variable may characterize the underlying evolution of the micro-structures, if the representative volume can be considered large enough to allow viewing the distribution of the internal flaws as homogeneous (Hamiel et al., 2004). Rather than looking at the evolution of internal flaws, which exist out of the core of the fault gouge zone, in using the theory of breakage mechanics we target the granulated gouge material (Fig. 1).

In cataclastic shear energy is dissipated, and it is important to quantify how much of the energy is dissipated in relation to the various mechanisms (Olgaard and Brace, 1983; Chester et al., 2005; Reches and Dewers, 2005). We examine this point from a mechanistic direction (Nguyen and Einav, 2009). In breakage mechanics, the breakage dissipation relates to the entire particle statistics. Results that come from statistical arguments often end up being less intuitive, albeit their tendency to be simpler in structure. The proposed breakage dissipation expression by Einav (2007b) is simple (the luxury of using statistical arguments), yet deserves clarification in terms of a visualisation example. For that purpose, we propose a simple mechanical analog. In particular, we uniquely

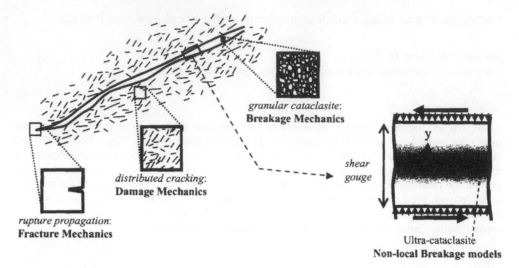

granular cataclasite:
Breakage Mechanics

shear gouge

y

distributed cracking:
Damage Mechanics

rupture propagation:
Fracture Mechanics

Ultra-cataclasite
Non-local Breakage models

Figure 1.   Fault gouge, cataclasis, ultra-cataclasis and the scope of breakage mechanics.

distinguish between two seemingly similar, but different entities: the dissipation from grain crushing, or as we call it 'breakage dissipation', and from the creation of new surface area, or 'surface area dissipation'. Breakage dissipation embodies not only the surface area dissipation, but also the dissipation to the surrounding from the re-distribution of locked-in strain energy; this latter factor, in fact, appears more significantly in value, though triggered by the same grain crushing event.

Finally, we demonstrate how breakage mechanics can be applied to gouge shear, with and without consideration of non-local effects, with the latter giving rise to formation of ultracalaclasite (Fig. 1).

2   DISSIPATION IN COMMINUTING GRANULAR MEDIA

An important consequence of the first and second laws of thermodynamics is the Clausius-Duhem dissipation inequality (Ziegler, 1982; Maugin, 1992):

$$\Phi_m + \Phi_q \geq 0 \tag{1}$$

where Φ_m and Φ_q are the mechanical and thermal dissipations. The thermal dissipation is expressed by:

$$\Phi_q = \theta q \cdot \nabla(1/\theta) \geq 0 \tag{2}$$

where θ is the temperature, $q = \{q_i\}$ is the heat flux vector, and $\nabla := \{\partial/\partial x_i; \ i = 1, 2, 3\}$ is the nabla operator. Since the directions of temperature gradient and heat flux are opposed, the thermal dissipation is always positive. A more stringent inequality is to assume that the mechanical dissipation is by itself non-negative:

$$\Phi_m \geq 0 \tag{3}$$

According to the internal variable formulation of thermodynamics, the mechanical dissipation expresses the integration of the dot products of the mechanical dissipative forces and the rate of their conjugated internal variables. We consider internal variables to be successful if they are measurable. Formulating a constitutive model based on measurable internal variables let those to be predicted, and the connection of their evolution be understood in terms of the postulated form of the dissipation (and the energy potential). Each of the various internal variables is associated with a separate mechanism. These mechanisms could be coupled in the sense that one leads to the other, but the internal variables should be independent. While successful internal variables could be measured objectively and the total mechanical dissipation be indirectly measured by infrared tomography (Maugin, 1992), the break up of the total mechanical dissipation into the various components would depend on the successfulness of the proposed constitutive models. Here we propose a simple analog to explore the conceptual functional form of the dissipation in relation to part of the various mechanisms.

2.1 *Mechanical analog of dissipation*

We start by a qualitative examination of the interplays between the various mechanisms, in relation to a single grain crushing event. Consider a single crushable particle and its first ring of neighbours (Fig. 2). Prior to crushing this particle transmits a given set of forces through its contacts. The directions and thickness of the lines in Figs. 2a and 2b represent the direction and magnitude of the corresponding contact forces. Immediately after a particle crushes, it dissipates energy in the sense of fracture mechanics (the so called "surface area dissipation"). The forces that were transmitted through this particle will be redistributed to its neighbours. In that respect its stored energy is redistributed somewhere else. We shall see that this redistribution gives rise to a new dissipation, which we call "energy redistribution dissipation". This term is a function of the surrounding topology. A further contribution comes from reorganization of particles (Fig. 2c), which in the simplest form could be effectively described by "plastic dissipation". Plastic dissipation is distinctively different from the other contributors, since the surface area and the energy redistribution dissipations are attainable even without topological rearrangement of surrounding particles (only a minute rearrangement of fragments is sufficient to totally alter the transmission of contact forces).

Let us now introduce a simple mechanical analog to portray the connection between the surface area and the energy redistribution dissipations (Nguyen and Einav, 2009). At this stage the analog ignores plastic dissipation, i.e., the topological reorganization of the particles. We skip much of the detail, which can be found in the appendix of Nguyen and Einav (2009).

Two sets of particle blocks are stacked alongside in a one dimensional configuration, separated by a single particle (blackened in Fig. 3). For simplicity we assume that each of the separated sets contain similar number of particles, each of which having a stiffness K_p. Again for simplicity, the particle that separates between these packs, i.e., the blackened particle, is assumed to be fully rigid but connected via two flexible bonds to the boundaries. Together, the particle and its bond are analogous to the blackened particle in Fig. 2. The bond breaking represents internal surface area creation,

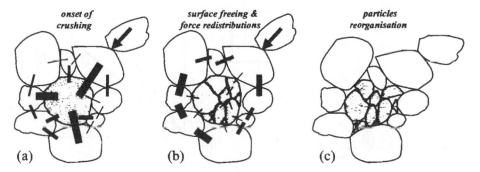

onset of crushing   *surface freeing & force redistributions*   *particles reorganisation*

(a)   (b)   (c)

Figure 2.   The dissipative events following the split of a particle: (a) a particle is about to crush, delivering a given network of contact forces represented by lines that are thickened according to the force magnitude; (b) new surface area is suddenly liberated, freeing surface energy, whereby force redistribution immediately follows that redistribute locked in strain energy; (c) reorganisation of the fragments and their neighbours, leading to plastic dissipation. We shall show that the first two modes can be lumped together. The third mode would obviously cause further redistribution of forces, which is why it is seen coupled to the other two modes.

and the neighbouring blocks represent the surrounding particles in Fig. 2. Each of the neighbouring blocks could by itself be connected using a bond-breaking connector to the ground (which could illustrates his potential to crush), but for simplicity this information is removed as we deal with only a single crushing event. The bond stiffness is assumed to be $K_b = aK_p$, where a is introduced as a non-dimensional parameter. Given N equally long particles in each of the blocks, with each particle having a length l, a single particle block has the stiffness $K = K_p/N$ and length $L = Nl$.

We start to load the analog, under sufficiently slow displacement controlled conditions. The whole system of particle blocks and bonds compress, as shown in Fig. 3b. When the bonds reach their ultimate shear force capacity T_u, and their ultimate elastic slip δ_u (see Fig. 3b and inset in Fig. 3d), the system reaches an ultimate state represented by $F = F_u$ and $\Delta = \Delta_u$ (see Figs. 3c and 3d), where $\Delta << L$ for small strain deformations. The bond breaking is abrupt (Fig. 3d), and therefore the elastically stored energy in the bonds ($K_b\delta_u^2 =$ double the shaded areas in Fig. 3d) is totally released for the creation of surface area, i.e., the increment of the surface area dissipation is:

$$\Delta\Phi_{\text{surface}} = K_b\delta_u^2 \tag{4}$$

After sufficient time, static equilibrium is approached towards a new minimum of the potential energy. The incremental change in the total energy dissipation per the single crushing event (denoted by Δ in front of the potential) is simply the difference between the elastic stored energy in ultimate-before Ψ_u and residual-after Ψ_r states:

$$\Delta\Phi_B = \Psi_u - \Psi_r = K_b\delta_u^2 + \frac{K_b}{K}K_b\delta_u^2 \tag{5}$$

The subscript 'B' was added to highlight that this dissipation relates to the total dissipation in the system directly related to the breakage (note also that since reorganization is prevented, the mechanical dissipation is entirely the breakage dissipation $\Delta\Phi_m = \Delta\Phi_B$). Considering eq. (4) we find that in this analog the increment in the breakage dissipation is the sum of two incremental dissipation terms, the surface area dissipation and energy redistribution dissipation:

$$\Delta\Phi_B = \Delta\Phi_{\text{surface}} + \Delta\Phi_{\text{redist}} = \Delta\Phi_{\text{surface}}(1 + aN) \tag{6}$$

where we designate the dissipation from the redistribution of initially locked-in strain energy:

$$\Delta\Phi_{\text{redist}} = aN\Delta\Phi_{\text{surface}} \tag{7}$$

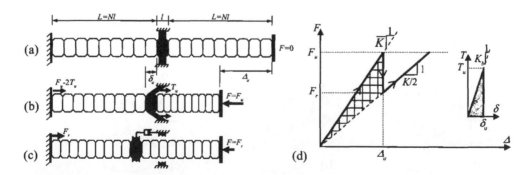

Figure 3. Mechanical analog: (a) initial configuration; (b) ultimate state before bond breaking; (c) bond breaking $\rightarrow$ release of locked-in strain energy & stress redistribution. The damper is showed only in (c), as it is irrelevant in (a) and (b) during quasi-static loading. (d) Load-displacement curves of the system response (hatched area designates the overall breakage dissipation, including both surface area dissipation and dissipation from redistribution of locked-in strain energy). Inset picture shows the response of an individual breakable bond (in this case the shaded area designates only the surface area dissipation of a single bond).

Since a is of the order of unity, this analog suggests that the energy-redistribution dissipation is about N times larger than the surface area dissipation, with N being of the order of the number of particles in a typical force chain in a granular media. For a growing number of particles N in the block, $\Delta\Phi_{redist} >> \Delta\Phi_{surface}$. As can be seen via eq. (6), the breakage energy dissipation (hatched area in Fig. 3d), which in this example is the total dissipation, contains the dissipations from both the creation of new surface area (shaded area in inset, Fig. 3d), and the dissipation from the redistribution of the locked-in strain energy.

The above analysis shows that under confining conditions there is a fundamental difference between the breakage energy dissipation and the surface area dissipation. It is the entire breakage energy dissipation, which was shown to be primarily related to the redistribution of locked-in strain energy, and not only the surface area dissipation, that represents the grain crushing. Therefore, estimates of dissipation energetics in cataclasis must account for the path dependency of the material behaviour, in addition to measurements of BET surface area. Further focus on this point is given in Nguyen and Einav (2009).

3   DISSIPATION IN BREAKAGE MECHANICS

Breakage mechanics is only briefly reviewed here: the end-product is presented in the form of a set of master equations. The derivation of the master equations of the homogeneous theory of breakage mechanics can give necessary mathematical and physical information (e.g., statistical micromechanics, energy balance, and particle scale mechanisms) and can be found elsewhere (Einav, 2007a, b).

3.1   *Summary of master equations*

The breakage mechanics equations of brittle granular systems may be collected:

$$p(D) = p_0(D)(1 - B) + p_u(D)B \tag{8}$$

$$\vartheta = 1 - <D^2>_u / <D^2>_0 \tag{9}$$

$$\Psi \equiv (1 - \vartheta B)\psi_r(\varepsilon) \tag{10}$$

$$E_B = -\partial\Psi/\partial B = \vartheta\psi_r(\varepsilon) = \vartheta\Psi/(1 - \vartheta B) \tag{11}$$

$$\sigma = \partial\Psi/\partial\varepsilon = (1 - \vartheta B)\psi_r'(\varepsilon) \tag{12}$$

$$E_B^* = E_B(1 - B) \tag{13}$$

where B, the breakage, weighs using Eq. (8) the relative distance of the current *gsd* $p(D)$ from the initial and ultimate *gsd* $p_0(D)$ and $p_u(D)$; Fig. 4a presents the measurable definition of the breakage; ϑ in Eq. (9) is referred to as the criticality proximity parameter and is measured using the initial and ultimate second order moments of the *gsd*, i.e., using $<D^2>_0$ and $<D^2>_u$, i.e., how far the initial distribution is from the ultimate distribution (which in many practical situations can be postulated as a fractal distribution); the macroscopic specific Helmholtz free energy potential Ψ is connected to the breakage and macroscopic strain tensor ε based on Eq. (10); the stress-like conjugate to the breakage is referred to as the breakage energy E_B, and is expressed in various ways using Eq. (11) by differentiating Ψ; the real Cauchy stress tensor σ is, on the other hand, the conjugate of the strain, derived using Eq. (12). It is then convenient to define the residual breakage energy E_B^* using Eq. (13), to express how much energy is left in the system for breaking particles, at any given moment, even after some breakage has already occurred.

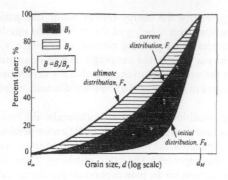 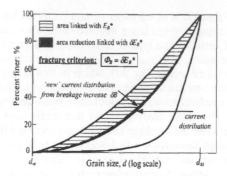

Figure 4.   The breakage measurement and evolution law (Einav, 2007a, b). The left figure (a) portrays the measurable definition of breakage, B, in terms of the initial, current and ultimate grain size distributions. The right diagram (b) presents the breakage propagation criterion for granular materials. Φ_B is the breakage dissipation, denoting the energy consumption from incremental increase of breakage. δE_B^* is the incremental reduction in the residual breakage energy. The energy balance equation, $\Phi_B = \delta E_B^*$, was shown to produce a physical evolution law for B.

3.2   *The fundamental structure of the breakage dissipation*

The loss in the residual breakage energy δE_B^*, resulting from the incremental breakage is postulated to equate to the increment of breakage dissipation Φ_B (Einav, 2007b):

$$\Phi_B = \delta E_B^* \tag{14}$$

Figure 4b shows the schematic representation of this energy balance equation. Combining Eqs. (13) and (14) and integrating the obtained equation gives the breaking yield condition y_B, where E_c is introduced as a constant of integration, referred to as the critical breakage energy constant (Einav, 2007b).

$$y_B = E_B(1 - B)^2 - E_c = 0 \tag{15}$$

where the equality could be introduced back into Eq. (14) to give an explicit expression of the breakage dissipation:

$$\Phi_B = \frac{\sqrt{2E_B E_c}}{(1 - B)} \delta B \geq 0 \tag{16}$$

The outcome in Eq. (15) reveals certain links with Griffith's energy method for cracks under tension (Einav, 2007b). For example, assume a breakage model with linear bulk elasticity, $\Psi = (1 - \vartheta B)K\varepsilon_v^2/2$, where ε_v denotes (elastic) volumetric strain. The effective mean stress is then $p = \partial\Psi/\partial\varepsilon_v = (1 - \vartheta B)K\varepsilon_v$, and the breakage energy $E_B = \vartheta K\varepsilon_v^2/2 = \vartheta p^2/2K(1 - \theta B)^2$. Combining this with Eq. (15), we get the critical pressure for the onset ($B = 0$) of comminution (Einav, 2007b):

$$p_{cr} = \sqrt{\frac{2KE_c}{\vartheta}} \tag{17}$$

3.3   *Plastic dissipation*

We identify two principal modes of plastic dissipations. The first occurs during shear and can be expressed by the usual plastic dissipation of a coulomb friction model:

$$\Phi_p^s = Mp\left|\delta\varepsilon_s^p\right| \geq 0 \tag{18}$$

It is important to realize that any crushing of a particle immediately results in another mode of plastic dissipation, arising from the configurational reorganization of the surrounding particles (Fig. 2c). This reorganization principally results from the new capability of the system to occupy new available volume, and is thus isotropic. Seeing the crushing as the active event and the reorganization as the passive mechanism, Einav (2007c) suggested that:

$$\Phi_p^v = \frac{p}{(1-B)}\sqrt{\frac{2E_c}{E_B}}\delta\varepsilon_v^p \geq 0 \tag{19}$$

So that Eq. (15) is preserved if Eq. (18) is ignored. To consider the coulomb friction model, we have adopted the following logic. We see the mechanisms of Figs. 2a and 2b as coupled with that of Fig. 2c, and that this latter volumetric rearrangement mechanism is in turn coupled with the plastic shear dissipation. Therefore we postulate that the total mechanical dissipation is expressed as follows (Einav, 2007a, c; Nguyen and Einav, 2009):

$$\Phi_m = \sqrt{\Phi_B^2 + \Phi_p^2} = \sqrt{\Phi_B^2 + \Phi_p^{v2} + \Phi_p^{s2}} \geq 0 \tag{20}$$

4   MODELLING CATACLASTIC SHEAR

4.1   *A model based on breakage mechanics*

Einav (2007c) made a full consideration of the thermodynamics of breakage mechanics to construct a consistent but simple elastic-plastic-breakage model. The model was referred to as a student model, given its simplicity and many novel features. The general yield function was derived by taking the Legendre transformation of the dissipation in Eq. (20):

$$y = \frac{E_B(1-B)^2}{E_c} + \left(\frac{q}{Mp}\right)^2 - 1 \leq 0 \tag{21}$$

The breakage and volumetric plastic dissipation terms (Eqs. (16) and (19)) both sums up to give the first term of the equation, while the plastic shear dissipation term (Eq. (18)) contributes via the second term of the yield function. The model was formulated in a general way, irrespective to the details of an assumed stored energy. An example was given in terms of a simple linear elastic strain energy function, which was sufficient to highlight some fundamental outcomes. Nguyen and Einav (2009) explored the capability of the model to predict compaction and cataclastic shear deformations, by developing a non-linear elastic law, introduced by appropriate strain energy potential.

4.2   *Regularization of the model using nonlocal theory: ultra-cataclastic shear*

Softening behaviour is well known in the literature as the cause of instability when solving boundary value problems using conventional continuum mechanics. A review on instability and bifurcation due to material softening has been well documented in the literature (e.g. Neilsen and Schreyer, 1993). Without enhancements (called regularization), either to the equilibrium equations or to the constitutive model, the numerical simulations of problems involving material softening usually results in discretization-dependent solutions (e.g. demonstration in Fig. 5a, using the model of the preceding section). Various regularization methods have been proposed and used effectively in the literature, including using rate-dependent effects and/or nonlocal enhancement to the constitutive modelling. In this study, nonlocal regularization of integral type is adopted for the local breakage model of the preceding section. We note that this is only a preliminary approach towards a complete rigorous solution. For that reason, nonlocal treatment is applied here to variables/terms directly controlling the softening behaviour of the model, rather than having to follow a consistent nonlocal thermo-mechanical formulation.

We recall the local breakage/yield criterion in Eq. (22). The way nonlocality is introduced to the constitutive equations is important, as illustrated in a paper by Jirasek (1998), in which different nonlocal treatments to a simple damage model were explored. Inappropriate treatment of nonlocal variable can lead to bad features (e.g. no regularization effect and/or instability of the numerical analysis) of the nonlocal model. From the yield function (21), it can be seen that the second term involving shear stress q becomes dominant when breakage B is large enough. Therefore applying nonlocality only to the breakage energy E_B (or either, the first term of the yield function) was experienced not to help prevent localization into infinitesimal zone in problems involving shearing. On the other hand, applying nonlocal averaging to both terms (effectively to the whole yield function in its current form) is also not a good way, this time because the current yield function was normalized to compare to 1. A slight rearrangement of the yield function would lead to a possible useful mathematical form for nonlocal averaging:

$$y = \frac{E_B}{E_c} + \left(\frac{q}{Mp(1-B)} \right)^2 - \frac{1}{(1-B)^2} \leq 0 \tag{22}$$

The nonlocal breakage/yield function, written for a material point at point **x**, is then:

$$y(\mathbf{x}) = \hat{C}(\mathbf{x}) - \frac{1}{(1-B(\mathbf{x}))^2} \leq 0 \tag{23}$$

where

$$\hat{C}(\mathbf{x}) = \frac{1}{G(\mathbf{x})} \int_{V_d} g(\|\mathbf{y} - \mathbf{x}\|) C(\mathbf{y}) dV(\mathbf{y}) \quad \text{and} \quad C = \frac{E_B}{E_c} + \left(\frac{q}{Mp(1-B)} \right)^2 \tag{24}$$

In the above expression V_d is the volume where the nonlocal averaging takes place; it is dependent on the type of weighting function $g(\|\mathbf{y} - \mathbf{x}\|) \geq 0$, e.g. Gaussian or bell-shaped distributions; and $G(\mathbf{x})$ is defined as a weight associated with the material point **x**, aiming at normalizing the weighting scheme:

$$G(\mathbf{x}) = \int_{V_d} g(\|\mathbf{x} - \mathbf{y}\|) dV(\mathbf{y}) \tag{25}$$

The flow rules of the model retain their local form. The stress update algorithm and numerical implementation of this nonlocal model is presented in a separate paper (Nguyen and Einav, 200?).

4.3   *Numerical example*

We use the nonlocal model in the preceding section to study the localization due to grain crushing in cataclasite zone (Fig. 1). Ignoring body force and inertia effects, the 2D equilibrium equations can be written as:

$$\frac{\partial \sigma_x}{\partial x} + \frac{\partial \tau_{xy}}{\partial y} = 0 \quad \text{and} \quad \frac{\partial \tau_{xy}}{\partial x} + \frac{\partial \sigma_y}{\partial y} = 0 \tag{26}$$

Further assuming that the variation in the stress condition along the fault line is negligible, and also the thickness of the fault layer is very small compared to its length (Fig. 1), we have the following constraints for the boundary conditions in horizontal direction:

$$\varepsilon_x = \varepsilon_z = 0; \quad \frac{\partial \sigma_x}{\partial x} = 0 \quad \text{and} \quad \frac{\partial \tau_{xy}}{\partial x} = 0 \tag{27}$$

The equilibrium equations (26) are then separated purely for the vertical direction:

$$\frac{\partial \tau_{xy}}{\partial y} = 0 \quad \text{and} \quad \frac{\partial \sigma_y}{\partial y} = 0 \tag{28}$$

which can be solved numerically using 1D finite elements. It should be noted in this case that the stress tensor is still that in 2D plane strain; its time variation is governed by the constitutive behaviour of the model under shearing at constant vertical stress σ_y. Only the spatial variations of σ_x and τ_{xy} in the horizontal direction are zero. Further note that the relation between τ_{xy} and σ_y is not seen directly via Eq. (28), but arises from the relevant coupling in the constitutive equations.

In our numerical simulation, the material in the cataclasite zone is first subject to a certain isotropic pressure p_0 corresponding to the depth of the fault. The shearing phase is then carried out under isochoric (undrained with infinite bulk modulus of water) condition. We use the following material properties and parameters in the numerical simulation: shear modulus $G = 5$ GPa; bulk modulus $K = 7.5$ GPa; crushing pressure $p_c = 90$ MPa; initial pressure $p_0 = 70$ MPa; $M = 1.5$; $\vartheta = 0.9$. The localization is triggered by weakening an element in the middle of the layer, using only 90% of the crushing pressure. The averaging scheme is based on the bell-shaped weighting function of the form:

$$g(\|\mathbf{y} - \mathbf{x}\|) = g(r) = \begin{cases} 0 & \text{if } r > R \\ \left(1 - \dfrac{r^2}{R^2}\right)^2 & \text{if } r \leq R \end{cases} \tag{29}$$

The spatial parameter in this case is the nonlocal interaction radius R ($R = 1.7$ mm in our numerical simulations), which controls the spatial interactions between material points. In this case V_d (eqs. 4 and 5) at a material point $\mathbf{x}$ is a sphere of radius R.

The modelled cataclasite zone is of 36 mm thickness, which is comparable to values reported in Chester et al. (2005), and used in DEM simulation by Guo and Morgan (2007). Due to symmetry, only half of the zone is modelled. During the numerical simulation, while the element behaviour evolves during the shearing process, the coordinates of all finite elements are unchanged, representing the flow of the material through a fixed frame. Different finite element meshes are used to illustrate the mesh-independence of the numerical results (Fig. 5b).

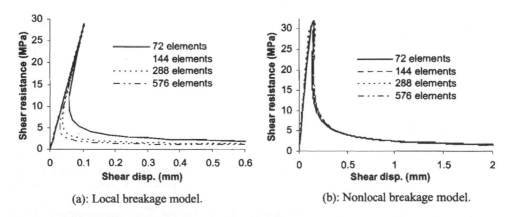

(a): Local breakage model. (b): Nonlocal breakage model.

Figure 5.   Mobilized shear resistance vs. shear displacement.

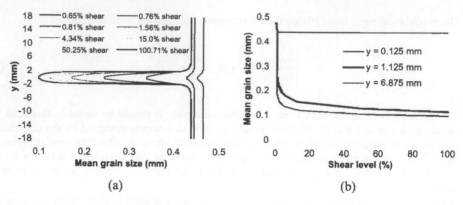

Figure 6.   Evolution of mean grain size.

The evolution of the mean grain size of the material in cataclasite zone can be calculated directly using the evolving breakage variable B. In doing that, we assume power law grain size distributions for both the initial and ultimate conditions:

$$p_i(D) = (3 - \alpha_i) \frac{D^{2-\alpha_i}}{D_M^{3-\alpha_i}} \tag{30}$$

where i stands for "u" (ultimate) or "0" (initial). We use an ultimate fractal dimension of $\alpha_u = 2.8$ and set α_0 according to prescribed $\vartheta = 0.9$. Using Eq. (8), the mean evolving grain size D is:

$$\bar{D} = \int_0^{D_M} D p(D) dD = (1 - B)\bar{D}_0 + B\bar{D}_u \tag{31}$$

where the maximum grain size is assumed constant throughout the deformations $D_M = 0.5$ mm. Fig. 6 plots the evolving mean grain size during the shearing process.

From Fig. 6a, it can be seen that the size of core of the cataclasite layer, i.e., the size of the ultra-cataclasite zone, is unchanged beyond a certain shear level. The size of this core in this example is about $2R$. Inside this core, the grains are continuously crushed towards the ultimate grain size (Fig. 6a), while the material outsize the core is unloaded elastically, and does not involves cataclastic deformation. The rate of crushing in the core is however decreasing (Fig. 6b) over the time. In other words the material gradually becomes harder to break.

The nonlocal regularization applied to breakage model has been numerically shown to bring stability to the model behaviour and well-posedness to the BVP. This is realized through the mesh-independence of the numerical solutions, and the localization of deformation into non-zero volume.

5   CONCLUSIONS

We have been studying the connection of dissipation to constitutive behaviour of brittle granular material. Various components of dissipations were identified, which in combination were shown to constitute the total mechanical dissipation. In relations to the recent development of breakage mechanics, and via the introduction of a simple mechanical analog, the concept of breakage dissipation was examined. The analog suggested that the breakage dissipation in a representative volume element may be viewed as the simple summation of both surface area dissipation, like in fracture mechanics, and redistribution of locked-in strain energy dissipation. The latter contribution to the overall mechanical dissipation is often neglected, but seems to dominate the breakage dissipation.

The overall mechanical dissipation is then seen to involve the coupling of the breakage dissipation and plastic dissipation, which arises from volumetric configurational changes and Coulomb friction.

Finally, we have demonstrated how the breakage mechanics theory can be applied to studying gouge shear, with and without consideration of non-local effects, with the latter introducing the formation of ultracalaclasite. Careful regularisation of the problem was executed, and the analysis demonstrated the formation of different gsd within and out of the very narrow core of the cataclasite zone.

REFERENCES

Chester, J.S., Chester, F.M. and Kronenberg, A.K. 2005. Fracture surface energy of the Punchbowl fault, San Andreas system, *Nature* 437(7055): 133–136.

Einav, I. 2007a. Breakage mechanics—Part I: Theory, *Journal of the Mechanics and Physics of Solids* 55 (6): 1274–1297. (2007a); Part II: Modelling granular materials, *Journal of the Mechanics and Physics of Solids* 55(6): 1298–1320.

Einav, I. 2007c. Fracture propagation in brittle granular matter, *Proceedings of the Royal Society A: Mathematical, Physical and Engineering Sciences* 463(2087): 3021–3035.

Einav, I. 2007d. Soil mechanics: breaking ground, *Philosophical Transactions of the Royal Society A: Mathematical, Physical and Engineering Sciences* 365(1861): 2985–3002.

Guo, Y. and Morgan, J.K. 2007. Fault gouge evolution and its dependence on normal stress and rock strength—Results of discrete element simulations: Gouge zone properties. *J. Geophys. Res.*, 112: B10403.

Hamiel, Y., Lyakhovsky, V. and Agnon, A. 2004. Coupled evolution of damage and porosity in poroelastic media: theory and applications to deformation of porous rocks, *Geophysical Journal International* 156: 701–713.

Jirásek, M. 1998. Nonlocal models for damage and fracture: comparison of approaches, *Int. J. Solids Structures* 35(31–32): 4133–4145.

Lyakhovsky, V., Ben-Zion, Y. and Agnon, A. 1997. Distributed damage, faulting, and friction. *J. Geophys. Res.*, 102(B12): 27, 635–27, 649.

Maugin, G.A. 1992. *The Thermodynamics of Plasticity and Fracture*, Cambridge University Press, Cambridge.

Neilsen, M.K. and Schreyer, H.L. 1991. Bifurcation in elastic-plastic materials. *Int. J. Solids Structures* 30(4): 521–544.

Nguyen, G.D. and Einav, I. 2009. Cataclasis and permeability reduction: an energetic approach based on breakage mechanics. *PAGEOPH Invited paper, future publication.*

Nguyen, G.D., Einav, I. 2009. The energetics of cataclasis based on breakage mechanics. *Pure and applied geophysics* 166: 1–32.

Nguyen, G.D., Einav, I. 200?. A stress return algorithm for nonlocal constitutive models of softening materials. *International Journal for Numerical Methods in Engineering.* under revision.

Olgaard, D.L. and Brace, W.F. 1983. The microstructure of gouge from a mining-induced seismic shear zone, *International Journal of Rock Mechanics and Mining Science & Geomechanics Abstracts* 20(1): 11–19.

Reches, Z. and Dewers, T.A. 2005. Gouge formation by dynamic pulverization during earthquake rupture, *Earth and Planetary Science Letters* 235(1–2): 361–374.

Ricard, Y. and Bercovici, D. 2003. Two-phase damage theory and crustal rock failure: the theoretical 'void' limit, and the prediction of experimental data, *Geophysical Journal International* 155: 1057–1064.

Ziegler, H. 1983. *An introduction to thermomechanics*, second ed. North Holland, Amsterdam.

III. *Experimental fault zone mechanics*

Strain localization in granular fault zones at laboratory and tectonic scales

Chris J. Marone
Department of Geosciences, Pennsylvania State University, University Park, PA, USA

Andrew Rathbun
Pennsylvania State University, University Park, PA, USA

ABSTRACT: We present results from laboratory experiments and a numerical model for frictional weakening and shear localization. Experiments document strain localization in sheared layers at normal stresses of 0.5 to 5 MPa, layer thicknesses of 3 to 10 mm, and imposed slip velocities of 10 to 100 µm/s. Passive strain markers and the response to load perturbations indicate that the degree of shear localization increases for shear strains γ of $0.15 < \gamma < 1$. Our numerical model employs rate-state friction and uses 1D elasto-frictional coupling with radiation damping. We interrogate the model frictional behavior by imposing perturbations in shearing rate at the fault zone boundary. The spatial distribution of shear strain depends strongly on frictional behavior of surfaces within the shear zone. We discuss the onset of strain localization and the width of active shear strain for conditions relevant to earthquake faulting and landslides.

1   INTRODUCTION

Laboratory and field evidence indicate that strain localization is accompanied by significant changes in hydraulic and mechanical properties of rocks (e.g., Wood 2002, Song et al., 2004, Rice 2006). Strain localization occurs at a broad range of scales and involves both formation of faults and, upon continued shear, confinement of shear to narrow bands within the wear and gouge materials that constitute the fault zone. Of particular interest is the connection between strain localization and the transition from stable to unstable frictional sliding within shear zones of finite width (e.g., Anand & Gu 2000, Rice & Cocco 2007).

2   SHEAR LOCALIZATION IN GRANULAR LAYERS

In this paper we focus on layers composed of granulated rock. Granular layers were sheared in a biaxial deformation apparatus using the double-direct shear configuration. Details of the testing apparatus and experimental procedures are reported in Rathbun et al. (2008). Layers were initially 3 to 10 mm-thick and we imposed slip velocities of 10 to 100 µm/s at the layer boundary. Normal stress was held constant during shear via a fast-acting servo-hydraulic control mechanism. We discuss experiments conducted at normal stresses in the range 0.5 to 5 MPa, which is high enough to result in inelastic yield at grain to grain contacts, on the upper end, and low enough to inhibit grain crushing, on the lower end.

2.1   *Dilation as a proxy for shear localization*

Previous studies of granular layers have established that upon shear loading, shear stress rises linearly before undergoing a progressive transition from elastic to inelastic behavior (e.g., Anthony & Marone 2005). Inelastic yield is associated with grain rearrangement, compaction, and bulk shear strain of the layer (e.g., Marone 1998). Figure 1 illustrates this behavior for a granular layer that was initially 10 mm thick and sheared at a normal stress of 1 MPa. Note the steep rise in shear stress followed by strain hardening and a transition to steady frictional sliding at a shear strain of ~0.5. Layers compact during the initial rise in shear stress and dilation beings at a normalized stress

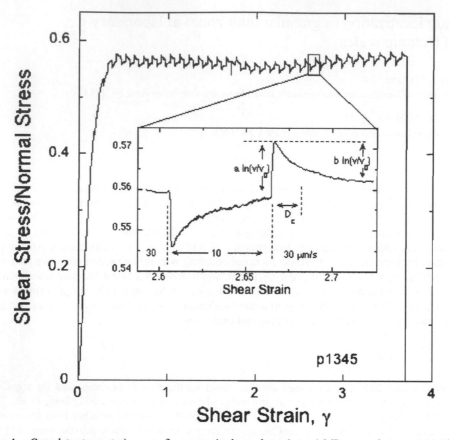

Figure 1.   Complete stress-strain curve for a granular layer sheared at a 1 MPa normal stress. Initial layer thickness was 10 mm. Normalized shear stress is plotted versus engineering shear strain computed from incremental displacement at the layer boundary divided by instantaneous layer thickness. Note the steep, quasi-linear, initial rise in shear stress followed by hardening and fully-mobilized shear at strains of 0.4 to 0.5. Beginning at a shear stain of 0.3, the shearing rate at the layer boundary was toggled between 10 and 30 μm/s. Inset shows detail of the friction response to step changes in slip velocity. The material is glacial till, (Caesar till, Ohio, USA) which is a mixed-size granular material (Particle size D: $D_{max} = 1$ mm, D50 ∼0.3 mm, 90% of the particles >0.1 mm) similar to natural fault gouge (after Rathbun et al., 2008).

(friction) level of 0.35 to 0.4 (Fig. 1). In this experiment we evaluated the effects of changes in slip velocity at the layer boundary. The systematic variation in friction seen throughout the experiment is associated with step changes in loading rate between 10 and 30 μm/s, as discussed further below.

In granular layers, the transition from initial strain hardening to steady-state (fully mobilized) frictional sliding is associated with the development of localized shear (e.g., Logan et al., 1979, Marone et al., 1992). Recently, Rathbun et al. (2008) showed that stress perturbations, and layer dilation, provide a more precise measure of the degree of shear localization. They showed that small perturbations in the shear stress level during creep friction tests provide a proxy for shear localization (Fig. 2). Creep friction tests were carried out at a constant shear stress level below that for stable sliding. Using this approach, the degree of shear localization can be assessed with the parameter Δh^*, which is the layer dilation for a unit increase in shear stress. We measured Δh^* as a function of shear strain for a range of conditions (Fig. 2). Dilation scales with layer thickness below a critical value of shear strain. For shear strains ≤ ∼0.5, thicker layers exhibit larger values of Δh^* than thinner layers (Fig. 2). However for shear strains greater than ∼1, dilation is independent of layer thickness. These data show that, initially, shear is distributed across the full thickness of the

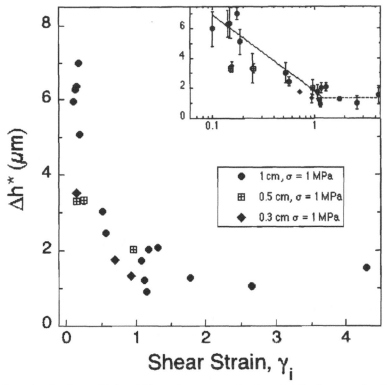

Figure 2. Data showing layer dilation Δh^*, used as a proxy for shear localization, as a function of shear strain. The parameter Δh^* is the layer dilation for an increase in shear stress equal to 5% of the strength during stable frictional sliding. Data are shown from multiple experiments with initial layer thicknesses ranging from 3 to 10 mm. Normal stress was 1 MPa in all cases. During steady-state frictional sliding, friction is typically 0.55 to 0.6 (e.g. Fig. 1); thus the shear stress perturbations were of order 0.03 MPa.

layer, but that shear becomes localized beyond a critical shear strain. The data of Figure 2 indicate that shear is fully localized by shear strains of roughly unity.

Figure 3 shows additional details of the relationship between shear stress, shear strain, and layer thickness. This figure shows stress-strain curves for representative experiments and one data set for changes in layer thickness as a function of shear strain (experiment p1025). Layer dilation occurs early in the strain history and then the layers compact slightly before reaching a steady level, consistent with a critical state, for shear strains of 0.3 and greater (Fig. 3).

2.2 *Rate/State friction and shear localization*

Slip velocity step tests have emerged as a powerful tool for interrogating friction constitutive behavior (Dieterich 1979, Ruina 1983, Scholz, 1998). A large body of literature shows that frictional strength of a wide range of materials exhibits two responses to a step increase in the imposed loading rate (e.g., Dieterich & Kilgore 1994, Tullis 1996, Marone 1998). First, there is an instantaneous change in frictional resistance of the same sign as the velocity change. This is referred to as the friction direct effect and it is described by the friction parameter a. Figure 4 defines the key parameters and outlines the rate and state friction equations. The direct effect is followed by a gradual evolution of strength, scaled by the friction parameter b (Fig. 4). The evolution effect is typically of the same as the change in velocity (Fig. 1). Existing studies show that the evolution effect occurs over a characteristic slip distance, D_c (sometimes referred to as L), for initially-bare solid surfaces or a characteristic strain for layers of granular/clay particles (e.g., Marone 1998).

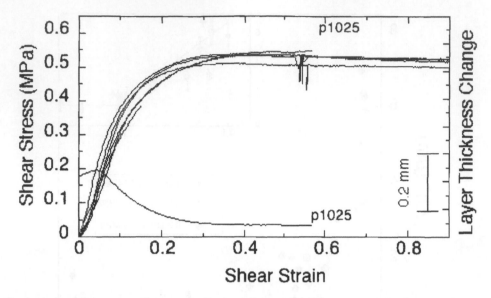

Figure 3.   Complete stress strain curves for five experiments along with a representative data set (experiment p1025) for changes in layer thickness as a function of shear strain. The layer thickness started at 10 mm in experiment p1025 and the thickness was measured continuously during shear with a DCDT. Note that dilation occurs during the initial increase in stress but that compaction begins prior to fully-mobilized shear within the granular layer.

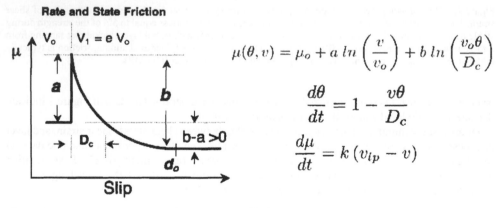

Figure 4.   The equations describing the rate and state friction constitutive law along with a schematic showing this behavior.

The values of D_c are typically larger for shear within a granular layer than for shear between solid surfaces (e.g., Marone & Kilgore 1993, Marone et al., 2009).

To the extent that the friction evolution effect is truly driven by shear displacement, step velocity increases and decreases are expected to yield symmetric behavior. Such symmetry was found by Ruina (1983) and Marone et al. (1990). Other studies have favored a model in which friction evolution occurs over a characteristic time (Beeler et al., 1994, Sleep 1997). Finally, a large number of works have evaluated only velocity increases or decreases, without considering the issue of symmetry (e.g., Marone & Kilgore 1993). Few studies have systematically evaluated symmetry of the friction response to changes in loading velocity.

2.3 *Mechanics of the critical slip distance for friction of granular materials*

For solid surfaces in contact, the critical slip distance for friction evolution can be thought of in terms of the asperity contact lifetime, given by the contact size divided by the average slip rate (Rabinowicz 1951, Dieterich 1979). When coupled with the adhesive theory of friction (e.g., Bowden & Tabor 1950), in which asperity strength (and size) is proportional to time of contact (lifetime), this model predicts that frictional strength during steady-sliding should decrease with increasing slip velocity, because contact lifetime (hence strength) is inversely proportional to sliding velocity. In the context of a velocity step test, the critical friction distance is the slip necessary to replace contacts with a lifetime given by the initial velocity by contacts corresponding to the final velocity (Dieterich 1979, Ruina 1983).

For granular materials the situation is slightly more complex. The model for solid friction can be applied directly for asperity contacts between grains. However, granular interactions and stress transmission via particle contacts lead to a second characteristic length scale, in addition to the asperity contact junction size. The work by Marone & Kilgore (1993) shows that the critical slip distance for granular shear depends on both the particle size and the shear localization dimension (Fig. 5).

The data of Figure 5 show measurements of the critical slip distance as a function of shear strain for granular layers sheared in the double-direct shear geometry. The average particle sizes ranged from 700 μm (Coarse) to 5 μm (Fine); fractal is a power-law size distribution between 45 and 720 μm. The critical slip distance is greatest for larger particles at all values of shear strain (Fig. 5). Fine particles, with an average size that is roughly 100 times smaller than the coarse

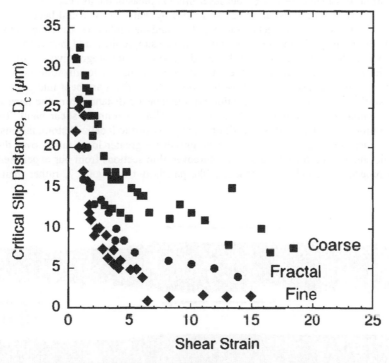

Figure 5. Measurements of the critical slip distance for granular layers as a function of shear strain. Data are shown for three particle size distributions. Note that the critical slip distance is greatest for larger particles at all values of shear strain. *Coarse*: Ottawa sand ASTM C-190 all particles are 600 to 800 μm. *Fine*: Silicosil 400 mesh (US Silica Co.) with median and maximum diameter of 1.4 and 10 μm, respectively. *Fractal*: given by $N(n) = bn^{-D}$, where $N(n)$ is the number of particles of size n, b is a constant and D is the fractal dimension 2.6, made using particles in the range <45 μm to 700 m. Data from Marone & Kilgore (1993).

particles, have a critical slip distance that is roughly 10 times smaller than the coarse particles, consistent with the expected scaling between contact junction dimension and particle diameter. It is important to note, however, that the difference in D_c values for coarse and fine particles is small compared to the observed evolution of D_c with shear strain (Fig. 5). The D_c values for layers of both coarse and fine particles decrease by more than 100% of the final value at shear strains greater than ~7. The decrease in D_c with shear strain is consistent with the effects of shear localization.

2.4   *Observations of shear localization*

Laboratory investigations of shear localization often include post-experiment examination of preserved microstructures (e.g., Mair & Marone 1999). In the experiments described here, we used passive markers in some experiments to record the strain distribution across layers (Figure 6). The markers were constructed with blue sand grains. Following the shear experiment, layers were impregnated with epoxy and then cut parallel to the shear direction. Figure 6 shows a layer that was sheared, top to the left, to a strain of 3.9. The layer had three markers that were initially vertical in the orientation of the photograph. The original image is shown below a copy (above) that has been marked to highlight the offset marker. Note that: 1) the marker is offset primarily along a zone near the center of the layer and 2) that the segments above and below the primary offset show distinct curvature. This curvature indicates a progressive localization process prior to development of the main shear zone in the center of the layer. By measuring offset of the top and bottom limbs of the marker, we calculated a shear strain of 3.25 along the shear zone at the center of the layer. Based on the total shear strain of 3.9, we estimate initiation of this shear band at a shear strain of 0.65, which is within the range indicated by our layer dilation measurements (Fig. 2).

Figure 7 is a schematic illustration of two modes by which shear could become localized. In panels a–c the marker is first subject to uniform strain and then cut by a shear zone. In this scenario, shear localizes abruptly. The markers are first rotated by simple shear and then cut and offset along a narrow zone at the center of the layer. Panels d–f show a more progressive localization process (Fig. 7). The markers are initially subject to uniform strain, but localization occurs gradually and on several surfaces near the center of the zone. The markers are bent into an arcuate shape by progressively greater strain concentration with increasing distance from the layer boundaries (Fig. 7). Eventually, the strained markers are offset along a primary shear band. This type of localization process, with a gradual transition from pervasive to localized strain is consistent with our dilation measurements, which indicate progressively greater localization over the range of macroscopic shear strain from 0.15 to 1.0. Moreover, thin sections from our experiments (Fig. 6) indicate a progressive localization process, like panels d–f of Figure 7, rather than an abrupt transition.

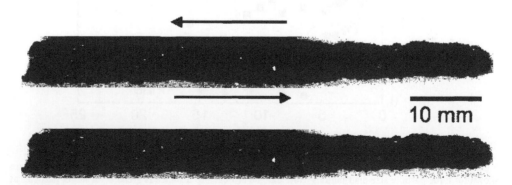

Figure 6.   Thin section of a granular layer that was subject to a shear strain of 3.9. The layer contains three passive markers that were initially vertical, in this orientation, formed by darker particles. Top image is annotated to show shear of the central marker. Lower image is unmarked photograph.

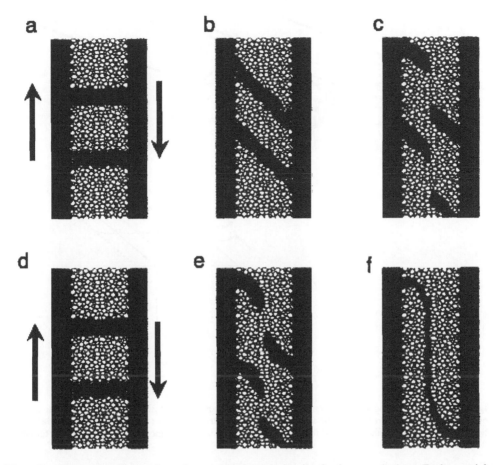

Figure 7. Schematic representations of two models for shear localization in a granular layer. Darker particles are passive markers. Panels a and d show initial, un-deformed state. Panels a–c show pervasive shear followed by abrupt shear localization. Panels d–f show progressive shear and localization. Strain is pervasive for the interval from panel d to panel e, but then shear localizes between e and f. Panel f shows one marker that is deformed and offset. Note similarity between panel f and micrograph in Figure 6.

3   A NUMERICAL MODEL FOR FRICTIONAL WEAKENING AND SHEAR LOCALIZATION

To address shear localization in tectonic fault zones and to improve our understanding of the scaling problem associated with applying laboratory observations to faults in Earth's crust, we employ a numerical model. The model describes frictional shear in a fault zone composed of multiple, parallel surfaces that obey rate and state friction (Fig. 8). The model used here is based on that described by Marone et al. (2009). We extend that model and focus on coupling between friction properties and shear localization.

3.1   *Elasto-frictional model for a fault zone of finite thickness*

A typical tectonic fault zone consists of a highly damaged zone surrounded by progressively less damaged country rock (e.g., Chester & Chester 1998). Thus, in the context of the seismic cycle, the zone of slip deformation is defined by a critical fracture density, above which slip and deformation occurs, and below which the rock behaves as intact material. One could imagine a model in which

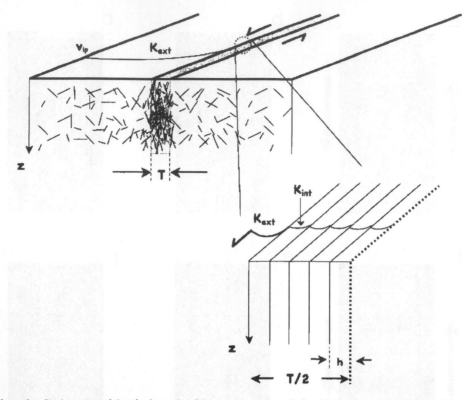

Figure 8.   Fault zone model and schematic of shear zone composed of multiple sub-parallel surfaces. The fault zone width is T. K_{ext} represents elastic stiffness of the crust surrounding the fault. K_{int} represents elastic coupling between surfaces in the model, which are separated by distance h. The model is symmetric about the center.

this critical fracture density depended on strain rate and other factors, such that the effective fault zone width varied throughout the seismic cycle, but we make the simplifying assumption of constant width T (Fig. 8).

Within the model fault zone, shear may occur on one or more sub-parallel surfaces (Fig. 8). Our aim is to investigate spatio-temporal complexity of shear localization in a fault zone that experiences a rapid change in imposed slip rate; for example due to earthquake propagation into the region of interest. Therefore we employ a simplistic geometry and elastic model. As an initial condition, we assume homogeneous creep within the fault zone, such that all surfaces are slipping at a background rate. We investigate the fault response to perturbations in slip rate imposed at the fault zone boundary at $\pm T/2$ relative to the center of the zone (Fig. 8).

Potential slip surfaces within the fault zone interact via elasto-frictional coupling. Stress is transmitted between surfaces only when: 1) the frictional strength of a surface exceeds the current stress level, or 2) a surface slips and its strength changes. In the models described here, stresses and frictional strengths are initially equal on all surfaces. Slip surfaces obey laboratory-based rate and state friction laws, and we focus here on the case of state evolution via the Ruina law (Dieterich 1979, Ruina 1983). A one-dimensional elastic model is used with radiation damping to solve the equations of motion.

Each surface i in the model shear zone obeys rate and state frictional behavior, such that friction μ_i is a function of state θ_i and slip velocity v_i according to:

$$\mu_i(\theta_i, v_i) = \mu_0 + a \ln\left(\frac{v_i}{v_0}\right) + b \ln\left(\frac{v_0 \theta_i}{L}\right) \tag{1}$$

Table 1. Model parameters. For all cases, $G = 30$ GPa, $\sigma = 100$ MPa, $K_{int} = G/h$; $K_{int}/K_{ext} = 10$; $v_0 = $ 1e-6 m. $n_s/2$ is the number of surfaces in the fault zone half width $T/2$.

a	b	L (m)	h (m)	K_{ext}/σ_n (m^{-1})	$n_s/2$	T (m)	v (m/s)
0.012	0.016	1e-5	6e-3	5e4	30	0.60	0.01

where μ_0 is a reference friction value at slip velocity v_0, and the parameters a, b, and L are empirically-derived friction constitutive parameters (e.g., Marone 1998). Note that we use L for the model critical slip distance, rather than D_c, which is the effective parameter measured from laboratory experiments. Tectonic fault zones are likely to include spatial variations of the friction constitutive parameters within the shear zone, and thus we allow such behaviors.

The model includes n_s parallel surfaces, where $i = 0$ is at the fault zone boundary. Surfaces are coupled elastically to their neighbors via stiffness K_{int}. We assume $K_{int} = G/h$, where G is shear modulus and h is layer spacing (Fig. 8) and use $G = 30$ GPa. We assume that remote tectonic loading of the shear zone boundary is compliant relative to K_{int} and take K_{int}/K_{ext} equal to 10.0. This is equivalent to assuming a constant spacing between surfaces and means that wider shear zones, with more internal surfaces, are effectively more compliant than narrower zones. Another approach would be to take K_{int}/K_{ext} equal to the number of surfaces in the shear zone. Details of the parameters used are report in Table 1.

We analyze friction state evolution according to:

$$\frac{d\theta_i}{dt} = -\frac{v_i\theta_i}{L} \ln\left(\frac{v_i\theta_i}{L}\right) \quad \text{(Ruina Law)} \tag{2}$$

Frictional slip on each surface satisfies the quasi-dynamic equation of motion with radiation damping (Rice 1993):

$$\mu_i = \frac{\tau_0}{\sigma_n} - \frac{G}{2\beta\sigma_n}(v_i - v_{pl}) + k(v_{pl}t - v_it) \tag{3}$$

where μ_i is the frictional stress, τ_0 is an initial stress, β is shear wave speed, σ_n is normal stress, k is stiffness divided by normal stress, and t is time. Differentiating Equations 1 and 3 with respect to time and solving for dv_i/dt yields:

$$\frac{dv_i}{dt} = \frac{k(v_{pl} - v_i) - \frac{b\frac{d\theta_i}{dt}}{\theta_i}}{\frac{a}{v_i} + \frac{G}{2\beta\sigma_n}}, \tag{4}$$

which applies for each surface within the shear zone. Our approach for including radiation damping is similar to that described in previous works (Perfettini & Avouac 2004, Ziv 2007).

We assume that the model begins with steady creep, and thus each surface of the fault zone undergoes steady state slip at velocity $v_i = v_0$ with $\mu_0 = 0.6$ and $\theta_{ss} = L/v_0$. The effective stiffness k_i between the load point and surface i within the fault zone is given by:

$$\frac{1}{k_i} = \frac{1}{K_{ext}} + \sum_{j=1}^{i} \frac{1}{K_{int_j}}. \tag{5}$$

To determine shear motion within the fault zone, we solve the coupled Equations 2, 4–5, using a 4th order Runga-Kutta numerical scheme. As noted above, perturbations in slip velocity are imposed at the shear zone boundary. This is assumed to occur via a remote loading stiffness K_{ext}.

Then, for each time step in the calculation, the surface with the lowest frictional strength is allowed to slip.

Our initial conditions are that shear and normal stress are the same on each surface. We ensure that time steps are small compared to the ratio of slip surface separation, h, to elastic wave speed. Thus, within a given time step, only one surface slips and it is coupled elastically to the remote loading velocity via the spring stiffness given in Equation 5.

3.2   *Frictional response to changes in imposed slip rate*

Figure 9 shows macroscopic shear strength of the fault zone as a function of slip at the fault zone boundary. Shear stresses are equal on all slip surfaces, however frictional strengths are not. Thus, Figure 9 shows friction of the weakest surface within the fault zone as a function of offset at the fault zone boundary. This case shows behavior of a fault zone that has homogeneous frictional properties.

The macroscopic frictional response of the fault zone differs from the constitutive response of the individual surfaces within it. In particular, the fault zone exhibits a protracted phase of strain hardening prior to reaching the maximum yield strength (Fig. 9). The peak strength is reached in a slip displacement of <5% of D_c for a single surface, whereas the fault as a whole requires slip equal to 200% of D_c before weakening begins. As a result, the effective critical friction distance for the fault zone significantly exceeds that for an individual slip surface (Fig. 9). The maximum yield strength of the fault zone, which is proportional to the friction parameter a, is nearly identical to that for an individual surface. Finally, the steady-state frictional strength is the same in both cases (Fig. 9).

The relationship between the intrinsic frictional behavior of a surface and the zone of active shear, as a whole, is important for several aspects of earthquake rupture and shear localization. Figure 10 shows this relationship for a series of model runs to different shear strains. In each case the intrinsic frictional response of a single surface is shown versus slip on that surface. In addition, the frictional strength for the shear zone is plotted versus boundary slip. The two curves are plotted on the same scale; but note that the single surface is subject to larger total slip displacement, so as to illustrate

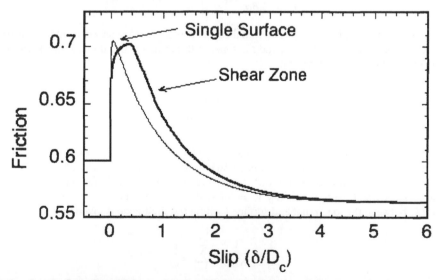

Figure 9.   Results form the model runs showing friction as a function of slip. For comparison, the intrinsic frictional response for a single surface is shown together with the frictional response of the shear zone. Note that the shear zone exhibits a prolonged phase of hardening, associated with the rate/state friction response of each layer, followed by weakening. See Table 1 for parameter values.

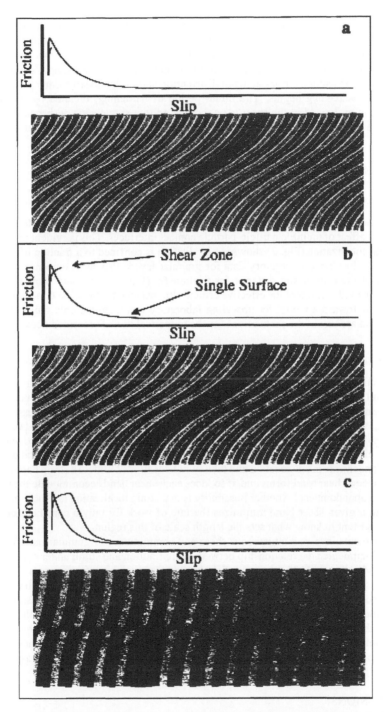

Figure 10. Three snap shots of the relation between frictional behavior and slip distribution within a model fault zone. In each case the response of a single surface is plotted together with the behavior for the complete shear zone. The single surface is the same in each panel. The shear zone response in panels a–c is shown for progressively greater boundary shear. Images below each plot show slip distribution, via offset of markers that were initially vertical in this orientation. Note that strain is initially pervasive but that localization occurs abruptly during frictional weakening. See Table 1 for parameter values.

the complete behavior. The panels of Figure 10 show three different amounts of shear applied at the boundary and below each plot is the spatial distribution of slip across the full shear zone.

The frictional model indicates that a perturbation in slip rate at the shear zone boundary results first in pervasive shear, up to a point, followed by localization along a single surface (Fig. 10). Comparison of the friction curves and slip distribution shows that localization occurs at the point that frictional weakening begins. The initial period of hardening, dictated by the friction rate parameter a, is prolonged in the shear zone, compared to a single surface, because each surface must proceed through this hardening phase before the zone as a whole can weaken.

4   DISCUSSION

4.1   *Shear localization and frictional behavior of granular layers*

Our observations indicate that the critical slip distance for friction of granular materials represents the combined effect of multiple particle-particle contact interfaces. This is evident in the laboratory data on shear localization (Fig. 2) showing that dilation is confined to a fraction of the layer once shear becomes localized. Laboratory data for granular layers also show evidence of localization in the form of the critical slip distance for friction D_c (Fig. 5). Our data show that laboratory measurements of D_c represent the effective critical slip distance for the zone of active shear, which points the way toward a model for upscaling laboratory results to tectonic faults. Indeed, these laboratory data are one of the motivations for the numerical model presented here.

4.2   *Spatio-temporal complexity of shear localization and delocalization*

One of the enduring puzzles of shear localization in granular fault zones is that of shear band migration and delocalization. A typical fault zone in nature, and in the laboratory, includes multiple zones of shear localization, rather than a single zone. This may indicate a progressive process of localization, where one type of feature is active for a limited time and then another takes over. Or, it may indicate that a set of shear zones operate simultaneously, to produce a penetrative shear fabric. However, in either case, the existence of multiple slip surfaces raises a fundamental question: why does shear concentrate in one location and then switch locations? Is there a strain hardening process that begins once a shear band forms and, if so, does each shear band accommodate the same critical strain prior to abandonment? Another possibility is that strain localization is a local process within the bulk, and a given shear band minimizes the rate of work for only a confined region. In this case, it is important to know what sets the length scale of this region.

Our experiments involve simultaneous shear and comminution of granular materials. Previous works have documented the relationship between comminution and strain under conditions similar to ours (e.g., Mandl et al., 1977, Marone & Scholz 1989). Although our experiments include detailed measurements of macroscopic stress and layer strain, these data are of limited value in answering the most important questions raised above. One would like to have independent assessment of the spatial distribution of shear strain as a function of imposed shear on the layer. While this is beyond the scope of the present data, our measurements of layer dilation as a proxy for shear localization offer some information, and the numerical model provides some insight about how the intrinsic frictional response of surfaces within a zone effect the overall response of a fault zone.

5   CONCLUSIONS

Laboratory friction experiments combined with constitutive modeling provide a powerful means of investigating problems in shear localization. Our laboratory data show that layer dilation, in response to small perturbations in creep stress rate or strain rate, can be used as a sensitive proxy for the degree of strain localization. For granular layers sheared at normal stresses up to a few MPa, shear strain becomes fully localized prior to engineering shear strains of 1. Thin section analysis shows that shear localization is a progressive, rather than abrupt, process within a granular layer. Our experiments included velocity step tests, which probe the friction constitutive behavior and its relation to shear localization. We present a numerical model for shear within fault zones composed of multiple slip surfaces and use the model to evaluate shear localization. The spatial distribution of fault zone shear depends strongly on the intrinsic frictional properties of the materials and on elasto-frictional interaction. The model shear distribution is strikingly similar to the slip distribution documented in thin sections from experiment. Frictional processes determine the onset of strain localization and the width of active shear strain in granular shear zones. Our work has important implications for a range of conditions relevant to earthquake faulting and landslides

ACKNOWLEDGMENTS

We thank Y.H. Hatzor and the other organizers of the Batshiva de Rothschild seminar on shear physics at the meso-scale in earthquake and landslide mechanics. The workshop was extremely stimulating and very enjoyable. We gratefully acknowledge support from the National Science Foundation under grant numbers ANT-0538195. EAR-0510182, and OCE-064833. J. Samuelson, A. Niemeijer, and B. Carpenter are thanked for stimulating discussions during the course of this work.

REFERENCES

Anand, L. & Gu, C. 2000. Granular materials: constitutive equations and strain localization. *Journal of Mechanics and Physics of Solids* 48: 1701–1733, 10.1016/S0022-5096(99)00066-6.

Anthony, J.L. & Marone, C. 2005. Influence of particle characteristics on granular friction. *Journal Geophysical Research* 110(B08409): 10.1029/2004JB003399.

Beeler, N.M., Tullis, T.E. & Weeks, J.D., 1994. The roles of time and displacement in evolution effect in rock friction. *Geophysical Research Letters* 21: 1987–1990.

Bowden, F.P. & Tabor, D. 1950. *The Friction and Lubrication of Solids.* Part I. Oxford: Clarenden Press.

Chester, F.M. & Chester, J.S. 1998. Ultracataclasite structure and friction processes of the Punchbowl fault, San Andreas system, California. *Tectonophysics* 295: 199–221.

Dieterich, J.H. 1979. Modeling of rock friction: 1. Experimental results and constitutive equations. *Journal Geophysical Research* 84: 2161–2168.

Dieterich, J.H. & Kilgore, B. 1994. Direct observation of frictional contacts: new insights for state-dependent properties. *Pure Applied Geophysics* 143: 283–302.

Gajo, A., Bigoni, D. & Wood, D.M. 2004. Multiple shear band development and related instabilities in granular materials. *Journal of Mechanics and Physics of Solids.* 52: 2683–2724, 10.1016/j.jmps.2004.05.010.

Karner, S.L., Chester, F.M. & Chester, J.S. 2005. Towards a general state-variable constitutive relation to describe granular deformation. *Earth and Planetary Science Letters* 237: 940–950, 10.1016/j.epsl.2005.06.056.

Lade, P.V. 2002. Instability, shear banding, and failure in granular materials. *International Journal of Solids and Structures* 39: 3337–3357.

Logan, J.M., Friedman, M., Higgs, N., Dengo, C. & Shimamoto, T. 1979. Experimental studies of simulated gouge and their application to studies of natural fault zones, Analyses of Actual Fault Zones in Bedrock. *U.S. Geol. Surv. Open File Rep.* 1239: 305–43.

Mair, K. & Marone, C. 1999. Friction of simulated fault gouge for a wide range of velocities and normal stresses. *Journal Geophysical Research* 104: 28, 899–28, 914.

Mandl, G., de Jong, N.L.J. & Maltha, A. 1977. Shear zones in granular material, an experimental study of their structure and mechanical genesis, *Rock Mech.*, 9, 95–144.

Marone, C. 1998. Laboratory-derived friction laws and their application to seismic faulting. *Annual Reviews of Earth & Planetary Science* 26: 643–696.

Marone, C., Cocco, M., Richardson, E. & Tinti, E. 2009. The critical slip distance for seismic and aseismic fault zones of finite width. In E. Fukuyama (ed.) *Fault-zone Properties and Earthquake Rupture Dynamics, International Geophysics Series* 94: 135–162, Elsevier.

Marone, C., Hobbs, B.E. & Ord, A. 1992. Coulomb constitutive laws for friction: contrasts in frictional behavior for distributed and localized shear. *Pure and Applied Geophysics.* 139: 195–214.

Marone, C. & Kilgore, B. 1993. Scaling of the critical slip distance for seismic faulting with shear strain in fault zones. *Nature* 362: 618–621.

Marone, C., Raleigh, C.B. & Scholz, C.H. 1990. Frictional behavior and constitutive modeling of simulated fault gouge. *Journal Geophysical Research* 95: 7007–7025.

Marone, C. & Scholz, C.H. 1989. Particle-size distribution and microstructures within simulated fault gouge. *Journal of Structural Geology* 11: 799–814.

Perfettini, H. & Avouac, J.-P. 2004. Stress transfer and strain rate variations during the seismic cycle. *Journal Geophysical Research* 109: 10.1029/2003JB002917.

Rabinowicz, E. 1951. The nature of static and kinetic coefficients of friction. *Journal Applied Physics* 22: 1373–1379.

Rathbun, A.P., Marone, C., Alley, R.B. & Anandakrishnan, S. 2008. Laboratory study of the frictional rheology of sheared till. *Journal Geophysical Research* 113(F02020): 10.1029/2007JF000815.

Rice, J.R. 1993. Spatio-temporal complexity of slip on a fault. *Journal Geophysical Research* 98: 9885–9907.

Rice, J.R. 2006. Heating and weakening of faults during earthquake slip. *Journal Geophysical Research* 111(B05311): 10.1029/2005JB004006.

Rice, J.R. & Cocco, M. 2007. Seismic fault rheology and earthquake dynamics. In Handy, M.R., Hirth, G. & Hovius, N. (eds.), *Tectonic Faults: Agents of Change on a Dynamic Earth; Dahlem Workshop 95, Berlin, January 2005, on The Dynamics of Fault Zones*. The MIT Press: Cambridge, MA, USA.

Richefeu, V., El Youssoufi, M.S. & Radjaï, F. 2006. Shear strength properties of wet granular materials. *Physical Review E* 73: 051304, 10.1103/PhysRevE.73.051304.

Ruina, A. 1983. Slip instability and state variable friction laws. *Journal Geophysical Research* 88: 10359–10370.

Scholz, C.H. 1998. Earthquakes and friction laws. *Nature* 391: 37–42.

Sleep, N.H. 1997. Application of a unified rate and state friction theory to the mechanics of fault zones with strain localization. *Journal Geophysical Research* 102: 2875–2895.

Song, I., Elphick, S.C., Odling, N., Main, I.G. & Ngwenya, B.T. 2004. Hydromechanical behaviour of fine-grained calcilutite and fault gouge from the Aigion Fault Zone, Greece. *Comptes Rendus Geosciences* 336: 445–454, 10.1016/j.crte.2003.11.019.

Tullis, T.E. 1996. Rock friction and its implications for earthquake prediction examined via models of parkfield earthqauakes. *Proc. Natl. Acad. Sci. USA* 93: 3803–3810.

Wood, D.M. 2002. Some observations of volumetric instabilities in soils. *International Journal of Solids and Structures* 39: 3429–3449.

Ziv, A. 2007. On the nucleation of creep and the interaction between creep and seismic slip on rate- and state-dependent faults. *Geophysical Research Letters* 34: 10.1029/2007GL030337.

Some new experimental observations on fracture roughness anisotropy

Giovanni Grasselli

Department of Civil Engineering, University of Toronto, Toronto, Ontario, Canada

ABSTRACT: A distinctive anisotropy in fracture roughness was observed in all surfaces created during Brazilian tests and fracture toughness tests using Cracked Chevron Notched Brazilian Disc (CCNBD) specimens suggesting the existence of a strong link between the fracture propagation mechanism and the resulting crack. Currently, it is hypothesized that the observed anisotropy in roughness could be related to the nature of energy dissipation during fracture propagation. The induced tensile loading causes a fracture front to propagate outward from the disc centre towards its outer edge where the compressional line loading is applied. It is thought that the fracture must dissipate more energy in this direction compared to the direction perpendicular to crack front propagation. It is observed that the advancing crack front often cuts through grains, while in the direction perpendicular to the propagation the fracture seems to preferentially follow grain-grain boundaries, resulting in a higher roughness.

1   INTRODUCTION

Fracture roughness describes the relief of a crack surface and it has been shown to highly contribute to the shear strength, frictional properties, and transport properties of fractures. For these reasons it has been of interest to researchers in geomechanical fields, ranging from oil, gas, and geothermal reservoirs to rock slope and dam foundation stability. The fracture geometry is the result of a failure process that drives the crack through the intact material. In the literature, most of the models and studies refer to fracture propagation as a tip that moves inside the material, creating a crack that is characterized by an irregular profile. However, in nature, the fracture is not a bi-dimensional feature, but it is a quasi-planar three-dimensional object. The presence of a third dimension introduces a whole new set of phenomena to have a role in the crack propagation. The crack front is not a point (crack-tip) any more, but it is characterized by many step lines, or an aggregation of segments which interconnects during the propagation of the crack. Thus, it is very likely that the crack front is not a continuous line whose real geometry is perturbed by imperfections and changes in material properties. Looking at average parameters, several researches agree that as fracture toughness increases, the surface roughness of the fracture generated, expressed as function of its fractal dimension, increases (Backers et al., 2003; Issa et al., 2003; Lange et al., 1993). Others directly relate the fracture energy to the roughness number, defined as the ratio between the real crack surface area and its nominal one measured as projection on the average plane of the crack. However, to the author's knowledge very few past studies accounted for the influence of the inherent heterogeneity of natural materials such as rocks (Ramanathan et al., 1997) and most of the published experiments and researches refer to homogeneous materials, disregarding the presence of flaws, voids, pores and micro-cracks. Only recently, experimental studies showed that preferentially oriented microstructural fabric and micro-cracks are largely responsible for rock anisotropic properties. For example, Nasseri et al. (2006; 2008), studying the relation between microstructural fabrics and fracture toughness anisotropy in granitic rocks, concluded that fracture toughness is highly sensitive to the presence of oriented micro-cracks and correlates well with measured differences in seismic velocities. Their experiments showed differences in fracture toughness ranging up to 2.4 times for the same rock type. This anisotropy has been linked directly to the microstructural anisotropy present in the rock, as represented by the size of dominant microstructural fabric, and its orientation with respect to the direction of propagation of the induced fracture (test-crack). They also noticed that, when a test-crack propagates at right angle to the dominant preferred microstructural fabric, it yields the highest fracture toughness value. The produced test-crack, when observed in a thin section, is characterized by increasing segmentation (sudden termination and reappearance of test-crack) and higher deflection (deviation from main direction) that result in a more tortuous profile. In the contrary when the test-crack is forced to propagate parallel to the dominant preferred microstructural fabric, yields the lowest fracture toughness value. The profile of such a test-crack, when observed under thin section, is smooth and does not show deflection. This paper summarizes the experimental observations made at the University of Toronto during the analysis of roughness distributions of mode I cracks created in granitic and sedimentary rock samples. In this study we will pay a special attention to the presence of anisotropy in roughness and its correlation with rock microstructure.

2   MATERIALS AND METHODS

2.1   *Creation of tensile fracture from intact specimens*

All granitic surfaces presented in this study were created in the laboratory under mode I tensile stress from intact rock samples. All granite surfaces were obtained from Chevron Cracked Notched Brazilian Disc (CCNBD) samples originally tested to evaluate fracture toughness (Figure 1), while limestone surfaces were produced during Brazilian indirect tensile tests.

2.2   *Microstructural investigation of the intact rocks*

Microstructural investigations of all granitic rock types presented in this article were carried out by analysing their thin sections. Microstructural mapping included mineral composition, grain orientation, micro- and macro-crack mapping, their density calculation and orientation with respect to the rift plan are reported in detail in a recent article by Nasseri and Mohanty (2008).

2.3   *Influence of thermal treatment on fracture roughness*

A series of 20 of Westerly granite were divided in five sets of four samples each. Different amounts of thermal damage were generated by heat treating four sets to 250°C, 450°C, 650°C, and 850°C, respectively, while leaving a fifth set, referred to as RT (room temperature) specimens, untreated. As described in detail by Tatone et al. (in press.) Westerly granite was selected for this part of the research because it is even textured, isotropic, and has been well studied in the past; thus, complexity related to pre-existing rock fabric anisotropy was avoided. The heat treatment process

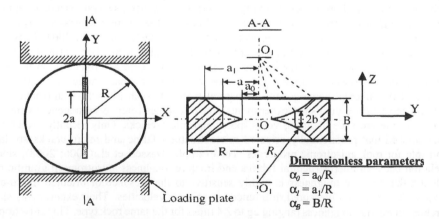

Figure 1.   Geometry of the CCNBD and related parameters: R = Radius of disc; B = Thickness of disc; R_S = Radius of saw; a = Length of crack; a_0 = Initial half length of chevron notch; a_1 = Final half length of chevron notch (ISRM, 1995).

involved slowly heating the specimens to the desired temperature and cooling them back to room temperature. A heating/cooling rate of roughly 1–2°C/min was maintained at all times, which was sufficiently slow to avoid cracking due to thermal shock. The 20 samples were then failed in a standard Mode I tensile test at room temperature. The surfaces of the produced cracks were then analyzed in order to asses the influence of thermal damage on fracture roughness.

2.4 *Fracture roughness: measurement techniques and procedure*

A 3D stereo-topometric measurement system, the Advanced Topometric Sensor (ATOS II) manufactured by GOM mbH, was adopted to digitize the fracture surfaces. The ATOS II system consists of a central projector unit which projects various white-light fringe patterns onto the surface being measured and two digital cameras which automatically capture images of these patterns on the surface (Figure 2). From these images the software computes three-dimensional coordinates with high accuracy for each pixel using triangulation methods and digital image processing (fringe projection and image shifting). Consistent exploitation of redundant information minimises the measurement errors. Due to the high data density resulting from the optical measurement process, details of the rough surface can be depicted precisely as point data. For the set-ups chosen for this study we measured points on a 250 μm xy and on a 44 μm xy grid. The accuracy of the point cloud along the z direction has been computed by the system to be ±20 μm for the 250 μm spacing, and ±2 μm for the 44 μm spacing setup. For the current research, the three-dimensional point clouds defining

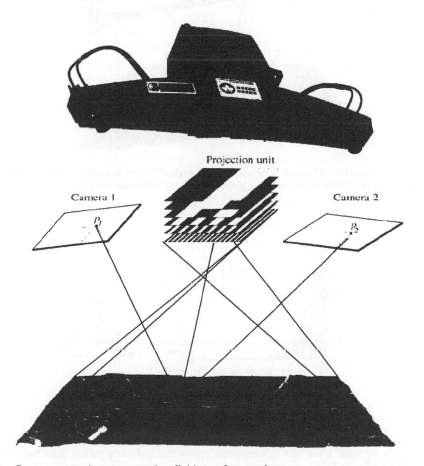

Figure 2. Stereo-topometric scanner used to digitize surface roughness.

the fractured surfaces were reconstructed using a Delaunay triangulation algorithm (.stl format) built into the ATOS software. This approach results in a discretization of the fracture surface into contiguous triangles, defined by vertices and by the orientation of the vector normal to the plane of the triangle. Looking at the reconstructed surface, each triangle orientation is uniquely identified by its azimuth angle (α) and dip angle (θ) (Figure 3). As presented in detail by Grasselli et al. (2002) and Grasselli (2006), it is possible to describe the fracture surface roughness as function of the apparent dip angle (θ^*), which quantifies "how steep" each triangle is with respect to the fracture average plane, measured along a chosen direction (Figure 3). For each surface it is possible to calculate the area of the joint that has an apparent dip angle equal or greater than a chosen threshold dip angle. The cumulative area is then normalized with respect to the total area of the fracture and it is termed A_θ^*. By varying the threshold dip value from 0 degrees to the maximum apparent dip angle (θ_{max}^*) measured on the surface, it is possible to plot the variation of A_θ^* as function of the

$$\tan\theta^* = -\tan\theta\cos\alpha$$

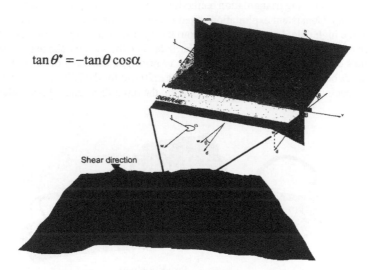

Figure 3.   Geometrical identification of the apparent dip angle, θ^*, measured along a chosen direction (shear direction) with respect to the fracture average plane (shear plane) (Grasselli et al. 2002).

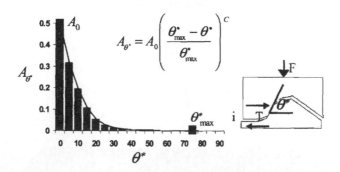

$$A_{\theta^*} = A_0\left(\frac{\theta_{max}^* - \theta^*}{\theta_{max}^*}\right)^C$$

Figure 4.   This plot describes, for a given shear direction, the distribution of the cumulative area of the surface A_θ^* characterized by an apparent dip angle (θ^*) equal or great than a chosen threshold value. θ^* is measured with respect to the average plane of the joint. As evident from the figure, increasing the threshold value, A_θ^* tends to rapidly decrease. It has been experimentally proven that such decay can be described by the above power law (Grasselli, 2006).

threshold dip angle (Figure 4). The relationship between the A_θ^* and the θ^* can be described by the following equation (see Grasselli et al. 2002 for more details):

$$A_\theta^* = A_0 \left(\frac{\theta_{max}^* - \theta^*}{\theta_{max}^*} \right)^C \tag{1}$$

where A_0 is the percentage of the fracture area having an apparent dip angle of 0 degrees or higher in the direction analyzed, θ_{max}^* is the maximum apparent dip angle in the direction analyzed, and C is a "roughness" parameter, calculated using a best-fit regression function, which characterizes the distribution of the apparent dip angles over the surface with respect to the direction analyzed (Grasselli et al., 2002).

As discussed in detail by Grasselli (2006), the parameters A_0, C and θ_{max}^* depend on the specified direction that is analyzed, as well as on the three-dimensional surface representation (i.e. triangulation algorithm and measurement resolution). In order to visualize the surface roughness anisotropy, the parameters A_0, C and θ_{max}^* are calculated along several possible directions, and the values of ratio θ_{max}^*/C, obtained for each of those, can be plotted in a polar diagram (Grasselli, 2006). The polar representation shows the anisotropy in roughness as a deviation of the plot from a circular shape with higher roughness to be associated with higher θ_{max}^*/C. Figure 5 illustrates an

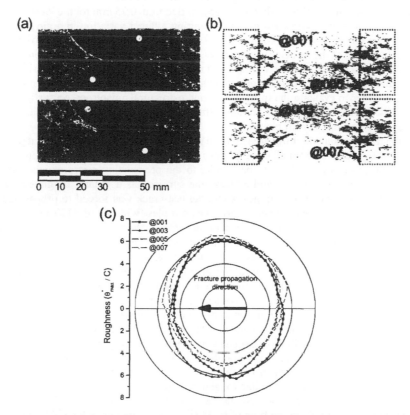

Figure 5. (a) Photograph of a Westerly granite sample, (b) three-dimensional surface reconstruction, and (c) polar distribution of roughness values (θ_{max}^*/C) for the fracture surfaces created during CCNBD tests of room temperature Westerly granite. Please note that in (c) the roughness values obtained from all four surfaces have been plotted such that the direction of fracture propagation is consistent. It should also be noted that @001 and @003 are mirror surfaces as are those denoted as @005 and @007.

example of the 3D reconstruction and resulting polar diagram describing the directional distribution of roughness of the four fracture surfaces of a CCNBD specimen obtained from a RT Westerly granite sample.

3 EXPERIMENTAL RESULTS

Surfaces of Barre (BG), Stanstead (STG), Bigwood (BWG), and Westerly (WG) granite specimens, created during toughness tests, and of Indiana Limestone (ILS), created during Brazilian tests, were analyzed with the purpose of estimating the surface roughness value and distribution. For each CCNBD specimen, only the areas of the fracture surfaces corresponding to unstable crack propagation were analyzed (Figure 5b). Surface the parameters A_0, C and θ^*_{max} were calculated along different directions all around the average-plane of the fracture in steps of 5 degrees. The values of θ^*_{max}/C, obtained for each plane, are plotted as polar diagrams (Figure 5) in order to readily visualize changes in roughness values as function of the direction analyzed. All measured fractures, independent of rock type, specific microstructure, and presence of micro-cracks, are characterized by an elliptical shape that reflects the directional dependence of roughness, with higher values, that correspond to greater roughness, always in the direction perpendicular to the propagation of the fracture (vertical axis in Figure 5). In our analysis we define "roughness anisotropy ratio" as the roughness perpendicular to the direction of fracture propagation (maximum value of the roughness) divided by the roughness parallel to propagation (minimum value).

The average grain size for the studied rock types varied from 0.75 mm for the Westerly to 1.3 mm for Stanstead granite, with Bigwood granite exhibiting a larger variation in grain size from 0.03 to 0.9 mm (Nasseri & Mohanty, 2008; Nasseri et al., 2007). The average micro-crack lengths were similar to those of the respective grain sizes, except for Bigwood granite, where much longer features designated as meso-cracks were present in addition. At micro-scale Nasseri and Mohanty found that all rock types but Westerly granite exhibit significant anisotropy in the microstructure. Therefore, in order to evaluate the influence of microstructural anisotropy on surface roughness, for Barre, Stanstead and Bigwood granites, fractures were generated and studied along their i) hard-way, ii) grain, and iii) rift planes.

The comparison among fractures created from the same sample but along different planes shows that average roughness value for all three granites changes as a function of the chosen plane. When the fracture propagates in a plane perpendicular to the pre-existing microstructural petrofabric orientation (i.e., samples BGKyx and STKzx) the surface has a greater roughness value with respect to the one measured on samples where the test-crack was forced to propagated parallel to the pre-existing microstructural petrofabric (i.e., samples BGKzy and STKyx). The difference in grain size between the four granites is also reflected in the magnitude of roughness values. For example, when comparing Barre tested fracture planes (Figure 6a) with those in Stanstead granite (Figure 6b), all created by forcing the test-crack to propagate at the same angle with respect to the pre-existing microstructural petrofabric, we notice that Stanstead fractured planes are generally characterized by slightly greater roughness than Barre granite ones. Observing the profile of the test-cracks on thin sections Nasseri et al. (in press) report that those that are forced to propagate perpendicularly to the pre-existent microstructural petrofabric, mostly follow grain boundaries, therefore, they tend to generate surfaces whose roughness is directly influenced by those fabrics. This could contribute to explain why, when comparing the four granites, we observe that the larger grain size for Stanstead granite is also reflected in generally greater roughness and in a higher relative increasing in roughness among the three tested planes (see comparison with Barre samples in Table 1). Similar roughness pattern has been also measured on Indiana limestone samples produced during Brazilian tests (Assane Oumarou et al., 2009). We can also notice that, independently from the type of rock, all surfaces are characterized by an anisotropic distribution of the roughness, with lower value always in the direction of propagation of the crack and higher values perpendicular to it. Interestingly, all tested samples, independently of the rock type, show anisotropy in roughness that ranges between 1.14 and 1.31 with an average of 1.22 (Table 1).

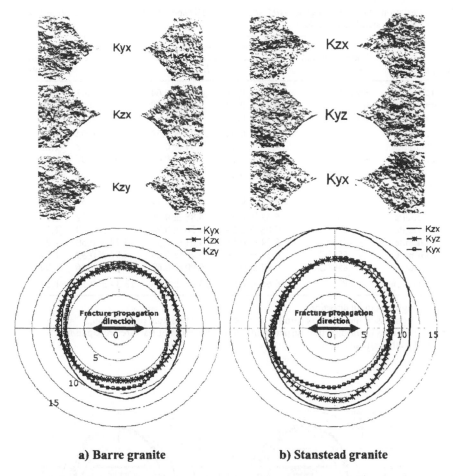

a) Barre granite   **b) Stanstead granite**

Figure 6.   a) Surface reconstruction and polar distribution of roughness values (θ^*_{max}/C) for three surfaces created during CCNBD tests on Barre granite samples. b) Surface reconstruction and polar distribution of roughness values (θ^*_{max}/C) for three surfaces created during CCNBD tests on Stanstead granite samples.

3.1   *Effect of temperature on fracture roughness anisotropy*

The fracture surfaces produced in thermally treated Westerly granite samples during CCNBD fracture toughness testing were digitized and analyzed with the aim of estimating the magnitude and directional distribution of surface roughness as a function of the temperature of thermal treatment. In total, the fracture surfaces of five CCNBD specimens (one for each treatment temperature) were analyzed. The resultant values of θ^*_{max}/C, for each direction are plotted in Figure 5c for the RT specimen and Figure 7 for the thermally treated specimens. To facilitate comparison of the roughness distributions for each thermal treatment temperature, the average roughness of the four surfaces of each CCNBD specimen were calculated and plotted in the same polar diagram (Figure 8a). For treatment temperatures up to 450°C, the distribution of roughness has an elliptical shape, with higher values perpendicular to the direction of fracture propagation (90–270 degrees) indicating there is a well-defined anisotropy in roughness. Beyond 450°C the polar distribution of roughness becomes nearly circular indicating that the anisotropy in roughness almost vanishes. To further illustrate this homogenization of roughness beyond 450°C, the average roughness values and their respective standard deviations in directions parallel and perpendicular to the direction of fracture propagation were plotted as a function of temperature (Figure 8b). The plot shows that

Table 1.   Variation of fracture roughness (θ_{max}^*/C) of CCNBD specimens of four types of granite analyzed in this study (mean $\pm$ standard deviation).

Rock type	Plane–Thermal treatment	Average roughness value (θ_{max}^*/C)			Roughness anisotropy ratio*
		a	b	c	
Barre Granite	Kyx - RT	10.21 ± 0.52	8.84 ± 0.64	11.83 ± 0.41	1.22
Barre Granite	Kzx - RT	10.32 ± 0.52	8.69 ± 0.70	11.69 ± 0.40	1.19
Barre Granite	Kzy - RT	9.10 ± 0.93	7.14 ± 1.04	10.72 ± 0.88	1.22
Stanstead Granite	Kzx - RT	12.62 ± 0.66	10.91 ± 0.59	14.99 ± 0.94	1.19
Stanstead Granite	Kyz - RT	9.93 ± 0.47	7.97 ± 0.63	11.43 ± 0.48	1.23
Stanstead Granite	Kyx - RT	9.84 ± 0.17	8.64 ± 0.42	11.00 ± 0.35	1.18
Bigwood Granite	Kzy - RT	9.88 ± 0.98	7.67 ± 1.18	12.20 ± 1.03	1.28
Bigwood Granite	Kzx - RT	7.97 ± 0.39	7.02 ± 0.37	8.84 ± 0.42	1.14
Bigwood Granite	Kyx - RT	8.23 ± 0.41	6.93 ± 0.11	10.53 ± 0.85	1.31
Westerly Granite	– - RT	5.33 ± 0.57	4.88 ± 0.26	5.85 ± 0.52	1.20
Westerly Granite	– - 250°C	5.06 ± 0.54	4.56 ± 0.24	5.77 ± 0.48	1.27
Westerly Granite	– - 450°C	5.24 ± 0.55	4.61 ± 0.35	5.78 ± 0.50	1.25
Westerly Granite	– - 650°C	6.36 ± 0.38	6.23 ± 0.23	6.51 ± 0.39	1.04
Westerly Granite	– - 850°C	6.97 ± 0.47	6.90 ± 0.62	7.17 ± 0.26	1.04
Indiana Limestone	– - RT	16.66 ± 0.07	14.95 ± 0.15	18.34 ± 0.11	1.21

a Average roughness in all directions,
b Average roughness in the direction parallel to the direction of fracture propagation,
c Average roughness in the direction perpendicular to the direction of fracture propagation,
* Calculated as the ratio of *c* to *b*.

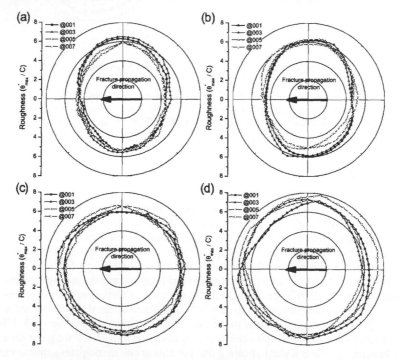

Figure 7.   Polar distribution of roughness values (θ_{max}^*/C) for the fracture surfaces created during cracked chevron notched Brazilian tests of thermally treated westerly granite: (a) 250°C, (b) 450°C, (c) 650°C, and (d) 850°C.

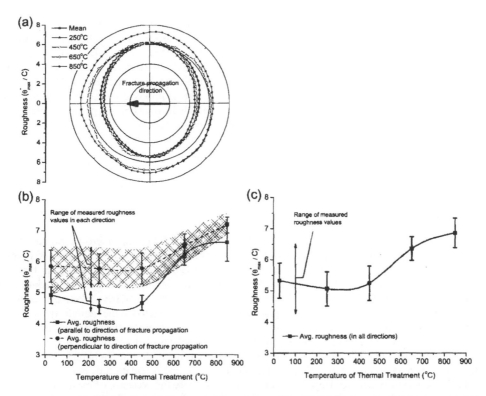

Figure 8. (a) Polar distribution of average roughness values (θ_{max}^*/C) for the fracture surfaces created during CCNBD tests of thermally treated westerly granite. (b) A comparison of the fracture roughness in directions parallel and perpendicular to fracture propagation as a function of the thermal treatment temperature. (c) A comparison of the average fracture roughness as a function of the thermal treatment temperature. Error bars represent ± one standard deviation.

up to 450°C the average roughness values measured in each direction differ by approximately 1.0 to 1.2 roughness units (18% to 23%), while for 650°C and 850°C the values differ by only 0.28 and 0.27 roughness units (4%), respectively. The thermally treated surfaces measured in this study display anisotropy ratios that range from 1.27 for 250°C to 1.04 for both 650°C and 850°C. A complete summary of the anisotropy ratios for each temperature as well as the corresponding values of roughness and their respective standard deviations are included in Table 1.

Comparison of the roughness values for specimens treated to various temperatures showed that, overall, the average roughness in all directions (Figure 8c), as well as the roughness measured parallel and perpendicular to the direction of fracture propagation (Figure 8b), increased with the temperature of thermal temperature. Considering Figure 8c, the change in roughness with increasing temperature up to 450°C is minimal (less than 4% change from RT to 450°C). However, between 450°C and 650°C there is a sizeable increase in roughness as the average roughness value increases from 5.2 to 6.4, an increase of more than 20%. This increase appears to correspond to the loss of anisotropy described above. Beyond 650°C the roughness continues to increase to an average value of 6.8 at 850°C.

4 DISCUSSION

All fracture surfaces measured in the present research, independent of rock type, show an anisotropic roughness (elliptical curves in Figure 6) with greater roughness values in the direction perpendicular to the fracture propagation. These experimental results suggest that the fracturing process and its direction of propagation could be correlated to the anisotropy in fracture roughness, leading to the creation of a preferential direction for shear displacements and fluid flow. Currently, we hypothesize that the observed anisotropy in roughness might be related to the nature of energy dissipation during fracture propagation in CCNBD and Brazilian discs specimens. The initiation and propagation direction of fractures in indirect tensile tests is imposed by the orientation of the applied loading. For example, for CCNBD samples, the tensile loading applied perpendicular to the plane containing the specimen notch causes a fracture front to propagate outward from the notch tip at the disc centre towards its outer edge where the compressional line loading is applied (along Y direction defined in Figure 1). The fracture must dissipate more energy in this direction compared to the direction perpendicular to crack front propagation (laterally along Z direction defined in Figure 1). Therefore, it is thought that additional energy present in the Y direction allows the advancing crack front to cut through grains, while the lesser energy in the direction perpendicular to the propagation (Z direction) causes the fracture to preferentially follow grain-grain boundaries (weakest bond).

SEM images show that the test-crack generally propagates following micro-fabric structures (Tatone et al., in press.). The fracture roughness values obtained from crcacks in Stanstead granite are higher than those obtained from fracture surfaces in Barre granite. This suggests that the difference in roughness between Barre and Stanstead granite is influenced by their variation in grain size. Interestingly, we observed that the roughness measured in BGKyx plane proves to be larger than STGKyz and STGKyx. Such a phenomenon could be related to the more frequent out of plane propagation of the test-crack as a result of its interaction with microstructural fabric oriented normal to propagation direction as evidenced in BGKyx. Analysis of Barre and Stanstead surfaces proves the fundamental link between surface roughness and microstructural characteristics of the intact rock such as grain size and distribution, and presence and orientation of micro- and meso-cracks. This is also evident for Bigwood granite samples where the presence of meso-cracks in the intact rock is key for the propagation of smoother cracks with respect to the other granites.

The last part of the study looked at how isotropic microsctructural features influence the roughness of tests-cracks. Westerly granite, an even textured, isotropic granite, was selected for this part of the research to avoid the effects of preferred microstructural orientation on the resultant fracture roughness. Using an isotropic granite, we have looked at how an increasing in internal micro-crack density influences the roughness of the test-crack. We have followed the evolution of the roughness value with respect to the amount of damage induced by thermal stresses. Granitic rocks are easily subject to significant amounts of thermally induced damage in the form of grain-grain boundary micro-cracking due to differential thermal expansion of their constituent minerals (Tatone et al., in press.). Weakening and opening of grain boundaries allows the test-crack to preferentially propagate along grain boundaries and thus the failed surfaces, at increasingly higher temperatures, portray the roughness of both expanded grains and grain-grain boundaries. Due to the lack of a pre-existing damaged area ahead of the propagating crack in samples treated to lower temperatures (RT and 250°C), test cracks are characterized by relatively smooth profiles as they cut through grains rather than follow expanded grain-grain boundaries, while for samples treated at higher temperatures (450°C, 650°C, and 850°C) the test-crack tends to follow the thermally damaged grain boundaries. This phenomenon can be observed in thin sections when comparing the test-crack path in the RT specimens to that of the 450°C specimen. Since the fractures increasingly follow grain-grain boundaries with temperature, the relative increase in fracture roughness with temperature is dictated by the shape, size, and expansion of the grains which comprise the rock. It is important to note that although a finer or coarser grained granitic rock would experience increases in grain-grain boundary micro-crack density when subjected to the same thermal treatment, the relative changes in fracture roughness between RT and 850°C specimens would likely differ significantly due to the different grain size.

A distinctive anisotropy in fracture roughness was observed in the fracture surfaces of the RT, 250°C, and 450°C specimens with increased roughness perpendicular to the direction of fracture propagation. Beyond 450°C this anisotropy greatly diminished as the roughness parallel and perpendicular to the fracture propagation direction became nearly equal. Interestingly, the largest increase in the measured surface roughness occurred between 450°C and 650°C which corresponded with a sharp decrease in both Vp, and K_{IC} (Tatone et al., in press.). A review of previous work on thermally treated granitic rocks found that several other physical/mechanical rock properties have been shown to undergo significant change when the temperature of thermal treatment extends over similar temperatures. Examples of such properties include: unconfined compressive strength, Young's modulus, tensile strength, and porosity and permeability. This drastic variation of mechanical/physical properties is explained by the $\alpha - \beta$ quartz phase transition which is known to occur at roughly 573°C under atmospheric conditions. This phase transition increases the differential expansion between the quartz grains and other constituent minerals leading to increased micro-cracking of the specimen. The quasi-total loss of the anisotropy observed between the 450°C and 650°C is again thought to be related to the $\alpha - \beta$ quartz transition. With the large increase in damage associated with the mineral phase transition, the advancing orthogonal crack front (along both Y and Z directions as shown in Figure 1) preferentially follows the path of least resistance which is characterized by open grain-grain boundaries. It should be emphasized that the above hypothesis is only one possible explanation for the deterioration of anisotropy in fracture roughness, which still remains measurable.

5 SUMMARY AND CONCLUSIONS

This article summarizes the research on rock fracture roughness carried at the University of Toronto in the past three years. The results presented in this contribution confirm the essential connection between petrofabric characteristics, evolution and extent of associated induced cracks along specific directions, and fracture roughness of the rocks studied. In particular, we have observed a direct correlation between fracture roughness values, rock grain size and crack orientations. All fracture surfaces were characterized by a remarkable anisotropy in the roughness value with consistently lower values in the direction we propagated the crack.

Investigating the effect of thermal damage on rocks, a distinctive anisotropy in fracture roughness was observed in the fracture surfaces of the RT, 250°C, and 450°C specimens, while beyond 450°C this anisotropy greatly diminished as the roughness along both directions within the fracture propagation plane became nearly equal. Reduction in the roughness anisotropy as a function of temperature is proposed to be related to the openness of grain-grain boundaries to which both the test crack propagation front and lateral front responded equally after $\alpha - \beta$ quartz phase transition is achieved.

The presence of anisotropy in roughness and its loss due to thermal treatment (isotropic damage of the rock) may have several important impacts on the engineering behavior of rock discontinuities. At low temperatures the anisotropy in fracture roughness may potentially lead to the creation of a preferential direction for shear displacements (anisotropic shear strength) and fluid flow (anisotropic permeability), while in rocks which have been exposed to high temperatures such properties will become isotropic. A better understanding of the effect of temperature on rock fracture genesis and related properties may lead to better understanding of volcanic activity, the movement of hydrothermal fluids in the subsurface, and crustal fault slip behaviour.

Another important note regarding the observed increase in fracture roughness with temperature has to do with discontinuity shear strength. Although increasing roughness of discontinuities in rock typically results in increased shear strength, the intact strength of the remaining asperities of the thermally damaged specimens is significantly reduced due to the presence of extensive grain-grain boundary micro-cracks. Therefore, despite increased roughness, the shear strength of thermally treated granite may be significantly decreased relative to RT specimens. Nonetheless, such a conclusion requires confirmation via a laboratory shear testing program of thermally treated specimens. In addition, a methodical investigation involving real fractures measured at multiple scales is needed to confirm and extend the present findings to large-scale geological features such as faults and joint-sets.

Finally, from a rock-physics point of view, we think that the data presented herein also suggest a strong correlation between the crack process and the produced three-dimensional crack surface. In particular we think that in rock the dissipation of fracture energy could be a strongly anisotropic process, possibly related to the heterogeneity of the material and to the particular rock microstructure. If this hypothesis will be proven correct, most fracture mechanics solutions proposed for metals and polymers, based on the assumption of isotropic materials, should be revisited and possibly modified before being applied to rocks and other heterogeneous geomaterials such as concrete.

ACKNOWLEDGMENT

This work has been supported by NSERC/Discovery Grant No. 341275 and NSERC/RTI Grant No. 345516. The author wish to thank Dr. J. Wirth and Mr. M. Braun, for their assistance with the initial surface scans, and to Mr. Tatone and Dr. Nasseri for providing part of the data herein presented and for their the precious help during laboratory tests and surface scans.

REFERENCES

Assane Oumarou, T., Cottrell, B.E., & Grasselli, G., 2009, Contribution of surface roughness on the shear strength of Indiana limestone cracks—An experimental study: *International Conference on Rock Joints and Jointed Rock Masses*.

Backers, T., Fardin, N., Dresen, G., & Stephansson, O., 2003, Effect of loading rate on Mode I fracture toughness, roughness and micromechanics of sandstone: *International Journal of Rock Mechanics and Mining Sciences*, v. 40, pp. 425–433.

Grasselli, G., 2006, Manuel Rocha Medal recipient—Shear strength of rock joints based on quantified surface description: *Rock Mechanics and Rock Engineering*, v. 39, pp. 295–314.

Grasselli, G., Wirth, J., & Egger, P., 2002, Quantitative three-dimensional description of a rough surface and parameter evolution with shearing: *International Journal of Rock Mechanics and Mining Sciences*, v. 39, pp. 789–800.

ISRM, 1995, Suggested method for determining mode I fracture toughness using cracked chevron notched Brazilian disc (CCNBD) specimens (Fowell, R.J.—Coordinator): *International Journal of Rock Mechanics and Mining Sciences*, v. 32, pp. 57–64.

Issa, M.A., Issa, M.A., Islam, M.S., & Chudnovsky, A., 2003, Fractal dimension—a measure of fracture roughness and toughness of concrete: *Engineering Fracture Mechanics*, v. 70, pp. 125–137.

Lange, D.A., Jennings, H.M., & Shah, S.P., 1993, Relationship between Fracture Surface-Roughness and Fracture-Behavior of Cement Paste and Mortar: *Journal of the American Ceramic Society*, v. 76, pp. 589–597.

Nasseri, M.H.B., Grasselli, G., & Mohanty, B., in press., Fracture toughness and fracture roughness in anisotropic granitic rocks: *Rock Mechanics and Rock Engineering*. Submitted on 19/3/2008.

Nasseri, M.H.B., & Mohanty, B., 2008, Fracture toughness anisotropy in granitic rocks: *International Journal of Rock Mechanics and Mining Sciences*, v. 45, pp. 167–193.

Nasseri, M.H.B., Mohanty, B., & Young, R.P., 2006, Fracture toughness measurements and acoustic emission activity in brittle rocks: *Pure and Applied Geophysics*, v. 163, pp. 917–945.

Nasseri, M.H.B., Schubnel, A., & Young, R.P., 2007, Coupled evolutions of fracture toughness and elastic wave velocities at high crack density in thermally treated Westerly Granite: *International Journal of Rock Mechanics and Mining Sciences*, v. 44, pp. 601–616.

Ramanathan, S., Ertas, D., & Fisher, D.S., 1997, Quasistatic Crack Propagation in Heterogeneous Media: *Physical Review Letters*, v. 79, pp. 873–876.

Tatone, B.S.A., Nasseri, M.H.B., Grasselli, G., & Young, P.R., 2009, Fracture toughness and fracture roughness interrelationship in thermally treated Westerly granite: *PAGEOPH*, v. 166, pp. 801–822.

Constraints on faulting mechanisms using 3D measurements of natural faults

Amir Sagy
Geological Survey of Israel, Jerusalem, Israel

Emily E. Brodsky
Department of Earth and Planetary Sciences, UC Santa Cruz, Santa Cruz, CA, USA

ABSTRACT: We use a combination of fault topography measurement with ground-based LiDAR and fault zone structure analyses to identify some major mechanical processes involved in faulting. Measurements on more than 30 normal and strike slip faults in different lithologies demonstrate that faults are not planar surfaces and roughness is strongly dependent on fault displacement. Thousands of profiles ranging from 10 μm to >100 m in length show that small-slip faults (slip <1 m) are rougher than large-slip faults (slip 10–100 m or more) parallel to the slip orientation. Surfaces of small-slip faults have asperities over the entire range of observed scales, while large-slip fault surfaces are polished on profiles as long as 1–2 m. In addition to the roughness exemplified by abrasive striations, we found quasi-elliptical topographical bumps with wavelengths of tens of meters in the slip orientation on faults with total displacements of tens to hundreds of meters. The internal structure of a bump on a large exposed fault surface southwest of Klamath Falls, Oregon was studied in detail. The surface reflects a variation of width of a cohesive grainy layer that borders the slip-surface. The layer contains evidence for particle rotation and flow, internal fracturing and slip. These observations together with the roughness measurements can be explained by wear on an initially rough surface. The on-going wear smoothes the surface, increases the localization and reduces the off-fault dissipation processes. Thus, a mature fault is likely to have absorbed larger slip events compared to a relatively less mature fault.

1 INTRODUCTION

The topography of fault surfaces is integral to all aspects of earthquake and fault mechanics (Chester & Chester, 2000; Brodsky & Kanamori, 2001; Adda-Bedia & Madariaga, 2008; Parsons, 2008). For example, one of the major determinants of slip distribution in an earthquake is the presence of asperities, which in some cases have their origins or termination in geometrical irregularities (Lay et al., 1982; Aki, 1984; Wesnousky, 1988). Furthermore, the variation of roughness with scale carries information about the dissipative processes on a fault (Power et al., 1988). Currently, it is unknown to what degree rock is ground down during earthquakes and, as a result, it is unknown what effect the brittle resistance plays in slowing down and stopping earthquakes. Since the difference between a large, societally damaging earthquake and a small, inconsequential one is simply the continued propagation of slip in the large earthquake, sorting out the role of resistance processes like wear is crucial to developing a physics-based model of earthquake damage. Here we combine multiscale fault surface geometry measurements with fault internal deformation structure. We use the data to measure the inter-relationship between surface-roughness and its near-by deformation, and for testing the role of wear as a mechanism for energy dissipation during earthquakes.

2   FAULT ROUGHNESS PARALLEL TO THE SLIP

We have previously measured the geometry of different exposed fault surfaces in the Western US and Italy using ground-based LiDAR and supplemented the data with laser profilometry on laboratory-scale hand-samples (Sagy et al., 2007). To quantify fault roughness we calculate the values of the power spectral densities of one-dimensional profiles parallel to the slip orientation (Fig. 1). The collected data has thus far yielded three major conclusions. a) Small-slip faults (slip <1 m) are rougher than large-slip faults (slip 10 to 100 m or more) on profiles parallel to the slip direction at the scale of slip during moderate earthquakes (a few meters). b) The power spectral density of small-slip faults perpendicular to the slip direction suggest that the roughness as a function of the measured profile can be modeled as a power-law along the measured scales. c) The slope of the power spectral density of large slip-faults parallel to the slip direction is close to logarithmic at the laboratory scales and is nearly self-affine at the larger field scales.

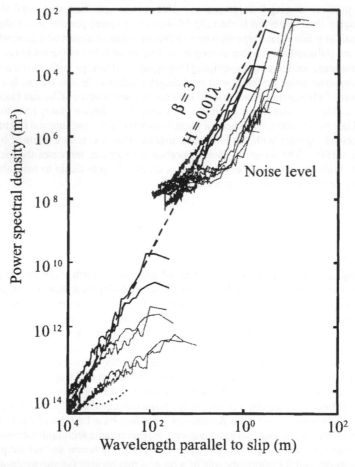

Figure 1.   Power spectral density calculated from sections from seven different fault surfaces that have been scanned using ground-based LiDAR, and from six hand samples scanned by a profilometer in the lab. Each curve includes 200–600 continuous individual profiles from the best part of the fault. Figure includes both LiDAR data (upper curves) and laboratory profilometer data (lower curves) from fault with slip of few cm to meters (black) and faults with total slip of tens of meters to hundreds of meters (grey). Doted lines represent scans of smooth, planar reference surfaces for the profilometer. Dashed black lines are slopes of $\beta = 3$. The graph is modified from Sagy et al. (2007).

3 DEFORMATION NEAR-BY THE SLIP SURFACE

Detailed structural analysis of a fault zone near Klamath Falls, Oregon combined with LiDAR measurements of the fault surface demonstrated the inter-relationship between the surface geometry and the deformation processes in the fault zone (Sagy & Brodsky, In print). The slip-surface is polished but contains elongated bumps with wavelengths of tens of meters and maximal amplitudes of ~0.5–1 m (Fig. 2a). Slip occurs along a band that contains principle slip surfaces with typical widths of 100 microns (Fig. 2b). The band structure present is evidence for extreme localization on slip surfaces accompanied by abrasional, fluid injection and granular flow structures (Fig. 2b–c).

The slip surfaces are parallel to the primary surface and are distinguished from the surroundings by particle size, shape and color (Fig. 2b). The sizes of the fragments in a single slip surface range between 10 μm–0.1μm. Fluidization and ductile deformation are also documented in this band associated with wavy layering and boudinage structures (Sagy and Brodsky, in print).

The slip band is hosted by a cohesive cataclasite layer. The layer contains aggregates of crystals comprised mainly of plagioclase rotated inside a matrix of finer particles that includes single crystals or fragments of plagioclase (Fig 2b–c). The grains generally have dimensions of a few millimeters to a few centimeters. One of the mechanisms to bring small grains into the cohesive layer is the injection of fine-grained material from the slip surfaces (Fig. 2b). Such injections of fluids mixed with ultrafine grains contribute to cementation. Consequently, we infer that the cohesiveness of the layer is a result of the on-going granular deformation, which increases the adhesive forces between comminuted grains as the surface area increases (Gilbert et al., 1991), repacks the rock volume to a denser configuration (Aydin, 1978), and promotes fluid transport into the layer. The combination of effects lithified the granular material. Our observations also suggest that bumps are not only remnants of previous large scale roughness but are manifestations of lenses of granular material which indicate volumetric flow is part of the process. Brittle deformation at a larger scale is also observed in the cataclasite layer. Small normal faults with displacements of a few millimeters appear mostly at acute angles to the main fault surface. In most cases, these faults have a sense that is sub-parallel to the slip orientation recorded by the striations and they typically cross the layer entirely, creating an S-shape in cross section.

The observed surface undulations are related to abrupt variations in this layer thickness. The layer widens under fault surface protrusions and becomes thin under surface depressions. Figure 2a presents examples of the thickening of the layer under the bumps and thinning of the layer under surface depression. The maximum exposed thickness of the layer can reach to more than a meter while under depression the width decreases down to 5 cm. Following this observation, we infer that the wavy appearance of the surface reflects variations of thickness of this cohesive layer.

Measurements on several fault strands in the same fault zone with total slips of 0.5–150 m reveal that the average thickness of the cohesive granular layer increases monotonically with slip (Sagy & Brodsky, in print). The change in thickness with increasing displacement (d (thickness)/dx) decreases as a function of the displacement.

4 MECHANICAL INTERPRETATION

The observations near the slip surface show that beside slip on a main surface the dissipation processes in the fault zone include fracturing, slip along secondary faults, comminution and particle rotation. These observations suggest that wear might be an essential process during faulting.

Experimental observations and related models demonstrated that wear volume during shear of two plates increases as a function of displacement by transient wear which is followed by steady state wear (Quineer et al., 1965; Yang, 2007). For the transient wear, the volume addition of wear material G in an increment of slip is a linear function of the volume V by which the surface departs from a perfectly flat surface

$$\frac{dG}{dx} = K_1 V$$

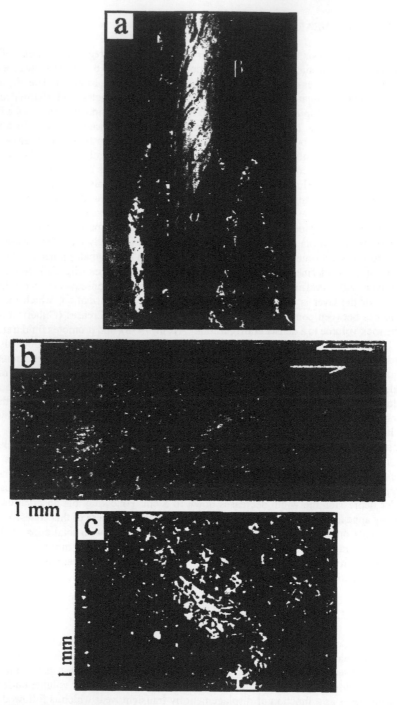

Figure 2.   Fault surface and related structure in Flower Pit fault. a) Oblique photo of a bump structure. The smooth surface (marked by I) is exposed above a cohesive grainy layer. The layer becomes thicker (0.5 m) under fault surface protrusions (marked by α) and thinner (5 cm) under surface depressions (marked by β). b) Thin section that shows slip surface (marked by the arrows) above the grainy cohesive layer. Fine grain material from the slip surface is injected to the grainy cohesive layer. The photo is rotated so that the fault surface is horizontal. c) Flow structure in the grainy cohesive layer that underlays the slip surface.

where x is the accumulated displacement, K_1 is a constant that depends on material properties and on a characteristic length of the specimen in a shear test, assuming a normal distribution of the roughness (Leong & Randolph, 1992).

In contrast to typical laboratory experiments, there is no obvious characteristic length in natural faults, and steady state wear might never be achieved (Power et al., 1988). Moreover, faults are generated in a natural anisotropic environment and develop under non-fixed stresses, at a variety of speeds and through large time scales. All these complications are expected to affect the geometrical surface evolution. Nevertheless, the observed smoothness of the fault surfaces parallel to the slip at <1 m scales (Fig. 1), together with the observation of bumps at scales of tens of meters (Fig. 2) are consistent with the preferential removal of the small-scale asperities by wear during slip.

These field measurements suggest that the effectiveness of the wear process in natural faults is a function of the displacement and of the size of the asperities. Small scale asperities are smeared out and disappear while large scale asperities are deformed but not removed. The measured roughness is thus expected to be depend on the ratio x/L, where x is the total displacement, and L is the measured profile parallel to the slip. For example, when $L >> x$ the wear effect is small and the roughness mostly influenced by the initial topography of the surface, and when $x >> L$ the observed surface is affected by the wear and the surface is predicted to be polished.

5   APPLICATIONS FOR ENERGY PARTITIONING DURING EARTHQUAKE

We suggest that wear processes and the geometric evolution of the slip-surface are essential for the energy dissipation processes. Energy dissipation in an ideal planar fault during an earthquake can be divided into radiant energy, creation of surface energy on the edge of the expanding fault, and work done on the fault plane during faulting (Kostrov, 1974). In a more realistic situation, a large part of the energy is absorbed by off-fault deformation (Shipton et al., 2006; Kanamori & Rivera, 2006). However, wear history decreases the roughness and in relatively mature faults less off-fault deformation is predicted to for a given slip amount. For example: Sagy & Brodsky, (in print) calculated the amount of slip along pre-existing slip-surface using a simple model of a planar surface with a sinusoidal perturbation of amplitude H and wavelength L (Chester & Chester, 2000). The calculations showed that smoother surfaces absorb larger amounts of slip before off-fault yielding occurs. These observations together with the roughness measurements (Fig. 1) indicate that less energy will dissipate in the fault zone as the fault matures. Thus, for a given amount of stress, slip along a mature fault is likely to be larger than slip on a relatively less mature fault.

REFERENCES

Adda-Bedia, M. & Madariaga, R. 2008. Seismic Radiation from a Kink on an Antiplane Fault, *Bull. Seismol. Soc. Am.* 98, 2291–2302.

Aki, K. 1984. Asperities, barriers, characteristic earthquakes and strong motion prediction, *J. Geophys. Res.*, 89, 5867–5872.

Aydin, A. 1978. Small faults formed as deformation bands in sandstone, *Pure Appl. Geophys.*, 116, 913–930.

Brodsky, E.E. & Kanamori, H. 2001. The elastohydrodynamic lubrication of faults, *J. Geophys. Res.*, 106, 16357–16374.

Chester, F.M. & Chester, J.S. 2000. Stress and deformation along wavy frictional faults, *J. Geophys. Res.*, 105, 23421–23430.

Gilbert, J.S., Lane, S.J. Sparks, R.S.J. & Koyaguchi, T. 1991. Charge measurements on particle fallout from a volcanic plume, *Nature*, 349, 598–600.

Jaeger, J.C., Cook, N.G.W. & Zimmerman, R.W. 2007. *Fundamentals of Rock Mechanics*, Blackwell Pub., 488 p.

Kanamori, H. & Rivera, L. 2006. Energy Partitioning During an Earthquake, *AGU Monograph Series* 170, Washington D.C. 3–14.

Kostrov, B.V. 1974. Seismic moment and energy of earthquakes, and seismic flow of rock (translated to English), *Izv. Earth Physics*, 1, 23–40.

Lay, T., Kanamori, H. & Ruff, L. 1982. The asperity model and the nature of large subduction zone earthquakes, *Earthquake Prediction Research*, 1, 3–71.

Leong, E.C. & Randolph, M.F. 1992. A model for rock interfacial behavior, *Rock Mech. Rock Eng.*, 25, 187–206.

Parsons, T. 2008. Persistent earthquake clusters and gaps from slip on irregular faults, *Nature Geoscience*, 1, 59–63.

Power, W.L., Tullis, T.E. & Weeks L.D. 1988. Roughness and wear during brittle faulting, *J. Geophys. Res.*, 93, 15,268–15,278.

Queener, C.A., Smith, T.C. & Mitchell W.L. 1965. Transient wear of machine parts. *Wear*, 8, 391–400.

Sagy A., Brodsky, E.E. & Axen, G.J. 2007. Evolution of fault-surface roughness with slip, *Geology*, 35, 283–286.

Sagy, A., & Brodsky E.E. 2008. Geometric and rheological asperities in an exposed fault zone, *J. Geophys. Res.*, doi: 10.1029/2008JB005701, in press.

Shipton, Z.K., Evans, J.P., Abercrombie, R.E. & Brodsky E.E. 2006. The missing sinks: slip localization in faults, damage zones, and the seismic energy budget. In Abercrombie, R. (Eds) *Earthquakes: Radiated Energy and the Physics of Faulting*, 217–222. Washington, DC. AGU.

Wesnousky, S.G. 1988. Seismological and structural evolution of strike-slip faults. *Nature*, 335, 340–343.

Yang, L.J. 2007. Modeling transient adhesive wear of tungsten carbide inserts tested with an angular setting, *Tribol. Inter.*, 40, 1075–1088.

Nonlinear elasticity and scalar damage rheology model for fractured rocks

Vladimir Lyakhovsky & Yariv Hamiel
Geological Survey of Israel, Jerusalem, Israel

ABSTRACT: Continuum damage mechanics models account for effects of distributed cracks. Crack density is expressed in terms of intensive scalar or tensorial damage parameters. A realistic damage rheology model should treat two aspects of the physics of fracturing: (1) the sensitivity of the macroscopic elastic moduli to distributed cracks and to the sense of loading, and (2) the evolution of damage (degradation-recovery of elasticity) in response to loading. The linear continuum damage mechanics of Kachanov with a scalar damage parameter accounts for these aspects but has several shortcomings, including constant Poisson ratio and thermodynamically-prohibited material healing. This led to the development of tensorial damage models that include many adjustable parameters which can hardly be constrained experimentally. Variations of Young modulus and Poisson ratio, as well as transition from material degradation to healing under different types of load, can be described in frame of non-linear continuum damage rheology model. We present here the theoretical basis for extension of the elastic potential which is fundamental for non-linear scalar damage rheology. This formulation is applied to various laboratory observations, including quasi-static modeling of composite material in order to reproduce damage accumulation and different effective moduli under tension and compression (e.g., Basaran & Nie, 2004); rock dilation under shear (e.g., Lockner & Stanchits, 2002); stress- and damage-induced seismic wave anisotropy observed during cycling load of Aue granite samples; and dynamic modeling of wave propagation to analyze the shift of the resonance frequency in rocks (e.g., Pasqualini et al., 2007). Comparison between numerical simulations and experimental results demonstrate that the nonlinear scalar damage rheology model can account for both quasi-static damage accumulation and nonlinear dynamic effects.

1 INTRODUCTION

Rock damage in the form of cracks, joints and other internal flaws is common in most rocks. It develops as part of rock formation and usually increases during tectonic loading. Damage accumulation under ongoing deformation can produce significant deviations from linear isotropic elasticity and may affect profoundly the elastic moduli especially just before macroscopic failure (Lockner and Byerlee, 1980; Lockner et al., 1992; Hamiel et al., 2004). Stress-strain relationships of a damaged rock are usually approximated by an elastic body with cracks or inclusions embedded inside an otherwise homogeneous matrix. For example, the elastic field of ellipsoidal inclusions (Eshelby, 1957) allows the construction of a model for a material with cracks. This approach was successfully applied to synthetic materials with known crack geometries and matrix elastic parameters (e.g. Christensen 1979). O'Connell & Budiansky (1974) and Budiansky & O'Connell (1976) proposed the self-consistent method for materials with random crack distributions. This approach was extended to the concept utilizing the crack-density tensor for finding the effective properties of a solid with arbitrary crack interactions (Kachanov 1987; Sayers & Kachanov 1991; Kachanov 1992) and was successfully applied for modeling mechanical properties and seismic wave propagation in anisotropic rock mass. However, the above approaches assume that cracks are open, with no contact between opposing faces, and that crack edges are blunt. Furthermore, cracks are assumed to be stationary and to have a constant aperture during deformation, that is, dilation and closure of cracks due to loading are ignored. The assumption of open stationary cracks forces the macroscopic elastic stress-strain relations of the host body to be linear (O'Connell & Budiansky 1974).

Various observations have demonstrated a non-linear elasticity in rocks and rock-like materials. Among them laboratory experiments by Nishihara (1957); Brace (1965); Zoback and Byerlee, (1975); Brady, (1969); Schock, (1977); Schmitt and Zoback, (1992); Weinberger et al. (1994); Johnson et al., (1996); Lockner and Stanchits (2002); Basaran and Nie (2004); Pasqualini et al. (2007) and others. Nonlinear elasticity is also observed in seismic records of ground motion in sediments and highly-damaged fault zone rocks (e.g., Field et al., 1997; Rubinstein and Beroza, 2004; Pavlenko and Irikura, 2003; Wu et al., 2008). Comparison of seismograms from weak and strong earthquakes indicates that attenuation becomes nonlinear at high amplitudes (e.g., Frankel et al., 2002; Beresnev, 2002; Hatzell et al., 2002, 2004; Bonilla et al., 2005; Tsuda et al., 2006, Sleep and Hagin, 2008).

In the present paper we provide theoretical developments based on a non-linear continuum model with a scalar damage state variable. Simultaneous change of both effective Young modulus and Poisson ratio with damage intensity as well as accounting for material degradation and healing can be accounted by extending the model to a nonlinear elastic theory incorporating an additional second-order term to the free energy of elastic solid. Simultaneous fit both the strain–stress curves and the measured seismic velocity values with the same set of model parameters demonstrate that the nonlinear damage model accounts for the different aspects of the stress–strain fields beyond linear elasticity.

2 THEORETICAL BACKGROUND

The linear continuum damage mechanics theory, originally framed by Kachanov (1986), introduces the scalar damage state variable characterizes a properly chosen volume of rock so that the density of the internal flaws within this volume may be considered uniform. Representative elementary volumes with a sufficiently large number of cracks corresponding to given values of the damage variable are assumed to be uniform and isotropic. In this framework the damaged material is a linear isotropic Hookean solid with the degraded Young modulus being proportional to a scalar damage parameter D defined as (Kachanov, 1986):

$$D = 1 - \frac{E}{E_0} \qquad (1)$$

where E is the instantaneous elastic modulus and E_0 is the initial undamaged value. In this approach, material with fixed damage (D = const.) remains linear with the same Young modulus under tension and compression. Such a damage parameter was determined experimentally by Basaran and Nie (2004) and compared with their thermodynamically-based damage evolution function. Strain-controlled tension-compression uniaxial tests that show clear correlation between increasing crack density and decreasing stiffness were done on a composite brittle material (lightly cross-linked poly-methyl methacrylate filled with alumina trihydrate). Predictions of the linear continuum damage models fit well tensile cycles of loading (Fig. 1), but show significant disagreement for material behavior under compression. Non-linear damage rheology approach allows a significantly improved fitting between results of the model predictions and the experimental data shown in Figure 1 by accounting for different effective elastic moduli of brittle material under tension and compaction.

The variations of Young modulus and Poisson ratio with damage intensity under different types of load in three dimensions can be described (e.g., Lyakhovsky et al., 1997a) by extending the free energy of the elastic solid to the form:

$$U = \frac{1}{\rho} \left(\frac{\lambda}{2} I_1^2 + \mu I_2 - \gamma I_1 \sqrt{I_2} \right) \qquad (2)$$

where $I_1 = \varepsilon_{ii}$ and $I_2 = \varepsilon_{ij}\varepsilon_{ij}$ are the first and second invariants of the elastic strain tensor and ρ is the mass density. The elastic energy potential (2) includes two quadratic Hookean terms of

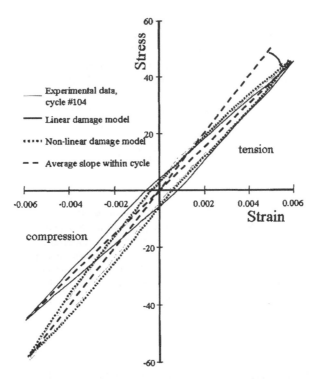

Figure 1. Stress-strain cycles from *Basaran and Nie* (2004); fit with linear damage model and non-linear damage model.

the elastic strain tensor ε_{ij} associated with the Lamé moduli λ and μ, and an additional non-linear second order term associated with a third modulus γ. The first two terms of (2) give the classical strain potential of linear elasticity. Theoretical derivations leading to this form of the elastic potential and comparison with available experimental results are discussed in this and following sections.

Lyakhovsky and Myasnikov (1984) and Lyakhovsky et al. (1997a) extended the self-consistent formulation of Budiansky & O'Connell (1976) for estimating effective elastic properties of the materials with non-interacting randomly distributed cracks. They account for the fact that cracks dilate and contract in response to the local tension and compression acting at the crack according to their orientation with respect to the principle axis of the stress tensor. They show that the non-analytical second order term $I_1\sqrt{I_2}$ should be included in the elastic potential together with the regular terms. They also demonstrated that the dependency of the elastic potential on the third strain invariant $I_3 = \det(\varepsilon_{ij})$ is very weak and may be neglected. The obtained potential form is the simplest mathematical expression for the elastic strain energy that generalizes the classical potential to a non-analytical second-order function of two strain invariants.

The same result can be obtained by considering a general function for the strain energy having any second-order term of the type $I_1^{2x} \cdot I_2^{(1-x)}$ with $0 < x < 1$, and eliminating unphysical values of the exponents. The limit values $x = 0$ and $x = 1$ are associated with the Hookean first two terms of (2). The relation between the mean stress (σ_{kk}) and volumetric deformation (I_1) for the assumed general form is

$$\sigma_{kk} = \rho \frac{\partial U}{\partial \varepsilon_{kk}} \sim 2x I_1^{(2x-1)} \cdot I_2^{(1-x)} + \frac{2}{3}(1-x)I_1^{(2x+1)} \cdot I_2^{-x} \tag{3}$$

The second term in eq. (3) is regular for every $0 < x < 1$, while the first term has a non-physical singularity for $0 < x < 1/2$. Only exponent values in the range $x \geq 1/2$ lead (in addition to $x = 0$)

to a non-singular stress–strain relation. However, for $x > 1/2$ the volumetric strain is zero ($I_1 = 0$) for zero mean stress ($\sigma_{kk} = 0$) and any non-zero shear loading ($I_2 \neq 0$). This is not compatible with material dilation under shear loading, which is widely observed in rock deformation experiments (e.g. Jaeger & Cook 1976). Thus the only exponent (other than the classical 0 and 1 values) associated with realistic rock deformation is the $x = 1/2$ value represented by the third term in eq. (2).

The resulting stress–strain relation associated with the strain energy function (2) is

$$\sigma_{ij} = \rho \frac{\partial U}{\partial \varepsilon_{ij}} = \left(\lambda - \frac{\gamma}{\xi} \right) I_1 \delta_{ij} + 2 \left(\mu - \frac{1}{2} \gamma \xi \right) \varepsilon_{ij} \tag{4}$$

where δ_{ij} is Kronecker delta and $\xi = I_1/\sqrt{I_2}$. The variable ξ is referred to as the strain invariants ratio and it ranges from $\xi = -\sqrt{3}$ for isotropic compaction to $\xi = +\sqrt{3}$ for isotropic dilation. Equation (4) reduces to linear Hookean elasticity for an undamaged solid ($\gamma = 0$). The cumulative effect of distributed micro-cracks and flaws in the elastic material leads to reduction of the effective elastic moduli and non-linear elasticity with asymmetric response to loading under tension and compression conditions observed in a four-point beam test (Weinberger et al., 1994). The stress–strain relations (4) with nonzero γ are reduced in a one-dimensional case to the bi-linear relations observed in the uniaxial tests (Fig. 1).

3 ROCK DILATION

The dilation of rock under shear gives rise to detectable effects both in laboratory experiments and in field observations. Such effects include hardening due to reduction in pore pressure and asymmetrical distribution of deformation following strike-slip earthquakes. Linear poro elasticity predicts that under undrained conditions the pore pressure change is proportional only to the change in the mean stress. However, based on experiments with clays, Skempton (1954) pointed out that the deviatoric stresses can also affect pore pressure under undrained conditions. Recently, Lockner and Stanchits (2002) experimentally related the change in pore pressure for axial loading under confining pressure with the mean stress, σ_m, and differential stress, τ, as:

$$dP = \frac{\partial P}{\partial \sigma_m} d\sigma_m + \frac{\partial P}{\partial \tau} d\tau = B d\sigma_m + \eta d\tau \tag{5}$$

where B is Skempton's proportionality coefficient, $\eta < 0$ and $|\eta|$ strongly increases with τ. Linear elasticity as well as linear poro-elasticity fails reproducing this observation, which requires non-linear stress-strain relations. The coupling between volumetric deformation and shear strain in (2) leads to the connection between mean stress and shear strain through the strain invariant I_2:

$$\sigma_{kk} = 3(\lambda I_1 - \gamma \sqrt{I_2}) + \left(2\mu - \gamma \frac{I_1}{\sqrt{I_2}} \right) I_1 \tag{6}$$

This coupling between volumetric deformation and shear strain leads to the connection between pore pressure and shear stress. Hamiel et al. (2005) calculated the poro-elastic coefficients B and η for the experimental conditions reported by Lockner and Stanchits (2002) and using the model parameters $\lambda = 5,000$ MPa, $\mu = 8,500$ MPa, $\gamma = 5,800$ MPa for the Berea sandstone and $\lambda = 8,000$ MPa, $\mu = 8,500$ MPa, $\gamma = 4,800$ MPa for the Navajo sandstone (Fig. 2). Good agreement between calculated and measured properties of the high porosity rocks confirm that the presented formulation provides an internally consistent solution for coupling of shear stress and pore pressure.

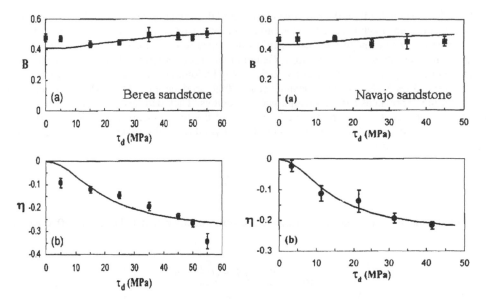

Figure 2. Measured poroelastic coefficients from Lockner and Stanchits (2002) for Berea and Navajo sandstone samples and simulated functional relation between the coefficients and shear stress.

4 STRESS- AND DAMAGE-INDUCED SEISMIC WAVE ANISOTROPY

The elastic energy potential (2) is formulated in terms of strain invariants without explicitly introducing material anisotropy. However, the potential includes non-linear elastic response that produces local anisotropy in a damaged volume. A propagating small amplitude seismic wave in the non-linear elastic media is associated with perturbations of the pre-defined state of stress corresponding to the initial strain value $\varepsilon_{ij}^{(0)}$. The relation between the wave-related perturbations of stress $\sigma_{ij}^{(1)}$ and strain $\varepsilon_{ij}^{(1)}$, in the vicinity of this pre-defined state of stress, is obtained by linearization of the non-linear stress-strain relation (4):

$$\sigma_{ij}^{(1)} = \left(\lambda \delta_{nm} - \gamma \frac{\varepsilon_{nm}^{(0)}}{\sqrt{I_2^{(0)}}} \right) \delta_{ij}\varepsilon_{nm}^{(1)} + \left(2\mu - \gamma \frac{I_1^{(0)}}{\sqrt{I_2^{(0)}}} \right) \delta_{in}\delta_{jm}\varepsilon_{nm}^{(1)}$$

$$- \gamma \left(\frac{\delta_{nm}}{\sqrt{I_2^{(0)}}} - \frac{I_1^{(0)}}{I_2^{(0)}\sqrt{I_2^{(0)}}}\varepsilon_{nm}^{(0)} \right) \varepsilon_{ij}^{(0)}\varepsilon_{nm}^{(1)} \tag{7}$$

The stress-strain relations (7) are equivalent to the usual stress-strain relations for an anisotropic elastic media medium, with a fourth-rank tensor C_{ijnm} of the strain-dependent effective elastic moduli:

$$C_{ijnm} = \left(\lambda \delta_{nm} - \gamma \frac{\varepsilon_{nm}^{(0)}}{\sqrt{I_2^{(0)}}} \right) \delta_{ij} + \left(2\mu - \gamma \frac{I_1^{(0)}}{\sqrt{I_2^{(0)}}} \right) \delta_{in}\delta_{jm} - \gamma \left(\frac{\delta_{nm}}{\sqrt{I_2^{(0)}}} - \frac{I_1^{(0)}}{I_2^{(0)}\sqrt{I_2^{(0)}}}\varepsilon_{nm}^{(0)} \right) \varepsilon_{ij}^{(0)}$$

$$\tag{8}$$

For a hydrostatic loading, the effective elastic moduli (8) can be reduced to the ordinary Lamé constants, or Young modulus and Poisson ratio of linear elasticity. However, under non-hydrostatic loading, stress- and damage-induced seismic wave anisotropy is expected due to crack opening in different stress-preferred orientations. General relations between the fourth-rank tensor of the elastic moduli C_{ijnm} and seismic wave velocities may be found in general references (e.g., Mavko et al., 1998; Aki & Richards, 2002). Hamiel et al. (2009) used the special form (8) of this tensor and re-derive the expressions for the seismic wave velocities. They demonstrated that three different types of waves exist in anisotropic media instead of the two standard P and S types of waves of isotropic solid referred to as quasi-longitudial, quasi-shear and pure-shear with three different seismic wave velocities:

$$V_s^2 = \frac{\mu^e}{\rho} \qquad (9)$$

Vs corresponds to a pure shear and isotropic wave with velocity calculated using the effective shear modulus.

$$V_p^2 = \frac{\mu^e + \frac{1}{2}A + \sqrt{\frac{A^2}{4} - B}}{\rho} \qquad (10)$$

Vp is compressional or quasi-longitudinal wave and another type of quasi-shear wave *Vqs* is given by

$$V_{qs}^2 = \frac{\mu^e + \frac{1}{2}A - \sqrt{\frac{A^2}{4} - B}}{\rho} \qquad (11)$$

where the coefficients A and B are

$$A = \lambda + \mu^e - 2\gamma e_{11} + \gamma \xi (e_{11}^2 + e_{12}^2 + e_{13}^2)$$
$$B = [(\lambda + \mu^e)\gamma \xi - \gamma^2](e_{12}^2 + e_{13}^2) \qquad (12)$$

Note that in the case of $\gamma = 0$, i.e. damage-free intact rock, the stress-strain relations reduce to Hookean elasticity and the wave velocities become the well known expressions from linear elasticity. Equations (10, 11) predict that the seismic wave velocities *Vp* and *Vqs* become anisotropic for a solid with pre-existing damage ($\gamma > 0$) under non-hydrostatic load.

Using the stress-strain measurements during three loading cycles in the laboratory experiment, Hamiel et al. (2009) constrained the parameters of the damage model that are appropriate for the Aue granite, Germany. The experimental setup has been described in detail in Stanchits et al. (2006). The detailed description of the main features of the visco-elastic damage rheology model including evolving elastic moduli and non-reversible strain accumulation may be found in Hamiel et al (this issue). Using the same set of parameters, the simulated stress-strains fit the overall features through all the loading cycles almost up to the final macroscopic failure (Fig. 3). The predicted P-wave anisotropic velocities and stress- and damage-induced anisotropy using the model parameters constrained by the stress-strain relation of the P-waves in the axial and transversal directions fit well with the measured velocity values (Fig. 3). Experimental estimation of velocities of different types of the S-waves is very complicated and is not presented in this study.

Thus, the employed damage model accounts quantitatively for the overall aspects of the stress-strain fields beyond linear elasticity, while simultaneously reproducing the main features of the damage- and stress-induced seismic elastic wave anisotropy measured during the experiments with tri-axial loading. Since the model is scalar, it can not treat anisotropy related to the internal rock structure associated with material layering, oriented crystals, or anisotropic damage in form of high density of cracks oriented in certain direction. In this case the seismic wave anisotropy is expected

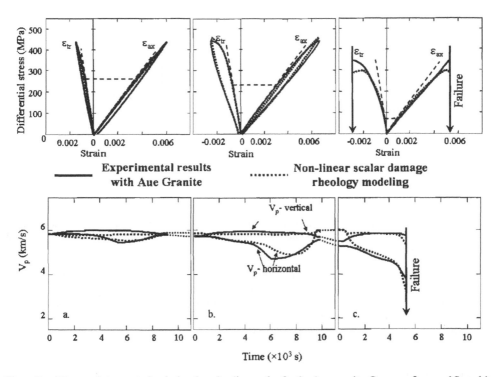

Figure 3. Measured stress-strains during three loading cycles for the Aue granite, Germany from and Stanchits et al. (2006) and simulated stress-strains using the damage rheology model (upper panel). Lower panel show measured and predicted P-wave velocities.

also under hydrostatic load. This is what would presumably require a tensor rather than a scalar damage measure.

5   NON-LINEAR WAVE RESONANCE

Another manifestation of nonlinear elastic behavior in rocks was reported by several authors who analyzed a set of resonant bar experiments with rock samples (e.g., Gordon and Davis, 1968; Winkler et al., 1979; Johnson et al., 1996; Guyer et al., 1999; Smith and TenCate, 2000; Pasqualini et al., 2007). Instead of the constant resonant frequency expected for linear elastic media, increased external forcing was shown to produce a shift of the spectral peak to lower frequencies and asymmetric shape of the resonance curves (Fig. 4a). The overall observed frequency shift in experiments with Lavoux sandstone (Fig. 4) and other rocks is of the order of a few percent. Several authors simulated a frequency shift at relatively low strains with a model that combines higher order terms in the free energy of a solid with an additional non-analytic term depending on both the strain and strain rate (e.g., Guyer et al., 1997; Guyer and Johnson, 1999; Guyer et al., 1999). Their model represents the bond system of a rock as an assemblage of hysteretic elastic elements that can only be in one of two states, open or closed. This basic model assumption is similar to those used to derive the elastic potential (2). The existence of these elements forms cusps in the hysteresis loop (discrete memory) and also predicts the existence of cusps in low-amplitude stress-strain curves. Pasqualini et al. (2007) criticized this model and argued that there is no experimental evidence for cusp behavior in low-amplitude stress-strain loops. They also showed that predictions of a simple Duffing oscillator with a cubic term are consistent (see also TenCate et al., 2004) with the data obtained at strains below a certain threshold. Lyakhovsky et al. (2009) presented theoretical results

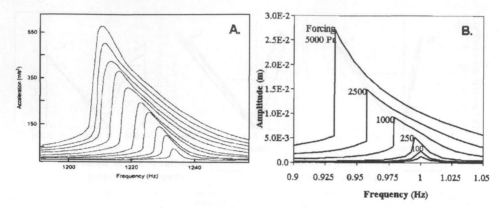

Figure 4.   A. Acceleration versus frequency for different excitation levels versus acceleration in Lavoux sandstone (from Johnson et al., 1996). B. Simulated resonance curves under periodic forcing with amplitude from 100 to 5,000 Pa.

and numerical simulations of wave propagation in elastic material governed by nonlinear damage rheology model and simulated the entire shape of the resonance curve under different forcing. In agreement with laboratory observations (Fig. 4a), the shape of the simulated resonant curves (Fig. 4b) is asymmetric with gradual decrease of the wave amplitude for frequencies higher than the resonance value and abrupt decreases for lower frequencies.

6   CONCLUSION

The non-linear elastic deformation of rocks is approached in this study by modifying the elastic potential. The energy expression includes a non-analytical, second-order term, $I_1\sqrt{I_2}$ in addition to the quadratic terms containing I_1^2 and I_2, of the Hookean elastic solid. The present model preserves its non-linear properties even under small deformations as the terms in (2) are of the same second order, leading to first-order terms in the stress-strain relation (3). This property differs from some non-linear elastic models, which use higher-order terms in the elastic energy and in the stress-strain relations. Models with high-order terms can be successful in large strain analysis of the Earth's interior, but they are out of range for small elastic deformation.

Several experimental observations can be understood in light of the non-linear stress-strain relations derived in this study. These observations include experimentally observed change in the effective elastic moduli under stress reversal; rock dilation due to deviatoric stresses; non-linear shift of the resonance frequency together with asymmetric shape of the entire resonance curve.

ACKNOWLEDGMENTS

We thank Terry Tullis and Yossef Hatzor for constructive review. The studies were supported by grant from the Israel Science Foundation (ISF 753/08).

REFERENCES

Aki, K. & Richards, P.G. 2002. *Quantitative Seismology* (second edition), University Science Books.
Basaran C. & Nie, S. 2004. An irreversible thermodynamics theory for damage mechanics of solids. *Int. J. Damage Mach.* 13: 205–223.
Beresnev, I.A. 2002. Nonlinearity at California generic soil sites from modeling recent strong-motion data. *Bull. Seismol. Soc. Am.*, 92, 863–870, doi:10.1785/0120000263.

Bonilla, L.F., Archuletta, R.J. & Lavallee, D. 2005. Hysteretic and dilatant behavior of cohesionless soils and their effects on nonlinear site response: Field data observations and modeling. *Bull. Seismol. Soc. Am.*, 95, 2273–2395, doi:10.1785/0120040128.

Brace, W.F. 1965. Some new measurements of the linear compressibility of rocks, *J. Geophys. Res.* 70, 391–398.

Brady, B.T. 1969. The non-linear behavior of brittle rock, *Int. J. Rock Mech. Min. Sci. Geomech. Abstr.* 6, 301–310.

Budiansky, B. & O'Connell, R.J. 1976. Elastic moduli of a cracked solid. *Int. J. Solids Struct.*, 12, 81–97.

Christensen, R.M. 1979. Mechanical of Composite Materials, Wiley-interscience, New York.

Eshelby, J.D. 1957. The determination of the elastic field of an ellipsoidal inclusion and related problems, Proc. R. Sot. Lond., A, 241, 376–396.

Field, E.H., Johnson, P.A., Beresnev, I.A. & Zeng, Y. 1997. Nonlinear ground-motion amplification by sediments during the 1994 Northridge earthquake, *Nature*, 390, 599–602.

Frankel, A.D., Carver, D.L. & Williams, R.A. 2002. Non-linear and linear site response and basin effects in Seattle for the M 6.8 Nisqually, Washington, earthquake. *Bull. Seismol. Soc. Am.*, 92, 2090–2109. doi:10.1785/0120010254.

Gordon, R.B., & Davis, L.A. 1968. Velocity and attenuation of seismic waves in imperfectly elastic rocks, *J. Geophys. Res.*, 73, 3917–3935.

Guyer et al., 1999.

Guyer, R.A., TenCate, J. & Johnson, P.A. 1999. Hysteresis and the dynamic elasticity of consolidated granular materials, *Phys. Rev. Lett.*, 82, 3280–3283.

Hamiel, Y., Lyakhovsky, V., Stanchits, S., Dresen, G. & Ben-Zion, Y. 2009. Brittle deformation and damage-induced seismic wave anisotropy in rocks. *J. Geophys. Int.* doi: 10.1111/j.1365-246X.2009.04200.x.

Hamiel, Y., Liu, Y., Lyakhovsky, V., Ben-Zion, Y. & Lockner, D. 2004. A visco-elastic damage model with applications to stable and unstable fracturing, *J. Geophys. Int.*, 159, 1155–1165.

Hamiel, Y., Lyakhovsky, V. & Agnon, A., 2005. Rock dilation, Nonlinear deformation, and pore pressure change under shear, *Earth Plan. Sci. Lett.*, 237, 577–589.

Hatzell, S., Leeds, A., Frankel, A., Williams, R.A., Odum, J., Stephenson, W. & Silva, W. 2002. Simulation of brad-band ground motion including nonlinear soil effects for a magnitude 6.5 earthquake on the Seattle fault, Seattle, Washington. *Bull. Seismol. Soc. Am.*, 92, 831–853, doi:10.1785/0120010114.

Hatzell, S., Bonilla, L.F. & Williams, R.A. 2004. Prediction of nonlinear soil effects, *Bull. Seismol. Soc. Am.*, 94, 1609–1629, doi:10.1785/012003256.

Jaeger, J.C. & Cook, N.G.W. 1976. *Fundamentals of Rock Mechanics*, Chapman & Hall, New York, NY.

Johnson, P.A., Zinszner, B. & Rasolofosaon, P.N.J. 1996. Resonance and elastic nonlinear phenomena in rock, *J. Geophys. Res.* 101, 11553–11564.

Kachanov, L.M. 1986. Introduction to Continuum Damage Mechanics, *Martinus Nijhoff Pub.*, 135 p.

Kachanov, M. 1987. Elastic solids with many cracks: a simple method of analysis, *Int. J. Solids Struct.*, 23, 23–43.

Kachanov, M. 1992. Effective elastic properties of cracked solids; critical review of some basic concepts, *Appl. Mech. Rev.*, 45, 304–335, 1992.

Lockner, D.A. & Byerlee, J.D. 1980. Development of fracture planes during creep in granite, in 2nd conference on acoustic emission/microseismic activity in geological structures and materials, pp. 11–25, eds Hardy, II.R. & Leighton, F.W., Trans-Tech. Publications, Clausthal-Zellerfeld, Germany.

Lockner, D.A. & Stanchits, S. 2002. Undrained poroelastic response of sandstones to deviatoric stress change, *J. Geophys. Res.* 107, doi:10.1029/2001JB001460.

Lockner, D.A., Byerlee, J.D., Kuksenko, V., Ponomarev, A. & Sidorin, A. 1992. Observations of quasi-static fault growth from acoustic emissions. in Fault mechanics and transport properties of rocks, International Geophysics Series, 51, 3–31, eds Evans, B. & Wong, T.-f., Academic Press, San Diego.

Lyakhovsky, V. & Myasnikov, V.P. 1984. On the behavior of elastic cracked solid, *Phys. Solid Earth*, 10, 71–75.

Lyakhovsky, V., Reches, Z., Weinberger., R. & Scott, T.E. 1997a. Nonlinear elastic behavior of damaged rocks, *Geophys. J. Int.*, 130, 157–166.

Lyakhovsky, V., Ben-Zion, Y. & Agnon, A. 1997b. Distributed damage, faulting, and friction, *J. Geophys. Res.*, 102, 27635–27649.

Lyakhovsky, V. Hamiel, Y. Ampuero, P. & Ben-Zion, Y. 2009. Nonlinear damage rheology and wave resonance in rocks. *Geophys. J. Int.* doi:10.1111/j.1365-246X.2009.04205.x.

Mavko, G., Mukerji, T. & Dvorkin, J. 1998. The Rock Physics Handbook, Cambridge Univ. Press, New York, 329 p.

Nishihara, M. 1957. Stress-strain relation of rocks, *Doshisha Eng. Rev.*, 8, 32–54.

O'Connell, R.J. & Budiansky, B. 1974. Seismic wave velocities in dry and saturated cracked solid, *J. Geophys. Res.*, 79, 5412–5426.

Pasqualini, D., Heitmann, K., TenCate, J.A., Habib, S., Higdon, D. & Johnson, P.A. 2007. Nonequilibrium and nonlinear dynamics in Berea and Fontainebleau sandstones: Low-strain regime. *J. Geophys. Res.*, 112, B01204, doi:10.1029/2006JB004264.

Pavlenko, O.V. & Irikura, K. 2003. Estimation of nonlinear time-dependent soil behavior in strong ground motion based on vertical array data, *Pure Appl. Geophys.*, 160, 2365–2379.

Rubinstein, J.L. & Beroza, G.C. 2004. Nonlinear strong ground motion in the ML 5.4 Chittenden earthquake: Evidence that preexisting damage increases susceptibility to further damage, *Geophys. Res. Lett.*, 31, L23614, doi:10.1029/2004GL021357.

Sayers, C. & Kachanov, M. 1991. A simple technique for funding effective elastic constants of cracked solids for arbitrary crack orientation statistics, *Int. J. Solids Structures*, 27, 671–680.

Schmitt, D.R. & Zoback, M.D. 1992. Diminished pore pressure in low-porosity crystalline rock under tensional failure: apparent strengthening by dilatancy, *J. Geophys. Res.* 97, 273–288.

Schock, R.N. 1977. The response of rocks to large stresses, in: Impact and Explosion Cratering, pp. 657–688, eds Roddy, D.L., Pepin, R.O. & Merrill, R.B., Pergamon press, New York.

Skempton, A.W. 1954. The pore-pressure coefficients A and B, *Geotechnique* 4, 143–147.

Sleep, N.H. & Hagin P. 2008. Nonlinear attenuation and rock damage during strong seismic ground motions, *Geochem. Geophys. Geosyst.*, 9, Q10015, doi:10.1029/2008GC002045.

Smith, E. & TenCate, J.A. 2000. Sensitive determination of the nonlinear properties of Berea sandstone at low strains, *Geophys. Res. Lett.*, 27, 1985–1988.

Stanchits, S., Vinciguerra, S. & Dresen, G. 2006. Ultrasonic velocities, acoustic emission characteristics and crack damage of basalt and granite, *Pure Appl. Geophys.*, 163, 975–994.

Tsuda, K., Steidl, J., Archuleta, R. & Assimaki, D. 2006. Site-response estimation for the 2003 Miyagi-Oki earthquake sequence consider nonlinear site response. *Bull. Seismol. Soc. Am.*, 96, 1474–1482, doi:10.1785/0120050160.

Weinberger, R., Reches, Z., Eidelman, A. & Scott, T.S. 1994. Tensile properties of rocks in four-point beam tests under confining pressure, in Proceedings first North American Rock Mechanics Symposium, pp. 435–442, eds Nelson, P. & Laubach, S.E., Austin, Texas.

Winkler, K., Nur, A. & Gladwin, M. 1979. Friction and seismic attenuation in rocks, *Nature*, 277, 528–531.

Wu, C. Peng, Z. & Ben-Zion, Y. 2008. Non-linearity and Temporal Changes of Fault Zone Site Response Associated with Strong Ground Motion, *Geophys. J. Int.*, (in press).

Zoback, M.D. & Byerlee, J.D. 1975. The effect of microcrack dilatancy on the permeability of Western granite, *J. Geophys. Res.*, 80, 752–755.

Damage rheology and stable versus unstable fracturing of rocks

Y. Hamiel, V. Lyakhovsky & O. Katz
Geological Survey of Israel, Jerusalem, Israel

Y. Fialko
Institute of Geophysics and Planetary Physics, Scripps Institution of Oceanography, University of California, San Diego, CA, USA

Z. Reches
School of Geology and Geophysics, University of Oklahoma, Oklahoma, USA

ABSTRACT: We address the relations between the rock rigidity and crack density by comparing predictions of a visco-elastic damage rheology model to laboratory data that include direct microscopic mapping of cracks. The damage rheology provides a generalization of Hookean elasticity to a non-linear continuum mechanics framework incorporating degradation and recovery of the effective elastic properties, transition from stable to unstable fracturing, and gradual accumulation of irreversible deformation. This approach is based on the assumption that the density of micro cracks is uniform over a length scale much larger than the length of a typical crack, yet much smaller than the size of the entire deforming domain. For a system with a sufficiently large number of cracks, one can define a representative volume in which the crack density is uniform and introduce an intensive damage variable for this volume. We tested our damage rheology against sets of laboratory experiments done with granite samples. Based on fitting the entire stress-strain records the damage variable is constrained, and found to be a linear function of the crack density. An advantage of these sets experiments is that they were preformed with different loading paths and explicitly demonstrated the existence of stable and unstable fracturing regimes. We demonstrate that the visco-elastic damage rheology provides an adequate quantitative description of the brittle rock deformation and simulates both the stable and unstable damage evolution under various loading conditions. Comparison between the presented data analysis of experiments with Mount Scott granite and results with Westerly granite and Berea sandstone indicates that granular or porous rocks accumulate relatively high irreversible strain during the loading cycle. This implies that the portion of elastic strain released during a seismic cycle as brittle deformation depends on the lithology of the region. Hence, upper crustal regions with thick sedimentary cover, or fault-zones with high degree of damage are expected to undergo a more significant inelastic deformation in the interseismic period compared to "intact" crystalline rocks.

1 INTRODUCTION

Laboratory investigations of rock fracturing indicate that damage starts to develop well before the rock fails. The evolution of damage profoundly affects the mechanical properties of rocks (e.g. Nishihara, 1957; Zoback & Byerlee, 1975; Schock, 1977; Lockner & Byerlee, 1980; Reches & Lockner, 1994; Pestman & Munster, 1996; Rubinstein & Beroza, 2004), and decreases their elastic moduli at relatively large stresses prior to failure (e.g. Lockner & Byerlee, 1980; Lockner et al., 1991, 1992; Fialko, 2002; Katz & Reches, 2004). In order to simulate the observed degradation of the effective elastic properties, a non-dimensional scalar or tensor damage variable is introduced in continuum damage rheology models. The damage variable characterizes a properly chosen volume of rock so that the density of internal flaws (e.g., microcracks in a laboratory specimen or small faults in the Earth's crust) within this volume may be considered uniform. This paper focuses on the application of damage rheology to the failure of brittle granite. According to damage mechanics, the change in the intensity of the damage variable is proportional to the change in the rigidity of the rock. However, the relation between rock rigidity and micro-crack density has not been clear, and our main objective here is to construct these relations.

Micro-crack density in brittle rocks was evaluated by acoustic emission analysis (e.g., Lockner et al., 1991; Zang et al., 1996; Janssen et al., 2001), and by direct microstructural analyses of micro-fractures and dilatational micro-cracks (e.g., Hadley, 1976; Tapponnier & Brace, 1976; Kranz, 1979; Moore & Lockner, 1995; Homand et al., 2000; Oda et al., 2002; Katz & Reches, 2004). Only a few attempts have been made to compare experimentally obtained micro-crack density with theoretical damage parameters. Katz & Reches (2004) presented experimental evidence for the relations between experimentally measured microcracks density and the reduction of the elastic moduli of granite rock samples. Hamiel et al. (2004) constrained the visco-elastic parameters of damage rheology by using triaxial laboratory experiments done on samples of Westerly granite and Berea sandstone. They found clear correlation between simulated damage and accumulated acoustic emission recorded during the loading of the samples. Based on these results we hypothesize that the micro-crack density is directly proportional to the damage variable in the damage rheology presented by Lyakhovsky et al. (1997a) and Hamiel et al. (2004). The present study quantitatively compares the predictions of the damage rheology with laboratory data that include stress-strain relations and direct microscopic mapping of microcracks.

2   THEORETICAL BACKGROUND: DAMAGE RHEOLOGY

The main features of the visco-elastic damage rheology are outlined below. Detailed theoretical background and comparisons with rock mechanics experiments may be found in Lyakhovsky et al. (1997a, b) and Hamiel et al. (2004). To evaluate the damage effects Lyakhovsky et al. (1997b) generalize the elastic strain energy potential of a deforming solid to the form:

$$U = \frac{1}{\rho} \left(\frac{\lambda}{2} I_1^2 + \mu I_2 - \gamma I_1 \sqrt{I_2} \right). \tag{1}$$

The elastic energy potential (1) includes two quadratic Hookean terms of the elastic strain tensor ε_{ij} with the Lamé moduli λ and μ and an additional non-linear second order term with additional modulus γ. $I_1 = \varepsilon_{ii}$ and $I_2 = \varepsilon_{ij}\varepsilon_{ij}$ are two independent invariants of the elastic strain tensor, and ρ is the rock density. Differentiation of the elastic energy (1) leads to constitutive stress-strain relation for the stress tensor, σ_{ij}

$$\sigma_{ij} = \rho \frac{\partial U}{\partial \varepsilon_{ij}} = \left(\lambda - \frac{\gamma}{\xi} \right) I_1 \delta_{ij} + 2 \left(\mu - \frac{1}{2} \gamma \xi \right) \varepsilon_{ij}, \tag{2}$$

where $\xi = I_1/\sqrt{I_2}$ is a strain invariant ratio ranging from $\xi = -\sqrt{3}$ for isotropic compaction to $\xi = \sqrt{3}$ for isotropic dilation. Equation (2) reduces to linear Hookean elasticity for an undamaged solid ($\gamma = 0$). The cumulative effect of distributed micro-cracks and flaws in the elastic material leads to reduction of the effective elastic moduli and non-linear elasticity with asymmetric response to loading under tension and compression conditions. Eq. 2 can be expressed though the dependence of the effective elastic moduli ($\lambda^e = \lambda - \gamma/\xi; \mu^e = \mu - \gamma\xi/2$) on the strain invariant ratio and their abrupt change with transition from compacting ($\xi < 0$) to dilating ($\xi > 0$) strains. Change in the effective elastic moduli under stress reversal in a four-point beam test (Weinberger et al., 1994), rock dilation due to deviatoric stresses (Hamiel et al., 2005) and other rock mechanics experiments (Lyakhovsky et al., 1997b) confirm the applicability of the nonlinear stress-strain relations (2) derived from the potential (1).

The effect of rock degradation is achieved by changing the elastic moduli functions of a scalar damage variable α, i.e., $\lambda(\alpha)$, $\mu(\alpha)$ and $\gamma(\alpha)$. α ranges between 0 and 1, where in undamaged material $\alpha = 0$, and failure occurs at critical α. Using the balance equations of energy and entropy, and accounting for irreversible changes related to viscous deformation and material damage, the equation of damage evolution has the form (Lyakhovsky et al., 1997a)

$$\frac{d\alpha}{dt} = -C\frac{\partial U}{\partial \alpha}, \tag{3}$$

where the positive constant or function of state variables C provides the non-negative local entropy production related to damage evolution. Equation (3) can describe not only damage increase or material degradation, but also the process of material recovery associated with healing of micro-cracks. The latter is favored by high confining pressure, low shear stress and high temperature. In the context of the laboratory experiments discussed in this study this process is not relevant. To account for possible stable weakening Hamiel et al. (2004) suggested power-law relations between the damage variable and elastic moduli

$$\lambda = const.; \quad \mu = \mu_0 - \mu_1\alpha; \quad \gamma = \gamma_1\frac{\alpha^{1+\beta}}{1+\beta}, \tag{4}$$

where μ_0, μ_1, γ_1 and β are constants for each material. Substituting (1) into (3) using (4), the damage evolution can be rewritten as

$$\frac{d\alpha}{dt} = C_d I_2(\alpha^\beta \xi - \xi_0). \tag{5}$$

where $C_d > 0$ describes the rate of damage evolution for a given deformation, and $\xi_0 = -\mu_1/\gamma_1$ is material property. The proposed power-law relation (4) between the damage variable and elastic modulus, γ, leads to a non-linear coupling between the rate of damage evolution and the damage variable itself. Equation (5) gives rise to three types of damage evolution (Hamiel et al., 2004): (I) healing or damage decrease for $\xi < \xi_{tran}(\alpha) = \xi_0/\alpha^\beta$; (II) stable damage growth or steady-state solution for damage existing for $\xi_{tran} \leq \xi < \xi_0$; and (III) unstable weakening for $\xi \geq \xi_0$. In the stable regime (II) damage grows asymptotically only to a certain level but not to a level of complete failure ($\alpha < 1$). The onset of damage in the model with the power-law relations between the damage variable and elastic moduli (4) depends not only on the material property, ξ_0, but also on the pre-existing level of damage. The material strength or differential stress at the transition from stable to unstable fracturing for a given confining pressure decreases with damage accumulation. The more damaged the sample is, the lower will be the stress at which failure occurs. This change in the rock strength will be discussed later in this paper.

Comparison between theoretical predictions and the observed deformation and acoustic emissions from laboratory experiments in granites and sandstones led Hamiel et al. (2004) to incorporate gradual accumulation of a damage-related non-reversible deformation. This irreversible (inelastic) strain,, starts to accumulate with the onset of acoustic emission and the rate of its accumulation is suggested to be proportional to the rate of damage increase

$$\frac{d\varepsilon_{ij}^v}{dt} = \begin{cases} C_v\frac{d\alpha}{dt}\sigma_{ij}^d, & \frac{d\alpha}{dt} > 0 \\ 0, & \frac{d\alpha}{dt} \leq 0 \end{cases}. \tag{6}$$

where C_v is suggested to be a material constant and σ_{ij}^d is the deviatoric stress tensor. The effective fluidity or inverse of viscosity ($C_v d\alpha/dt$) relates the deviatoric stress to the rate of irreversible strain accumulation. Following Maxwell visco-elastic rheology model the total strain tensor, ξ_{ij}^{tot}, is assumed to be a sum of the elastic strain tensor and the irreversible viscous component of deformation, i.e., $\xi_{ij}^{tot} = \varepsilon_{ij} + \varepsilon_{ij}^v$. This model assumption means that the total irreversible strain accumulated during the loading should be proportional to the overall damage increase in the tested rock sample.

3   EXPERIMENTAL SETTING AND RESULTS

The experiments and microstructural analysis of Katz et al. (2001) and Katz & Reches (2004) provide an outstanding database to verify and constrain the damage model. These authors used 25.4 mm diameter cylinders of the medium-grained (0.9 ± 0.2 mm) Mount Scott Granite to perform two series of triaxial tests. In the first series, the samples were loaded to failure from uniaxial to 66 MPa confining pressures (Katz et al., 2001). In the second series of experiments, the samples were held at a constant stress after loading, a procedure termed load-hold tests (Katz & Reches, 2004). In these tests, the confining pressure was 41 MPa, for which the ultimate strength, $U_S = (\sigma_1 - \sigma_2)$ is 586 ± 16 MPa, estimated from several loaded to failure experiments. Each load-hold test consists of four steps: (a) Confining pressure loading at a constant rate of 0.023 MPa/s; (b) Axial loading to the pre-selected stress that ranges from $0.54 \cdot U_S$ to $1.05 \cdot U_S$. Axial shortening was at a strain rate of $\sim 10^{-5}$ s^{-1}; (c) Once the pre-selected stress was achieved, the specimen was held at a constant stroke for up to six hours; (d) After the hold period, the samples were unloaded by first reducing the axial stress to the confining pressure and then decreasing both stresses at the loading rates. During the experiments, load, axial and transversal strains were continuously monitored. One of the advantages of these sets of experiments is the different loading path. The experimental loading path for the different tests is schematically shown in Fig. 1. After unloading four selected samples were analyzed for microstructural damage induced by loading (Fig. 2 and Table 1). One non-loaded sample (sample 123) was also analyzed in order to obtain the initial micro-crack density of the initial rock. Together with the non-loaded sample, all five analyzed samples were selected

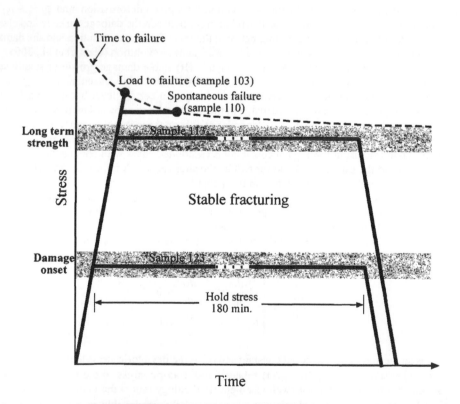

Figure 1.   Schematic diagram of the loading and types of experiments. Dark gray areas represent some variability in the onset of damage and the long term strength as deducted from the experiments. Light gray area represents the region where the damage evolution has steady-state solution, and dashed line represents the time-to-failure curve. The sample numbers denote experiments performed at different conditions.

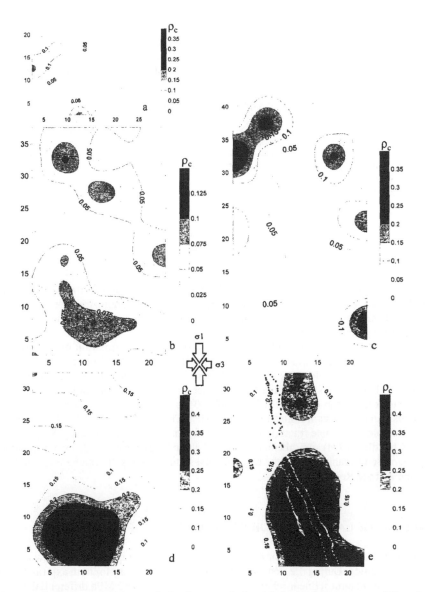

Figure 2. Local values of the microcrack density, ρ_c, calculated from the fracture maps of Katz & Reches (2004) using Eq. 7. a–e are samples 157, 123, 114, 113 and 110, respectively. Contours indicate the values of ρ_c (note that not all the samples are on the same scale), and diamonds (black and white) in sample 110 (e) are the fault trace (not included in the calculations of microcrack density); black and white colors of diamonds are used for clarity. Maps show the x ($\sigma 3$)–z ($\sigma 1$) plane (in mm) of the samples except for the not loaded sample 157 (a) which shows x–y plane.

in a way that they will span the whole loading path, from the initial damage in the non-loaded sample (sample 157) via onset of damage accumulation (sample 157) to total failure (sample 110). This microstructural mapping was conducted by simultaneous examination of the fracture on a petrographic microscope and mapping on the digitized image of the scanned thin section (for more details see Katz & Reches, 2004). The digitized microstructural data was also used to generate fracture maps and to calculate a dimensionless micro-crack density. Following Kachanov (1992), the micro-crack density, ρ_c, is defined as

Table 1. Experimental conditions and calculated micro-crack density of the analyzed sample.

Test	Test type	Hold stress[a] MPa/fraction	Hold time[b] minutes	Micro-crack density
157	Not loaded	0/0	0	0.037 ± 0.034
123	Load–Hold	334/0.5	180	0.049 ± 0.005
114	Load–Hold	518/0.88	180	0.068 ± 0.015
113	Load–Hold	563/0.96	180	0.145 ± 0.048
110	Failure	564/0.96	0.03	0.152 ± 0.071

[a] Maximum differential stress/fraction of the ultimate strength (Us = 586 MPa) at the start of holding.
[b] Time elapsed from start of stroke holding to unload or failure.

$$\rho_c = \frac{1}{A} \sum_n L_n^2. \tag{7}$$

where A is the representative area, L_n is a half-length of the crack number n. Spatial distribution of ρ_c calculated for five analyzed samples using Eq. 7 and the micro-fracture maps of Katz & Reches (2004) are shown in Fig. 2. The average microcracks density and standard deviation for each sample are given in Table 1.

The different types of experiments allow investigating rock deformation under different types of loading and regimes of fracturing. All samples loaded below 95% of the rock strength remained stable during the entire hold stage in spite of increase of the micro-crack density and material degradation. Most of the samples loaded to higher stresses failed, but not instantaneously. For example, it took as long as 61 minutes for sample 104 (see Table 1 in Katz & Reches, 2004) to fail spontaneously under constant hold stress of 613 MPa. These results explicitly demonstrate a transition from stable to unstable fracturing previously discussed by Kranz et al. (1982), Lawn (1993), Martin & Chandler (1994) and others in context of static fatigue tests. Additional mechanical data on the Mount Scott granite and on the experimental setting and results can be found in Katz et al. (2001), and Katz & Reches (2004).

4 ANALYSIS OF LABORATORY OBSERVATIONS

The measured stress-strain curves for the Mount Scott granite samples clearly show two different stages of deformation. During the first stage, the stress-strain relations for both axial and transversal components exhibit almost linear relations, until approximately 250 MPa differential stress, while during the next stage these relations significantly deviate from the straight line (dash line in Figs. 3a, 4a, and 5a). We use the linear part of the stress-strain curves to evaluate the initial elastic moduli ($\lambda = 2 \times 10^4$ MPa, $\mu_0 = 3 \times 10^4$ MPa) for all the samples. The point where the stress-strain curve deviates from linear relation allows one to estimate the yield stress and calculate the transitional strain invariant ratio (ξ_{tran}) at the onset of damage. For given values of initial damage (α) and power β (eq. 5), the critical strain invariant ratio is calculated directly ($\xi_0 = \xi_{tran}\alpha^\beta$). Estimation of the model coefficients controlling the rate of damage accumulation, β and C_d, is required for the analysis of the entire stress-strain curves from the onset of damage to unloading or failure (for more details see Hamiel et al., 2006). In all calculations we chose the same relatively low level of the initial damage $\alpha = 0.15$. In this study we adopt $\beta = 0.4$, that together with other coefficients of the visco-elastic damage rheology model presented in Table 2 provides a good fit to the measured stress-strain curves. Figs. 3a, 4a, and 5a show the measured and calculated axial and transversal stress-strain curves for samples 109, 115 and 103, respectively. The model results for all the samples (Figs. 3, 4, and 5) confirm that the visco-elastic damage rheology with power

Table 2. Experimental conditions and model coefficients obtained from fitting the measured stress-strain data (for all tests $\lambda = 2 \times 10^4$ MPa, $\mu_0 = 3 \times 10^4$ MPa).

Test	Hold stress MPa	Failure stress MPa	ξ_0	C_d 1/s	C_v 1/MPa	$R = C_v \mu_0$
102	601		−0.45	10	1.0×10^{-5}	0.3
103		595	−0.6	40		
109	546		−0.5	20	1.5×10^{-5}	0.45
110	564	564	−0.6	50		
113	563		−0.5	20	1.0×10^{-5}	0.3
114	518		−0.45	20	1.5×10^{-5}	0.45
115	534		−0.5	20	2.0×10^{-5}	0.6
117	318		−0.5	20		
123	334		−0.6	20		

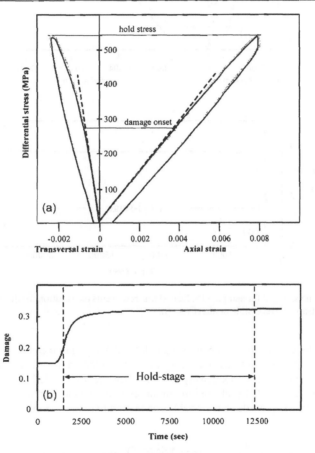

Figure 3. (a) Measured stress-strain curves, axial and transversal, for sample 109 (gray line) compared to the model prediction (black line). Dashed line represents the solution assuming linear elasticity. The values of the hold-stress and onset of damage are shown by thin black lines. (b) Simulated evolution of damage (α) during the experiment. See Table 2 for model coefficients.

law relation between damage variable and modulus γ (4) adequately represents the laboratory data, and reproduces the different fracturing regimes revealed by experiments with Mount Scott granite.

Modeling of sample loading with coefficients obtained from fitting the stress-strain data for these samples (Table 2) allows calculation of the damage evolution and estimation of the final damage accumulated during the whole cycle including loading, hold-stress and unloading stages.

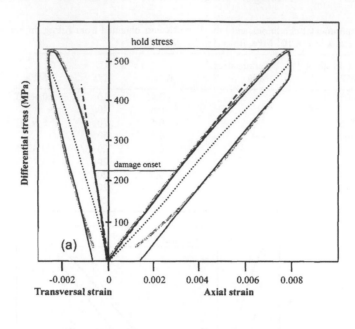

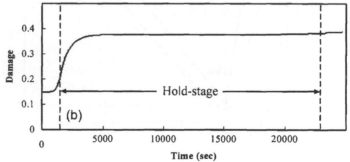

Figure 4.   The same as Fig. 3 for sample 115. Dotted line represents the solution for the elastic damage model, without accumulation of irreversible strain ($C_v = 0$).

Figure 6 shows the connection between the calculated damage and measured microcrack density, ρ_c, for the five analyzed samples (including the non-loaded sample 157; for more details see Table 1). The value of microcrack density measured for the non-loaded sample 157 corresponds to the initial value of $\alpha = 0.15$ adopted in our simulations for the starting material. As shown in Fig. 6 linear regression provides a reasonable connection between the microcrack density and the damage variable

$$\alpha = 4.5(\pm 0.5) \cdot \rho_c. \tag{8}$$

This relation means that the shear elastic modulus μ decreases linearly with the increase of microcracks density. However, increasing the elastic modulus γ, from zero for a linear-elastic damage-free material to its maximum value at the critical damage, amplifies the material non-linearity with damage accumulation.

The laboratory experiments preformed by Katz et al. (2001) show an increase in the strength of Mount Scott granite samples with confining pressure (Fig. 7). The maximum differential stresses for all the hold-tests are also shown in Fig. 7. Following Lyakhovsky et al. (2005) we use a pressure-dependent C_d that decays exponentially with characteristic pressure-scale of 14 MPa.

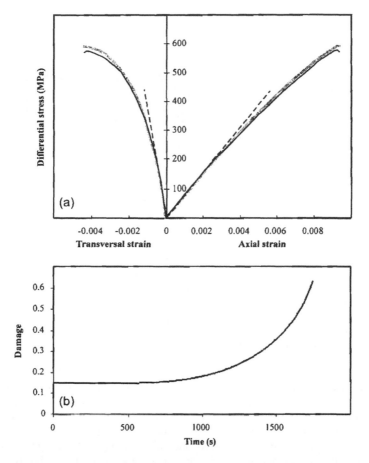

Figure 5. The same as Fig. 3 for sample 103. The duration of the load stage is 1749 seconds, while the duration of the hold stage is only 1.8 seconds and therefore too short to be seen in this plot. Note the different damage evolution here compared to the load hold tests 109 (Fig. 3) and 115 (Fig. 4).

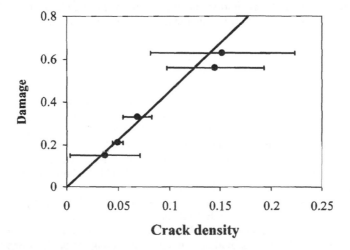

Figure 6. The measured crack density, ρ_c, versus the calculated damage variable, α. The inferred linear relation $\alpha = 4.5\rho_c$ is shown by black line.

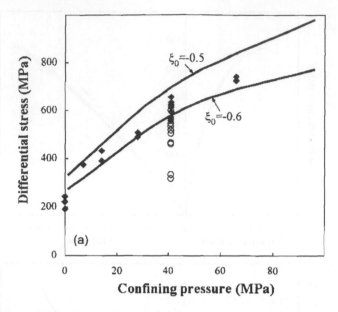

Figure 7.   The measured failure stress versus confining pressure (gray diamonds) compared with calculated failure curves for $\xi_0 = -0.5$ and $\xi_0 = -0.6$ ($C_v = 1.5 \times 10^{-5}$ MPa^{-1}). The stress at which the load-hold experiments were held (without failure of the sample) is shown by empty circles.

Two calculated yield curves, i.e., strength versus confining pressure, are presented in Fig. 7, one with $\xi_0 = -0.6$ and the other with $\xi_0 = -0.5$. The value of the critical strain invariant ratio for most of the samples (Table 2) falls into this interval corresponding to the variation of the strength value reflecting the inherent heterogeneity of the granite. The change of ξ_0 from −0.6 to −0.5 leads to onset of damage at higher strain (or stress), and therefore, higher failure stress. Thus, the yield curve for $\xi_0 = -0.5$ is at higher stresses than the same for $\xi_0 = -0.6$. As shown in Fig. 7 all the measured data, except for zero confining pressure, fall between these two curves. The low value of the measured rock strength at zero confining pressure indicates that the damage rate in the vicinity of zero confining pressure is probably even higher than those constrained here. It should be noted that the inferred coefficient C_d is approximately constant for pressures above 30 MPa, indicating that constant value may be adopted for simulations of fracture processes in the seismogenic zone.

5   DISCUSSION AND CONCLUDING REMARKS

We tested the visco-elastic damage rheology introduced by Hamiel et al. (2004) against two new sets of laboratory experiments with Mount Scott granite. Thus we extended previous results that were based on experiments with Westerly granite and Berea sandstone samples (Hamiel et al., 2004) and provided additional constraints on the coefficients of damage rheology. An advantage of the new experiments with Mount Scott granite is that they have been performed with different loading paths and explicitly demonstrated the existence of stable and unstable fracturing regimes.

In the present study the damage variable is constrained based on fitting the entire stress-strain records, and found to be a linear function of the microcrack density ρ_c (Fig. 6). The linear relation between α and ρ_c (8) is in agreement with the theoretical prediction of Kachanov (1992) and with a previous comparison between the calculated rate of damage accumulation and measured acoustic emission (Hamiel et al., 2004). It should be noted that the damage variable, α and the microcrack density, ρ_c, characterize a properly chosen volume of rock with a large number of internal flaws (microcracks in a laboratory specimen or small faults in the Earth's crust) and are not related to any intrinsic length scale. Therefore we suggest that the linear relation between α and ρ_c (8) should be scale independent and hold on a scale of the thickness of the brittle crust.

The series of load-hold tests (Katz & Reches, 2004) analyzed here allow direct estimation of the strain partitioning between elastic and inelastic components based on the laboratory data. Ben-Zion & Lyakhovsky (2006) connected the rate of irreversible strain accumulation with partitioning between seismic and aseismic deformation in the seismogenic zone, or seismic coupling. They use a non-dimensional value $R = \mu_0 \cdot C_v$ and showed that the fraction of elastic strain released during a seismic cycle, i.e., the seismic coupling, can be estimated as $\chi = 1/(1 + R)$. The R-value represents the ratio between time scale of damage accumulation and the time scale of the damage-related irreversible strain accumulation under a given loading conditions and is the major factor controlling the aftershock productivity and the rate of aftershock decay. Ben-Zion & Lyakhovsky (2006) also demonstrated that long aftershock sequences fitted well by Omori law are expected in regions with $R < 1$, or large seismic coupling, $\chi > 50\%$. This theoretical prediction is supported by previous estimates of R value for Westerly Granite (Hamiel et al., 2004) and estimates for Mount Scott granite presented in this study (Table 2). For all granite samples the R value falls between 0.3 and 0.6 corresponding to seismic coupling $60\% < \chi < 80\%$. Previous analyses of experimental results with Berea sandstone indicate that the latter accumulates more irreversible strain, i.e., $R = 1.4$ corresponding to $\chi = 42\%$ (Hamiel et al., 2004). A comparison between different types of rocks, i.e., granites and sandstones, enables one to relate the seismic coupling to the lithology. We suggest that granular rocks or rocks with higher porosity have lower seismic coupling. This implies that the portion of elastic strain released during a seismic cycle as brittle deformation depends on the lithology of the region. For example, fault-zones that comprise significant accumulation of gouge or regions that have thick sedimentary cover are expected to release more energy by aseismic slip. The dependence of χ on the lithology might provide an explanation for the shallow slip deficit reported by Fialko et al. (2005) for some large earthquakes. However, to date there exist only few observational constraints on the depth distribution of fault slip averaged over multiple earthquake cycles, and in different host rocks. The lack of such constraints impedes robust conclusions about the effect of the fault zone lithology on the degree of seismic coupling.

ACKNOWLEDGMENTS

We thank Terry Tullis and Yossef Hatzor for constructive review. This study was supported by grant from the Israel Science Foundation (ISF 753/08).

REFERENCES

Ben-Zion, Y. & Lyakhovsky, V., 2006. Analysis of aftershocks in a lithospheric model with seismogenic zone governed by damage rheology, Geophys. J. Int., 165, 197–210.

Fialko Y., Sandwell, D., Agnew, D., Simons, M., Shearer P. & Minster, B., 2002. Deformation on nearby faults induced by the 1999 Hector Mine earthquake, Science, 297, 1858–1862.

Fialko, Y., Sandwell, D., Simons, M. & Rosen, P., 2005. Three-dimensional deformation caused by the Bam, Iran, earthquake and the origin of shallow slip deficit, Nature, 435, 295–299.

Hadley, K., 1976. Comparison of calculated and observed crack densities and seismic velocities in Westerly granite, J. Geophys. Res., 81, 3484–3494.

Hamiel, Y., Liu, Y., Lyakhovsky, V., Ben-Zion, Y. & Lockner, D., 2004. A visco-elastic damage model with applications to stable and unstable fracturing, J. Geophys. Int., 159, 1155–1165.

Hamiel, Y., Lyakhovsky, V. & Agnon, A., 2005. Rock dilation, Nonlinear deformation, and pore pressure change under shear, Earth Plan. Sci. Lett., 237, 577–589.

Hamiel, Y., Katz, O., Lyakhovsky, V., Reches, Z. & Fialko Y., 2006. Stable and unstable damage evolution in rocks with implications to fracturing of granite, Geophys. J. Int., 167, 1005–1016.

Homand, F., Hoxha, D., Belem, T., Pons, M.-N. & Hoteit N., 2000. Geometric analysis of damaged microcracking in granites, Mech. Mat., 32, 361–376.

Janssen, C., Wagner, F.C., Zang, A. & Dresen, G., 2001. Fracture process zone in granite: a microstructural analysis, Int. J. Earth Sci., 90, 46–59.

Kachanov, M., 1992. Effective elastic properties of cracked solids; critical review of some basic concepts, Appl. Mech. Rev., 45, 304–335, 1992.

Katz, O., Gilbert, M.C., Reches, Z. & Roegiers, J.C., 2001. Mechanical properties of Mount Scott granite, Wichita Mountains, Oklahoma, Oklahoma Geology Notes, 61, 28–34.

Katz, O. & Reches, Z., 2004. Microfracturing, damage, and failure of brittle granites, J. Geophys. Res., 109, B01206, doi:10.1029/2002JB001961.

Kranz, L.K., 1979. Crack growth and development during creep in Westerly granite, Int. J. Rock. Mech. Min Sci, 16, 23–36.

Kranz, R.L., Harris, W.J. & Carter, N.L., 1982. Static fatigue of granite at 2000C, J. Geophys. Res., 9, 1–4.

Lawn, B., 1993. Fracture of Brittle Solids, second edition, pp. 378, Cambridge University Press, Cambridge, UK.

Lockner, D.A. & Byerlee, J.D., 1980. Development of fracture planes during creep in granite, in 2nd conference on acoustic emission/microseismic activity in geological structures and materials, pp. 11–25, eds Hardy, H.R. & Leighton, F.W., Trans-Tech. Publications, Clausthal-Zellerfeld, Germany.

Lockner, D.A., Byerlee, J.D., Kuksenko, V., Ponomarev, A. & Sidorin, A., 1991. Quasi-static fault growth and shear fracture energy in granite, Nature, 350, 39–42.

Lockner, D.A., Byerlee, J.D., Kuksenko, V., Ponomarev, A. & Sidorin, A., 1992. Observations of quasi-static fault growth from acoustic emissions. in Fault mechanics and transport properties of rocks, International Geophysics Series, 51, 3–31, eds Evans, B. & Wong, T.-F., Academic Press, San Diego.

Lyakhovsky, V., Ben-Zion, Y. & Agnon, A., 1997a. Distributed damage, faulting, and friction, J. Geophys. Res., 102, 27635–27649.

Lyakhovsky, V., Reches, Z., Weinberger., R. & Scott, T.E., 1997b. Nonlinear elastic behavior of damaged rocks, Geophys. J. Int., 130, 157–166.

Lyakhovsky, V., Ben-Zion, Y. & Agnon, A., 2005. A viscoelastic damage rheology and rate- and state-dependent friction, Geophys. J. Int., 161, 179–190.

Martin, C.D. & Chandler, N.A., 1994. The progressive fracture of Lac du Bonnet granite, Int. J. Rock Mech. Min. Sci & Geomech. Abstr., 31, 643–659.

Moore, D.E., & Lockner, D.A., 1995. The role of microcracking in shear-fracture propagation in granite, J. Struct. Geol., 17, 95–114.

Nishihara, M., 1957. Stress-strain relation of rocks, Doshisha Eng. Rev., 8, 32–54.

Oda, M., Katsube, T. & Takemura T., 2002. Microcrack evolution and brittle failue of Inada granite in triaxial compression tests at 140 MPa, J. Geophys. Res., 107, 2233, doi:2210.1029/2001JB000272.

Pestman, B.J. & Munster, J.G., 1996. An acoustic emission study of damage development and stress—memory effects in sandstone, Int. J. Rock Mech. Min. Sci. & Geomech. Abstr., 33, 585–593.

Reches, Z. & Lockner, D.A., 1994. Nucleation and growth of faults in brittle rocks, J. Geophys. Res., 99, 18159–18173.

Rubinstein, J.L. & Beroza, G.C., 2004. Nonlinear strong ground motion in the ML 5.4 Chittenden earthquake: Evidence that preexisting damage increases susceptibility to further damage, Geophys. Res. Lett., 31, L23614, doi:10.1029/2004GL021357.

Schock, R.N., 1977. The response of rocks to large stresses, in: Impact and Explosion Cratering, pp. 657–688, eds Roddy, D.L., Pepin, R.O. & Merrill, R.B., Pergamon press, New York.

Tapponnier, P., & Brace, W.F., 1976. Development of stress-induced microcracks in Westerly granite, Int. J. Rock Mech Min Sci & Geomech Abstr., 13, 103–112.

Weinberger, R., Reches, Z. Eidelman, A. & Scott, T.S., 1994. Tensile properties of rocks in four-point beam tests under confining pressure, in Proceedings first North American Rock Mechanics Symposium, pp. 435–442, eds Nelson, P. & Laubach, S.E., Austin, Texas.

Zang, A., Wagner, C.F. & Dresen, G., 1996. Acoustic emission, microstructure, and damage model of dry and wet sandstone stressed to failure, J. Geophys. Res., 101, 17, 507–17,521.

Zoback, M.D. & Byerlee, J.D. 1975. The effect of microcrack dilatancy on the permeability of Western granite, J. Geophys. Res., 80, 752–755.

Micro-scale roughness effects on the friction coefficient of granite surfaces under varying levels of normal stress

Omer Biran, Yossef H. Hatzor & Alon Ziv

Department of Geological and Environmental Sciences, Ben-Gurion University of the Negev, Beer-Sheva, Israel

ABSTRACT: In this paper we explore roughness and normal stress effects on the steady-state friction coefficient of granite. Our tests were performed on a single direct shear, servo controlled, apparatus. To check the validity of the direct shear apparatus, we begin with a series of classic rate and state experiments. After establishing a good agreement with published results, we proceed with measurements of micro-scale roughness effects on friction using four levels of micro roughness on polished and ground surfaces, under normal stresses between 2.5 MPa and 15 MPa. For rough sliding surfaces we observe a second order normal stress effect on friction, where under normal stresses up to 5 MPa the friction coefficient decreases with increasing normal stress; above that normal stress level and up to 12 MPa the friction coefficient increases with normal stress, beyond which the friction coefficient exhibits a constant value similar to what is observed in standard direct shear tests. We find that behavior to be roughness dependent. It diminishes with decreasing roughness, and is not at all observed if the sliding surface is extremely smooth.

1   INTRODUCTION

The slip of solid materials along a pre-existing surface is resisted by friction. The first documented study on friction was performed by Leonardo da Vinci. His finding was confirmed later by Amontons (1699), who found that the frictional force is independent of the geometrical contact area, and is equal to the ratio between the shear and normal stresses. Bowden & Tabor (1954) suggested an explanation for the independence of the friction coefficient on the geometrical contact area. In their work on metals, they noted that the surfaces are in contact only at discrete locations, or "asperities", and therefore the "real" contact area is much smaller than the geometrical area. While Amonton's law provides a good description for friction for many applications, it is not suitable for the description of unstable friction phenomena such as earthquakes (Scholz 2002). In order to model and better describe dynamic friction phenomena, rate and state friction laws were developed by Dieterich (1979) based on experimental observations. Dietrich's laws describe the friction coefficient as a function of the slip rate and a state variable (Dieterich 1979, Ruina 1983). Rate and state friction laws provide powerful tools for investigating the mechanics of earthquakes and faulting, and the incorporation of the state variable provides a means to account for complex friction memory effects and history dependence. In the past few years, concentrated efforts were invested in the improvement of rate and state friction laws by adding more variables that influence the friction coefficient, such as normal stress variations (Linker & Dieterich 1992, Richardson & Marone 1999, Boettcher & Marone 2004, Hong & Marone 2005). In this paper we present preliminary experimental results that show how roughness and normal stress effects on the steady-state friction coefficient.

2   BACKGROUND

Rate and state friction was originally proposed by Dieterich (1979) on an empirical basis using an extensive body of shear tests. Experimental results show that the friction coefficient, μ, is a function of slip rate, V, and a state variable, θ, as follows (Dieterich 1979, Ruina 1983):

$$\mu = \mu^* + A \ln\left(\frac{V}{V^*} + 1\right) + B \ln\left(\frac{\theta V^*}{D_c} + 1\right) \tag{1}$$

where A and B are dimensionless empirical fitting parameters, D_c is a characteristic sliding distance from one steady state to another, V^* is the reference velocity, and μ^* is the coefficient of friction when the contact surface slips under a constant slip rate V^*. Various evolution laws have been proposed for the state variable. The two most commonly used state evolution laws are the 'Dieterich law' (Dieterich 1979):

$$\frac{d\theta}{dt} = 1 - \frac{V\theta}{D_c} \tag{2}$$

and the 'Ruina law' (Ruina 1983):

$$\frac{d\theta}{dt} = -\frac{V\theta}{D_c} \ln\left(\frac{V\theta}{D_c}\right) \tag{3}$$

These laws were originally formulated for conditions of constant normal stress and were later extended by Linker & Dieterich (1992) to account for variable normal stress. In equation 2, state, and thus friction, evolve even for $V = 0$. At hold time, $d\theta/dt = 1$, therefore the state variable during the hold time is:

$$\theta = \theta_0 + \Delta t \tag{4}$$

Substituting (4) in (1) shows that coefficient of friction increases with the logarithm of the hold time, i.e.:

$$\mu \propto B \ln(t) \tag{5}$$

In Ruina's evolution law any change in friction, including strengthening during quasi-stationary contact, requires slip. In either case, the condition for steady state coefficient of friction is $V\theta/D_c = 1$ and for both laws, the steady state velocity dependence is:

$$\mu_{ss} = \mu^* + (A - B) \ln\left(\frac{V_{ss}}{V^*} + 1\right) \tag{6}$$

Thus, steady state friction exhibits velocity weakening if B is greater than A, and velocity strengthening otherwise. The rate and state friction laws have been successfully used to address various geophysical problems, including simulation of crustal deformation along faults (Rice 1993, Ziv & Cochard 2006), and earthquake modeling (Marone et al., 1995, Tullis 1996, Scholz 1998).

3 EXPERIMENTAL METHODS

3.1 *Testing apparatus*

Our tests were performed on a single direct shear, hydraulic, servo controlled apparatus (Figure 1A, 1B). Normal and shear load cell capacities are 1000 kN and 300 kN, respectively. Shear box size is 180 mm × 180 mm × 140 mm. The sample is cemented into the shear box using Portland 350 cement. Shear and dilatational displacements are monitored by 6 LVDT type displacement transducers, with a maximum range of 50 mm and 0.25% linearity full scale. Additional 4 LVDT transducers, used to monitor vertical (dilatational) displacements were mounted on four corners of the shear box; 2 LVDT transducers, used to monitor horizontal (shear) displacement were mounted on two opposite sides of the tested interface (Figure 1C). In order to obtain optimal feedback signals and control, all six transducers were mounted as closely as possible to the sliding interface. Using a computer control interface, output signals from all channels can be used as servo control variables,

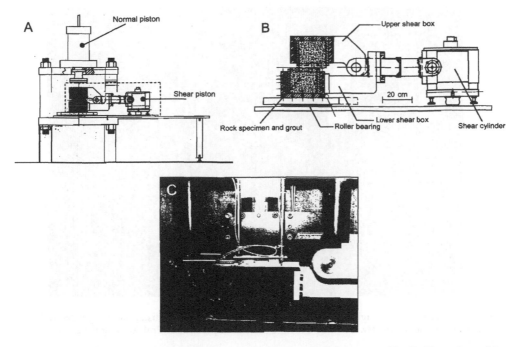

Figure 1. The direct shear system assembly. A. The complete load frame assembly. B. Close view of the shear box, shear piston and shear load frame. C. Close view on the horizontal (shear) and vertical (dilatational) displacement transducers (LVDT).

currently at an acquisition rate of 50 Hz. We used the average of the two shear displacement transducers output to control the sliding velocity, and used the output from the normal piston load cell to control the normal stress during sliding.

3.2 Tested material

Direct shear tests were performed on machined Timna granite blocks, typically with a 10 cm by 10 cm nominal (i.e. geometric) contact area. In order to ensure that the nominal contact area remains constant during sliding, the upper block was machined to slightly smaller dimensions than the lower block.

Four surface roughness levels were examined. These were obtained with three different preparation methods: 1. Rough (SC)—roughness that results from sawing the sample with standard rock saw, 2. Smooth (SG)—roughness obtained with a standard surface grinder; 3. Surface polished by hand-lapping using a 180 Silicone Carbide powder (#180), 4. Surface polished by hand-lapping using a 220 Silicone Carbide powder (#220).

Surface roughness was measured using an optical profilometer, Zygo New View 5000 white-light interferometer. The optical output is processed, statistically analyzed and graphically displayed using a post processing software package (MetroPro) (Figure 2B, 2C, 2D). Three statistical parameters of surface topography are used for characterising the different finished surfaces. 1. RMS—the root mean square, 2. R_a—the average distance of all points from an arbitrary plane, and 3. PV—the maximum peak to valley height over the sample area. To assess roughness anisotropy, measurements were taken parallel and perpendicular to slip direction.

Statistical analyses of the obtained roughness profiles indicate that there are significant differences between the studied surfaces. While the polished surfaces are nearly isotropic, the saw-cut surfaces are clearly not (Figure 3). Direct shear tests were always performed while keeping the shear direction parallel to the direction of maximum roughness.

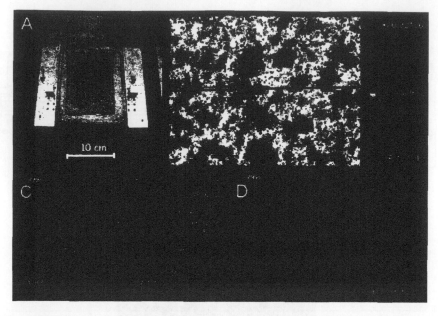

Figure 2.   An example of roughness graphic display. A. Interface smoothed using surface grinder. B. Roughness distribution in plane view. C. Three dimensional view of panel B. D. Profile along the cross section in panel B.

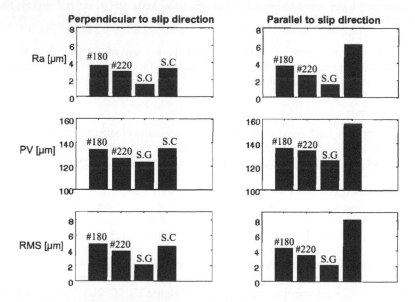

Figure 3.   Statistics of roughness parameters measured using the Zygo NewView 5000 interferometer prior to the direct shear tests. Note roughness anisotropy in the saw cut samples.

3.3 *Preliminary smoothing segment*

In order to suppress stick-slip episodes, and to reach the displacement magnitude required for sliding at steady state, all experiments were first began with a smoothing cycle, consisting of a forward shearing segment at a constant sliding rate of 10 μm/S to a distance of 5 mm under

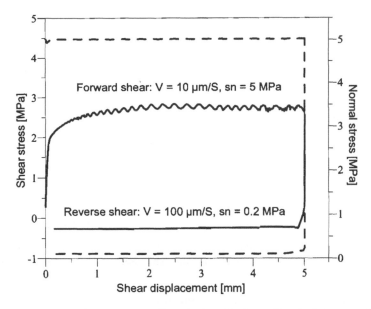

Figure 4. Preliminary smoothing segment performed prior to each experiment. Shear stress (solid line, left vertical axis) and normal stress (dashed line, right vertical axis) as a function of shear displacement.

a constant normal stress of 5 MPa followed by a backward shearing segment, under a constant normal stress of 0.2 MPa at a sliding velocity of 100 μm/S (Figure 4).

4 REPRODUCING THE CLASSICAL RATE AND STATE EXPERIMENTS USING A DIRECT SHEAR APPPRATUS

Previous experiments that examined the effects of slip rate and contact age were performed using a double shear apparatus (Dieterich 1979, Linker & Dieterich 1992, Dieterich 1972, Kilgore et al., 1993). In the double shear configuration the net torque in the system may be negligible, whereas in the direct shear system it is not. We first reproduce some typical rate and state experiments to verify that the differences between the two testing configurations do not influence the results. Below we present three types of rate and state experiments: slide-hold-slide, velocity stepping and normal stress stepping, which were successfully reproduced with the direct shear apparatus.

4.1 *Slide-Hold-Slide experiments*

In order to examine the effect of the hold duration on the static friction coefficient Slide-Hold-Slide (SHS) experiments were preformed, similar to the procedure described by Dieterich (1972) and Dieterich & Kilgore (1994). During these experiments the normal stress was held constant at 5 MPa. Constant slip rates of 100 and 50 μm/S were imposed on the sample until a steady state friction coefficient was attained. Once steady state sliding was attained, the shear piston displacement was stopped for specified time intervals, after which shear sliding commenced at the original slip rate before the hold segment. SHS segments were repeated several times, with increasing hold interval each time. A typical result for a SHS experiment is displayed in Figure 5A for a #180 SiC surface. Inspection of Figure 5B reveals that the peak (or static) friction coefficient and consequently $\Delta\mu$ (where $\Delta\mu$ is the difference between static and initial friction coefficients), increase with hold time and slip rate, consistent with previous findings (Dieterich 1972, Dieterich & Kilgore 1994, Marone 1998). This data together with equation 5 were used to obtain the rate-and-state B parameter (see Table 1).

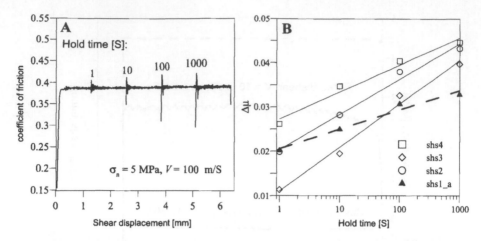

Figure 5.   Result of SHS tests. A. A typical result for a single run. B. Peak less residual friction coefficient as a function of hold time (See Table 1 for regression coefficients). Solid lines $V = 100$ mm/S and dashed line $V = 50$ mm/S.

Table 1.   Summary of velocity stepping and slide hold slide tests.

Experiment	Normal stress [MPa]	Sliding rate [μm/S]	B	A–B	A	R^2
shs1_a	5	50	0.002	–	–	0.973
shs2	5	100	0.004	–	–	0.988
shs3	5	100	0.004	–	–	0.986
shs4	5	100	0.003	–	–	0.975
vs6_g1	7.5	0.0125–1.6	–	–0.006	0.006	0.984
vs4_g1	7.5	0.2–10	–	–0.006	0.005	0.991
vs3_g1	5	0.1–100	–	–0.01	0.005	0.959
vs4b_g1	2.5	0.1–100	–	–0.008	0.005	0.865
vs5b_g1	5	0.1–100	–	–0.009	0.008	0.839

shs—Slide hold slide experiment, vs—Velocity stepping experiment.

4.2   *Velocity stepping experiments*

We performed velocity stepping experiments, similar to those performed in the past by other researchers (Dieterich 1979, Kilgore et al. 1993, Tullis & Weeks 1986). These experiments were preformed on rough surfaces (saw cat) at slip rates between 0.0125 and 100 μm/S, and under a constant normal stress of 5 MPa for slip rats between 1 and 100 μm/S) and under 7.5 MPa for slip rates between 0.0125 and 10 μm/S. The imposed sliding velocity was increased by a factor of 10 between adjacent segments during sliding velocities between 1 and 100 μm/S and by a factor of 2 during sliding velocities between 1 and 0.0125 μm/S. We found that in response to step decrease in slip rate the coefficient of friction rapidly decreased and than rose to a new higher steady value as sliding commenced, in agreement with the previous works cited above (see Figure 6A). Furthermore, we found that the rise to a new steady state friction coefficient value roughly followed an exponential curve with a characteristic displacement magnitude D_c ranging from 1 to 1.5 μm in those tests for which D_c was estimated. The *A* parameter was retrieved directly from the experimental output and the value of *A–B* was estimated by using Equation 4 (see Figure 6B); the quantitative estimates of *A*, *B*, and *A–B* for all the tests are listed in Table 1.

4.3   *Rapid normal stress stepping*

Normal stress stepping tests were performed on a rough saw-cut surface, in a manner similar to experiments reported by previous researchers (Linker & Dieterich 1992, Hong & Marone 2005).

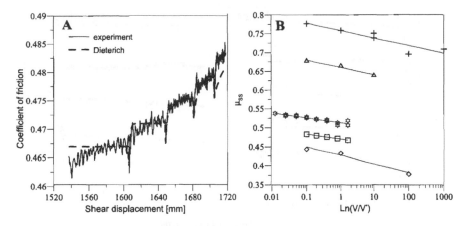

Figure 6.   Result of velocity stepping experiments. A. Friction coefficient vs. shear displacement, dashed line based on Dieterich constitutive law with: $A = 0.0049$, $A–B = -0.0057$ and $Dc = 0.001$ mm. B. Friction coefficient at steady state as a function of shear rate. See Table 1 for the slopes of the regression lines $(A–B)$.

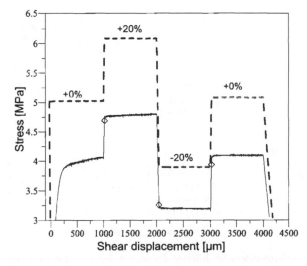

Figure 7.   Normal (dashed) and shear (solid) stresses as a function of shear displacement during normal stress stepping. The diamonds indicate the end of instantaneous response.

Rapid normal stress changes were imposed on the sample during steady-state sliding at a sliding velocity of 10 μm/S. The test was performed under an initial constant normal stress value of 5 MPa, and when the sample attained steady-state the normal stress level was changed by ±20% from the nominal value, at a very fast rate. The frictional response of the tested interface to changing normal stress is shown in Figure 7. In agreement with previously published results (Linker & Dieterich 1992, Hong & Marone 2005), we find that the change in normal stress was simultaneously accompanied by a linear change in shear stress, followed by a delayed non linear response, most likely due to plastic yield at contact points along the tested interface.

5   STEADY-STATE FRICTION DEPENDENCE ON SURFACE ROUGHNESS

Does steady state friction depend on surface roughness? To address this question the results of direct shear tests preformed on SC (rough) and SG (smooth) surfaces are compared. These testes

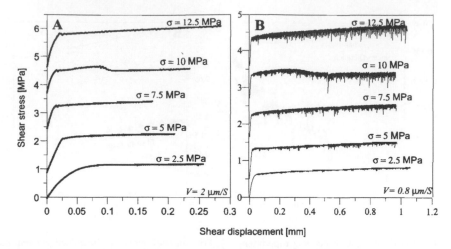

Figure 8.   Shear stress vs. shear displacement under an imposed constant normal stress condition. A. $V =$ 2 μm/S. B. $V = 0.8$ μm/S.

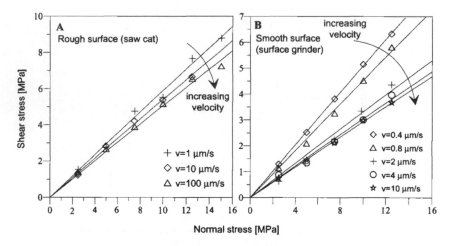

Figure 9.   Obtained Coulomb criteria for different slip rates. A. Rough surfaces (saw cut). B. Smooth surfaces (surface grinder). Linear regression coefficients for each sliding velocity are listed in Table 2.

Table 2.   The friction coefficient that is obtained from Coulomb failure criteria.

Sliding rate [μm/S]	Roughness	Friction coefficient	R^2
1	SC	0.591	0.998
10	SC	0.542	0.999
100	SC	0.495	0.999
0.4	SG	0.509	0.999
0.8	SG	0.453	0.999
2	SG	0.331	0.994
4	SG	0.305	0.997
10	SG	0.293	0.999

SC—Saw cat, SG—Surface grinder.

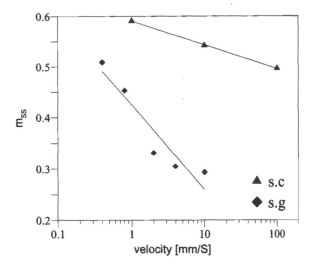

Figure 10.   Friction coefficient at steady state as a function of shear rate.

were carried out at various sliding velocities and normal stress. For each normal stress level the corresponding shear stress at steady state was determined graphically (Figure 8) and the results were plotted on a τ–σ space (Figure 9).

The friction coefficient for each surface and sliding velocity are listed in Table 2, along with the obtained values of the linear regression coefficient R^2. The experimental results show that the steady-state friction for either surface is velocity weakening, and that it is more velocity weakening for smooth surface then it is for rough surface (Figure 10).

6   SECOND ORDER NORMAL STRESS EFFECTS ON STEADY-STATE FRICTION

Through measurements of shear stress response to changing the normal stress during sliding under constant rate we have identified a second order effect of normal stress on the steady state friction coefficient. First steady-state sliding was obtained under a constant sliding velocity of 1 μm/S and an imposed constant normal stress. Next the normal stress was increased or decreased to a specified new target using normal stress output as the servo control parameter. Seeking the new normal stress target was performed while shear sliding continued under a displacement control mode imposed on the shear piston. Two sets of tests were thus performed. While in one, the starting normal stress was set to 15 MPa and has been decreased by 10% each time down to 2.5 MPa, in the other the starting normal stress was set to 2.5 MPa and has been increased by 10% each time up to 15 MPa. Stepping the normal stress was done at a rate of 0.05 MPa/S. Finally four types of surface finish were studied: 1) Saw Cat (SC.), 2) #180 grit of SiC, 3) #220 grit of SiC and 4) Surface grinder (SG), whose roughness statistics were reported in section 3.2. Typical result for shear stress response to descending and ascending normal stress are shown in Figures 11A and 11B respectively for saw cat surfaces.

For reasons that are not presently clear to us, friction coefficients for the stress-ascending experiment (Figure 11B) failed to reach steady-state. Since here we are interested in roughness and normal stress effects on steady-state friction, the result of stress-ascending experiments will not be considered further (Figure 11B). Note the normal stress dependence of the steady-state friction coefficient of rough samples (SC). Specifically, we find that the steady-state friction coefficient decreases with increasing normal stresses for normal stresses between 2.5 and 5.22 MPa, but increases for normal stresses between 5.22 and 12.5 MPa (Figure 12A). We are unable to draw conclusions for normal stresses greater than 12.5 MPa, since for these stress levels the system did not

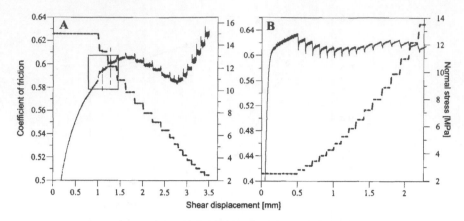

Figure 11.   Friction coefficient (solid curves and left vertical axes) and normal stresses (dashed curves and right vertical axes) as a function of shear displacement for decreasing (A) and increasing (B) normal stress level during sliding at a constant rate. The box indicates a segment that did not reach steady state.

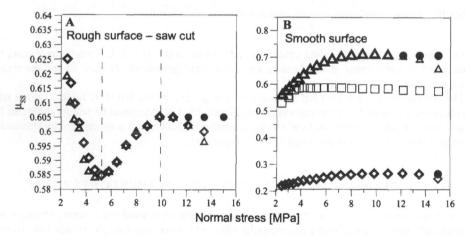

Figure 12.   Steady-state friction as a function of normal stress level. (A) obtained for the two rough surfaces (saw-cut). Both tests were performed under the same conditions and similar results are obtained. Strengthening effect of asperities decreases with increasing normal stress levels up to 5.2 MPa, beyond which increase in strength is detected with increasing normal stress up to a level of 12 MPa, at which stage a constant level of friction coefficient is attained. Note that the last two segments ($\sigma n = 12.5$ MPa, 15 MPa) did not reach steady state during the test (see boxes in Figure 11), the inferred friction coefficient values for these two tests are plotted in solid symbols. (B) obtained from three different surfaces: surface grinder (SG—open diamonds), polished with #220 grit (open squares), polished with #180 grit (open triangles). An extremely low friction coefficient of $\mu = 0.2$ is obtained with the SG samples. A pronounced increase in friction coefficient is obtained with the polished surfaces to $\mu = 0.5 - 0.6$. In all cases the strengthening is observed with increasing normal stress. Inferred μ values for segments where steady state was not reached during the test are plotted as solid symbols.

reach steady-state (box in Figure 11A). Interestingly, the result for smooth surfaces differs radically from that for rough surfaces. Steady-state friction for smooth surfaces, increases with increasing normal stresses (Figure 12B). The differences in the friction coefficient that were obtained for each normal stress level are clear and significant (Figure 13A–C).

We rationalize our results as follows: Beginning with the rough surfaces, under the lowest tested normal stress (2.5 MPa) it seems that the influence of roughness is most pronounced with a maximum value of coefficient of friction of 0.625 at that stress level. With increasing normal stress levels the obtained coefficient of friction for steady state decreases until a minimum value

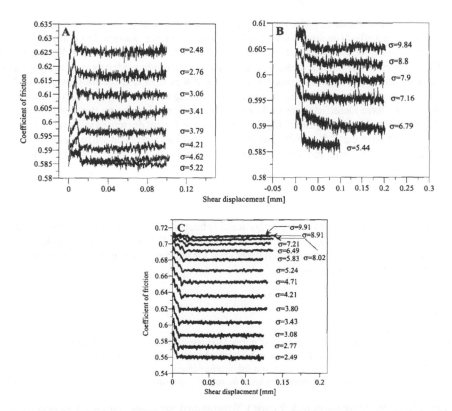

Figure 13. Obtained friction coefficients at steady state for each individual segment. A + B. Test on rough surface that presented in figure 11A. C. Test on smooth surface (#180) that presented in figure 11B.

of 0.583 is obtained. This value represents the lowest friction coefficient obtainable for Saw Cut surfaces under the studied normal stresses and sliding velocities. We believe that this minimum value marks the end of the role initial asperities play in the frictional response of the tested SC surfaces. When the normal stress is further increased the effect of plastic flow and healing at contact areas becomes more pronounced and therefore shear strengthening is obtained. With the smooth surfaces, however the behaviour is different. It seems that in all smooth surfaces the first shear softening stage is missing and we only get shear hardening as normal stress is increased, indicating that the role of plastic flow and healing around contact points is the dominant mechanism in initially smooth surfaces. The only difference between the samples polished by hand lapping in SiC powder (#180, #220) and the surfaces polished with mechanical surface grinder (SG) is the minimum value of the friction coefficient obtained: whilst in the surfaces polished with SiC powder the minimum value is 0.53 the surface polished with surface grinder exhibits a friction coefficient of 0.2. Note that the sample that was prepared with #180 SiC powder (Figure 12B—open triangles) exhibits shear hardening as normal stress is increased by as much as 20% with respect to the initial shear stress, representing an increase in coefficient of friction from 0.55 for low normal stresses to 0.7 for high normal stresses. This significant amount of shear hardening can not be considered a second order effect, and is in fact in contradiction with Coulomb law. This interesting preliminary observation requires further research.

7 SUMMARY

Using direct shear apparatus, we have reproduced some of the rate and state friction experiments including the slide-hold-slide, velocity stepping and normal stress stepping. We find that the differences between the direct and the double shear testing configurations do not influence the results. Roughness effects on steady state friction coefficient were examined by comparing constant slip rate and normal stress shear tests on surfaces of different roughness levels. We have showed, using Coulomb criterion, that the steady-state friction for all roughness levels is velocity weakening, and that it is more velocity weakening for smooth surfaces than it is for rough surfaces. Finally, through measurements of shear stress response to changing the normal stress during sliding under constant rate we have identified a second order effect of normal stress on the steady state friction coefficient. On rough surfaces, the steady state friction coefficient decreases with increasing normal stress at low normal stress (2.5–5 MPa), but increases at normal stress that are higher than 5 MPa. On smooth surfaces on the other hand, steady-state friction increases with increasing normal stresses.

ACKNOWLEDGEMENT

This research was funded by Ministry of National Infrastructures, state of Israel, through contract ES-19-2007/35517 28-17-023. The authors wish to thank Dr S. Biderman of Rotem Industries Ltd for providing access to the optical profilometer used for microscopic roughness measurements and for assistance with processing the data.

REFERENCES

Boettcher, M.S. & Marone, C. 2004, "Effects of normal stress variation on the strength and stability of creeping faults", *Journal of Geophysical Research,* vol. 109, B03406, doi:10.1029/2003JB002824.
Bowden, F.P. & Tabor, D. 1964, *The friction and lubrication of solids,* Oxford University Press.
Dieterich, J.H. 1972, "Time dependent friction in rocks", *Journal of Geophysical Research,* vol. 77, pp. 3690–3697.
Dieterich, J.H. 1979, "Modeling of rock friction 1. Experimental results and constitutive equations", *Journal of Geophysical Research,* vol. 84, pp. 2161–2168.
Dieterich, J.H. & Kilgore, B.D. 1994, "Direct observation of frictional contacts: New insights for state-dependent properties", *Pure and Applied Geophysics,* vol. 143, pp. 283–302.
Hong, T.C. & Marone, C. 2005, "Effects of normal stress perturbations on the frictional properties of simulated faults", *Geochemistry Geophysics Geosystems,* vol. 6, Q03012, doi:10.1029/2004GC00082.
Kilgore, B.D., Blanpied, M.L. & Dieterich, J.H. 1993, "Velocity dependent friction of granite over a wide range of conditions", *Geophysical Research Letters,* vol. 20, pp. 903–906.
Linker, M.F. & Dieterich, J.H. 1992, "Effects of Variable Normal Stress on Rock Friction—Observations and Constitutive-Equations", *Journal of Geophysical Research,* vol. 97, pp. 4923–4940.
Marone, C., Vidale, J.E. & Ellsworth, W.L. 1995, "Fault healing inferred from time dependent variations in source properties of repeating earthquakes", *Geophysical Research Letters,* vol. 22, pp. 3095–3098.
Marone, C. 1998, "The effect of loading rate on static friction and the rate of fault healing during the earthquake cycle", *Nature,* vol. 391, pp. 69–72.
Rice, J.R. 1993, "Spatio-temporal complexity of slip on a fault", *Journal of Geophysical Research,* vol. 98, pp. 9885–9907.
Richardson, E. & Marone, C. 1999, "Effects of normal stress vibrations on frictional healing", *Journal of Geophysical Research,* vol. 104, pp. 28859–28878.
Ruina, A. 1983, "Slip instability and state variable friction laws", *Journal of Geophysical Research,* vol. 88, pp. 10,359–10,370.
Scholz, C.H. 1998, "Earthquakes and friction laws", *Nature,* vol. 391, pp. 37–42.
Scholz, C.H. 2002, *The mechanics of earthquakes and faulting,* Cambridge University Press.
Tullis, T.E. & Weeks, J.D. 1986, "Constitutive behavior and stability of frictional sliding of granite", *Pure and Applied Geophysics,* vol. 124, pp. 383–414.
Tullis, T.E. 1996, "Rock friction and its implications for earthquake prediction examined via models of Parkfield earthquakes", *Proceedings of the National Academy of Sciences,* vol. 93, pp. 3803–3810.
Ziv, A. & Cochard, A. 2006, "Quasi-dynamic modeling of seismicity on fault with depth variable rate- and state-dependent friction", *Journal of Geophysical Research,* vol. 111, doi:10.1029/2005JB004189.

IV. *Granular shear and liquefaction*

Friction in granular media

J.C. Santamarina & H. Shin
Civil and Environmental Engineering, Georgia Institute of Technology, Atlanta, GA, USA

ABSTRACT: Fine-grained and coarse grained granular materials exhibit normal-stress dependent frictional shear strength. The mineral-to-mineral friction mobilized at interparticle contacts emerges at the macroscale through a complex sequence of competing particle level processes. The observed frictional response of the soil mass varies with strain level; we can distinguish the constant volume friction angle, dilation angle, peak friction angle, residual friction angle after grain alignment, and post-granular-segregation friction angle. Compiled experimental data and particle-level simulations help identify the most relevant soil parameters that affect the frictional response in each case (including interfacial friction at boundaries). Other sediment conditions that affect frictional strength include: grain crushing, inherent anisotropy, intermediate stress, temperature, strain rate, vibration, pore fluid and contact-level adhesive forces. The measurement of soil friction is a boundary-value problem; information-intensive measurement methods may help overcome measurement limitations related to incomplete knowledge of boundary conditions.

1 INTRODUCTION

Friction denotes the normal-stress dependent energy loss when a medium is subjected to shear. The history of friction is briefly reviewed in Table 1, with emphasis on solid-to-solid friction.

The understanding of friction in granular materials remains challenging due to complexities in interparticle solid-to-solid friction, the role of surface topography in granular materials, the participation of multiple coexisting phenomena in upscaling interparticle friction to the macroscale of the granular medium, and the emergence of new phenomena. Specially, the grain size appears as an inherent length scale that prompts us to distinguish the macroscale shear response of fine and coarse grained materials.

Table 1. History of friction.

Reducing friction		
<3500 bc	Mesopotamia	The wheel
~2750 bc	Egyptians	Recognized differences between sliding on sand and on wet silt
Early theoretical developments		
1452–1519	Leonardo da Vinci	The shear force T is independent of the apparent contact area The shear force T doubles when N doubles
1663–1705	G Amontons	Re-discovered da Vinci's frictional laws. Suggested that friction is due to roughness and the overriding of asperities
1638–1744	JT Desaguliers	More polished surfaces exhibit higher friction Hinted on adhesion
1736–1806	CA Coulomb	Referred to Amontons observations. Kinematic friction is independent of sliding velocity. "Cohesion is zero . . . for newly-turned soils"
1766–1832	J Leslie	While some asperities climb, others fall. Wondered about the source of energy dissipation
1885	O Reynolds	Demonstrated the concepts of dilatancy (rubber bag filled with sand and water and connected glass tube).

(Continued)

Table 1.   (*Continued*)

Lubrication and adhesion		
>XIX		Hydrodynamic Lubrication—μ = f (sliding velocity)
1920's	K. Terzaghi	Shear strength in soils—Adhesion theory
1922	W Hardy, I Doubleday	Boundary Lubrication: Lubricant binds onto surface
1939	IV Kragelsky	Molecular-mechanical theory of friction (USSR). It does not necessarily imply adhesion or chemical bonding.
1948	DW Taylor	Confining stress dependent dilatancy
1950	P. Bowden, D. Tabor	Adhesion theory of friction (UK)
Fundamental understanding		
1950's ...	BJ Alder, TE Wainwright	Molecular dynamics (study of μ in the 1990's)
1957	JF Archard	Considered the distribution of asperity heights to link Hertz $f(N^{2/3})$ contact response with Amonton's f(N) observations.
1964	AW Skempton	Studied the role of grain characteristics on ϕ residual
1970's	P Cundall	Discrete element modeling—BALL
1986	G Binnig, CF Quate, C Gerber	Atomic force microscope
1990 ... Current		Noise-friction interaction Energy coupling
		Friction control/engineering Strain rate and other effects
		Measurement

In this manuscript we review the nature of friction in granular materials, identify differences in friction between fine and coarse grained sediments, and explore emergent phenomena in the load-deformation response of granular materials.

2   FRICTION BETWEEN TWO LARGE MINERAL SURFACES

Friction between two mineral surfaces involves multiple phenomena that take place at the thin interface between the two solid bodies, hence, it is affected by the physical-chemical characteristics of the external layer on minerals. These processes are explored next.

2.1   *Surface topography*

Surface topography plays a secondary role when large surfaces come into contact. Consider two rough surfaces approach each other (Fig. 1a). The tallest asperities interact first, *deform elastically* and the contact area is a non-linear function of the applied load (contact mechanics—e.g., Hertzian contact). Then, if shear resistance is a linear function of the contact area, the linearity between the shear force T and the normal force N in Amontons' law is satisfied when asperity heights follow an exponential or Gaussian distributions (Archard 1957, Greenwood & Williamson 1966, Chapter 1 in Kragelsky & Alisin 2001). As the normal load increases, asperities *deform plastically*, the contact area is a linear function of the normal force acting on the asperity, and friction is a linear function of the applied load regardless of surface topography. In both cases, *the precise distribution of surface topography has a secondary role on the sliding resistance between the two large surfaces.* (Note: elasto-plastic contact response is explored using finite element simulations in Vijaywargiya & Green 2007; topography and elasto-plastic asperity effects are discussed in Kogut & Etsion 2004).

2.2   *Dry-Friction—transient effects*

The shear resistance T is a function of the true contact area A and the yield strength at contacts σ_y, in agreement with the adhesion theory of friction: $\mu = T/N = A\sigma_y/N$ (Bowden & Tabor 1950).

Figure 1. Effect of surface topography. (a) Large surfaces: the true area of contact is independent of surface topography when contacts deform either elastically (exponential height distribution) or plastically. (b) Grains: the same scale of surface topography may cause interlocking between grains.

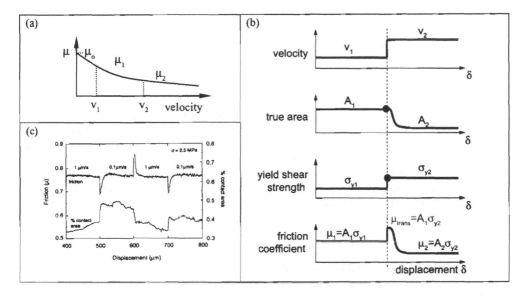

Figure 2. Mineral-to-mineral friction. (a) Velocity dependency. (b) Explanation of friction transient consistent with adhesion theory of friction. (c) Experimental data obtained with glass and acrylic plastic (Dieterich & Kilgore 1994).

Both components depend on shear velocity (Dieterich 1979—part I—see constitutive equation in Ruina 1983, Dieterich & Kilgore 1994, Sleep 1997).

The true contact area A increases with duration of the interaction between asperities, $A = a_1 + b_1 \log (t/sec)$; this behavior is a consequence of creep at contacts, displacement of adsorbed layers, time for the interpenetration of asperities, and/or lower contact oscillations at low sliding velocity among others. The most immediate consequence of contact creep is the increase in mineral-to-mineral friction coefficient at low sliding velocity, $\partial \mu / \partial v < 0$, reaching the static friction coefficient μ_0 as $v \to 0$ (Fig. 2a). A decrease in friction with increasing velocity leads to stick-slip behavior, particularly in media with low stiffness and when the shear velocity is low. On the other hand, the yield stress increases with strain rate $\dot{\varepsilon}$ in most minerals $\sigma_y = a_2 + b_2 \log (\dot{\varepsilon}.sec)$.

Time-dependent true contact area A(t) and strain rate dependent strength $\sigma_y(\dot{\varepsilon})$ combine to cause frictional transients when the sliding velocity changes. Consider a sliding block being displaced at velocity v_1 and then at $v_2 > v_1$ (Fig. 2b):

- when the velocity is v_1, the true contact area is A_1, the yield stress is σ_{y1} and the friction coefficient $\mu_1 = T_1/N = A_1 \sigma_{y1}/N$.
- when the displacement velocity increases to $v_2 > v_1$, the steady state true contact area decreases $A_2 < A_1$, the yield stress increases $\sigma_{y2} > \sigma_{y1}$ (applies to most minerals), and the friction coefficient decreases to $\mu_2 = T_2/N = A_2 \sigma_{y2}/N < \mu_1$ because the effect of contact area prevails.
- however, immediately following the transition $v_1 \to v_2$, the contact area is still A_1, yet its is now being sheared at a higher velocity and yield stress is σ_{y2}. Therefore, there is a transient

$\mu_{\text{trans}} = A_1 \sigma_{y2}/N$ so that a higher friction is measured immediately following the increase in velocity. The memory of A_1 is short term and the friction coefficient soon converges to its steady state value μ_2.

Conversely, a negative transient is observed when the velocity decreases. The transient length is about 10 μm in the experimental results shown in Figure 2c; limited data suggest an increase in transient length with asperity height (Dieterich 1979, Ruina 1983).

2.3   *Fluids—Shear velocity*

The presence of fluids adds velocity-dependency to the friction coefficient. If the sliding *velocity is sufficiently high*, a fluid layer forms at the interface. The fluid layer partially supports the external load through viscous resistance, preventing solid-to-solid contact, and leading to a reduction in rubbing friction. In this regime, resistance to sliding increases with velocity due to the hydrodynamic contribution (Fig. 3).

The fluid layer is squeezed by the normal load, and the thickness decreases at *low sliding velocity*, allowing for more pronounced solid-to-solid rubbing. In addition, the fluid viscosity increases when the fluid layer becomes thinner than ~10 Å (molecular interactions are discussed in the next section). Both effects contribute to the increase in friction coefficient as the sliding velocity decreases in the boundary lubrication regime (Fig. 3), $\partial\mu/\partial v < 0$, and stick-slip behavior emerges. Finally, if the fluid does not bind to the mineral surface, the two surfaces may experience solid-solid contact and a marked increase in the friction coefficient follows (Fig. 3).

2.4   *Energy loss*

Roughness, either geometrical or electrical, cannot explain losses as the energy gained by an asperity going down a peak is utilized by other asperities to climb other peaks (as noted by J. Leslie [1766–1832]). So, how is energy lost when shear takes place between two interacting mineral surfaces? Little energy is stored as strain energy when either smooth or rough mineral surfaces are sheared. Some energy remains as increased surface energy in newly created surfaces. The remaining mechanical energy is converted into some other form of energy through asperity crushing, wear, inelastic indentation, plastic work (deformation, adhesion-and-shear, material transfer from one surface to the other, ploughing), viscous losses and vibration (macroscale and molecular scale), the emission of elastic waves (from acoustic emissions to seismic activity), shear-induced polarization of double layers followed by ohmic losses and the emission of electromagnetic waves (i.e., seismo-electric coupling). All losses eventually end in heat.

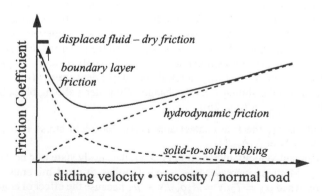

Figure 3.   The effect of fluids on friction—Stribeck curve (Taylor & Pollet 2000, Okawara & Mitachi 2003). Typically, the separation between surfaces decreases at low sliding velocities.

3   GRAIN SIZE AS A DETERMINING SCALE

Surface undulations on otherwise planar surfaces determine the contact points between them (Fig. 1a), but have a minor effect on the frictional resistance between the two large surfaces. This is not the case in granular materials: surface topography not only defines contact points, but it may also cause interlocking and hinder grain rotation (Fig. 1b). In this section we recognize grains size as a fundamental length scale.

3.1   *Scales in granular materials*

Three geometric scales emerge in the study of friction in granular materials:

- The grain size d: this is the inherent scale of the medium and it has immediate implications on particle-level forces and the relevance of surface topography.
- The size of interparticle contacts d_c: Assuming a Hertzian contact and a normal load estimated as $N = \sigma'd^2$ (first order approximation for a simple cubic packing subjected to normal stress σ'):

$$\frac{d_c}{d} = \sqrt[3]{\frac{3(1 - v_g)}{2}\frac{\sigma'}{G_g}} \qquad (1)$$

where G_g is the shear stiffness of the mineral that makes the grains. The size of contacts is typically $d_c/d = 1/20\text{-to-}1/80$ ($G_g = 10\text{-to-}50$ GPa; $\sigma' = 100$ kPa to 1 MPa). Then, a convenient definition for the length of roughness is $\sim d/100$.

- The Debye-Hückel length ζ: This is the thickness of the diffuse double layer that forms on mineral surfaces and it depends on fluid permittivity and ionic conductivity; typically, $\zeta = 20\text{-to-}200$ Å.

These scales have profound implications in interparticle friction.

3.2   *Particle size and shape*

There is a fundamental change in the nature of particle formation and the ensuing particle shape when the particle size varies across the $d = 10\text{-to-}50\,\mu$m region. Submicron particles ($d < 10\,\mu$m) form through chemical synthesis and precipitation, and are often platy and made of phillo-silicates clay minerals (rod and spherical shapes are also found). However, when the particle size exceeds the micron-scale $d > 50\,\mu$m, particle tends to be more rotund, and particle shape is a consequence of mechanical actions such as breakage, abrasion, and collisions.

The previous scales allow us to identify intermediate scales that define the particle shape and affect the frictional behavior of granular materials. Consider a sinusoidal oscillation of wavelength λ riding on top of a circular particle of perimeter πd; then (refer to Fig. 4):

- Eccentricity $\lambda \sim \pi d/2$: promotes particle alignment during shear and the development of residual shear strength
- Angularity $\lambda \sim d/10$: promotes interlocking and hinders particle rotation
- Roughness $\lambda \sim d/100$ (based on the observation that the size of contacts is $d_c/d = 1/20\text{-to-}1/80$): involved in the generation surface friction μ, as described in the previous section (large surfaces).

3.3   *Size, forces and fabric*

Consider the particle self weight $W = \pi d^3 \rho g/6$, the mean skeletal force $N = \sigma'd^2$, and electrical DLVO forces (van der Walls attraction and double layer repulsion). Skeletal and self weight forces prevail for particles $d > 10\text{-to-}50\,\mu$m. However, electrical interaction controls behavior for submicron size grains. If the medium is unsaturated (mixed fluid phase), capillary forces emerge and can play a dominant role when particles are smaller than 1 mm (Santamarina 2002).

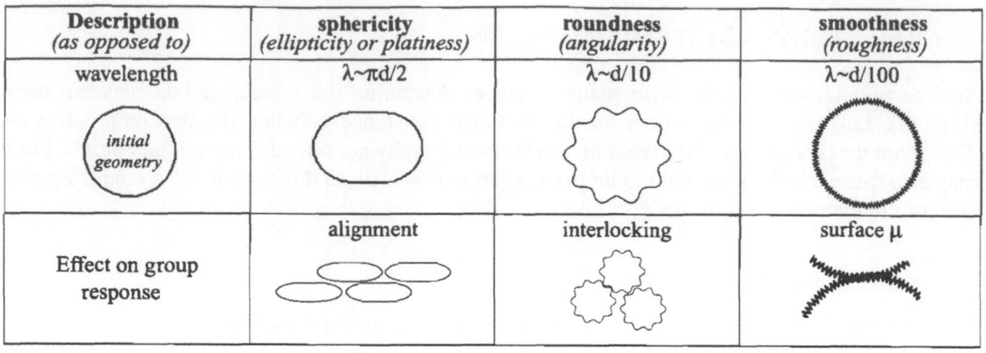

Description *(as opposed to)*	sphericity *(ellipticity or platiness)*	roundness *(angularity)*	smoothness *(roughness)*
wavelength	$\lambda \sim \pi d/2$	$\lambda \sim d/10$	$\lambda \sim d/100$
initial geometry			
Effect on group response	alignment	interlocking	surface μ

Figure 4.   Scales in particle shape: eccentricity, angularity and roughness—Implications in upscaling contact-level particle interaction to the soil mass.

The 10-to-50 μm size boundary-also encountered above in the context of formation and shape-allows us to separate granular materials into *coarse grained* (larger than ~50 μm) and *fine-grained* (smaller than 10 μm). Fabric formation in coarse grained sediments is determined by the relative size of particles (measured by the coefficient of uniformity) and particle shape (Fig. 5a). Well graded sediments made of rounded particles tend to pack at a higher density than poorly graded sediments made of angular particles (Fraser 1935, Youd 1973, Shimobe & Moroto 1995, Miura et al. 1998, Cubrinovski & Ishihara 2002, Cho et al. 2006).

Fabric formation in fine grained sediments is determined by particle shape and interparticle electrical interactions that are a function of pH-dependent surface charge, differences in edge and face charges, electrical interparticle forces, and the ionic concentration in the pore fluid (Fig. 5b). Furthermore, individual grains may form aggregations that effectively behave as coarser particles. Therefore, a wide range of fabrics can develop in fine grained sediments.

3.4   Roughness and the diffuse counterion cloud

Water hydrates counter ions on mineral surfaces and precipitated salts that rest in the pore space. The interaction between thermal activity and electrical interactions results in the diffuse counterion cloud around mineral surfaces. The relative size between asperity height and adsorbed layer thickness determines different frictional regimes in sediments. (Note: the discussion of roughness in Figure 4 was in terms of the topography wavelength; in this section we address topography height).

Let's consider the Deby-Huckel length ζ as the characteristic distance for the electrical interaction between particles. Surface roughness will exert full control on friction when the asperity height h exceeds the thickness of the diffuse layer $h > \zeta$ (Fig. 6a). If the relative scale that defines roughness is established in terms of the particle diameter, say $h/d \sim 1/100$, and we assume a nominal value $\zeta \sim 10$ nm, then, surface roughness becomes important when particles exceed the micron size: $d > 100\,h \sim 1$ μm. Conversely, the diffuse layer becomes important to sliding resistance when roughness decreases below the interaction distance $h < \zeta$ (Fig. 6b).

3.5   Electrical roughness—friction in fine-grained sediments

So far, roughness refers to a topographic characteristic of mineral surfaces. "Effective roughness" is also encountered at the atomic scale in otherwise topographically smooth mineral surfaces, such as in fine-grained clayey sediments: a test charge pulled parallel to the mineral surface experiences the energy wells associated to surface charges (Fig. 7a). The concept of electrical roughness can be extended from the atomic scale of surface charges to the particle scale: a test charge feels the undulating potential energy surface associated to the counter-ion clouds that surround particles (Fig. 7b). Resistance to shear is also caused by the hindered mobility of water molecules and

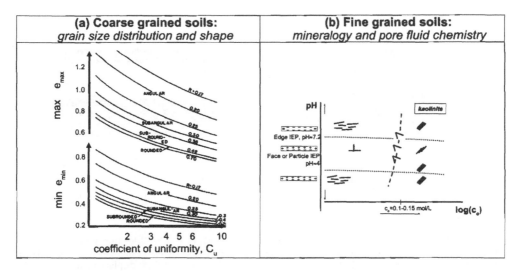

Figure 5. Fabric formation. (a) Coarse grained sediments (after Youd 1973). (b) Fine grained sediments (Palomino & Santamarina 2005).

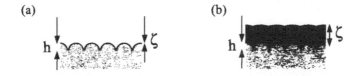

Figure 6. Mineral roughness (gray) and the thickness of adsorbed layers (black). (a) Rough surface $h \gg \zeta$: the adsorbed layer has limited effect. (b) Smooth surface $h > \zeta$: the diffuse layer exerts a strong effect on the normal interaction between sliding surfaces and the resulting shear resistance.

counterions in thin layers (Fig. 7c—the oscillatory nature of the hydration force during normal displacement is a related effect). In fact, when the platy surfaces are displaced relative to each other, molecules go through a sequence of ordered-disordered, solid-fluid islands (Fig. 7d—molecular dynamic simulations reported in Persson, 1998 and references therein).

When two mineral grains are sheared past each other, molecules and counterions are displaced relative to each other until the energy barrier is overcome, slip occurs and a new equilibrium configuration is reached (Fig. 7e—The schematic representation in Figure 7f is analogous to shear in metals—see Tabor 1992). Acoustic emissions are detected during shear even in fine grained bentonites (Matsui et al. 1980).

Adsorbed water layers become thinner as the effective stress increases. All processes in Figure 7 are accentuated in thinner layers: deeper energy wells, higher concentration of counterions, lower water mobility and higher shear strength of the water film (e.g., data in Israelachvili et al. 1988). In other words, the shear resistance between two wet mineral surfaces increases as the effective normal stress increases; hence, the shear resistance at the particle level in wet phillosilicate clay minerals has the characteristics of boundary layer type frictional response, shown in Figure 3.

3.6 *Summary: The relevance of grain size*

The previous discussion has identified critical changes in sediment characteristics and behavior between sediments with grain size smaller than 10 μm and those larger than 50 μm, on the bases of particle formation processes and particle shape, mineralogy, controlling particle level-forces,

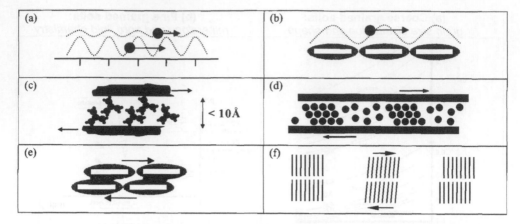

Figure 7.   Friction-related phenomena in fine-grained sediments. (a and b) "Electrical roughness" due to atomic-scale surface charges and particle-scale counterion clouds. (c) Bonds in thin counterion clouds and hindered mobility. (d) Solid-fluid islands during shear. (e & f) Evolution of shear displacement.

the development of fabric, and the relevance of roughness at contacts. Differences in the frictional response of fine-grained and coarse-grained sediments are identified in the following sections.

4   FRICTION IN SEDIMENTS

Contact-level shear resistance upscales through multiple particle interactions in the granular structure to eventually convey frictional strength to the granular medium. These mechanisms are explored in this section. First, we start with a particle-level estimation of the macroscale strain required to mobilize friction.

4.1   *Strain level for frictional losses*

A sediment deforms elastically at very small strains. Slippage at interparticle contacts advances gradually with the increase in shear force (Fig. 8—Refer to Kogut & Etsion 2004, Vijaywargiya & Green 2007 and references therein for more detailed discussions).

Full contact slippage takes place when the slip distance δ exceeds a threshold value δ^* (e.g., the junction size in metals—see implications in Rabinowicz 1995). For a Hertzian-Mindlin contact between particles size d, the equivalent macroscale threshold strain $\gamma_t \approx \delta^*/d$ when full contact slippage is reached can be estimated as (d_c is the diameter of a Hetzian contact)

$$\gamma_t \approx \frac{\delta^*}{d} = \frac{3}{4}\mu(2 - v_g)\frac{\sigma'}{G_g}\frac{d}{d_c} = 1.26\mu\left(\frac{\sigma'}{G_g}\right)^{2/3} \quad \text{for } v_g = 0.3 \quad (2)$$

where v_g, G_g are the Poisson's ratio and shear modulus of the mineral that makes the grains. This equivalent macroscale threshold strain defines the onset of large frictional losses in the sediment (see Dobry et al. 1982), which is accompanied by pronounced fabric changes. Note that the threshold strain for frictional losses is proportional to the effective confining stress $\sigma'^{2/3}$.

4.2   *Evolution of the granular skeleton during loading*

The boundary stress applied to a granular medium is not supported uniformly by all grains, and the distribution of interparticle forces can be matched with Weibull or exponential functions (Dantu 1968, Gherbi et al. 1993, Jaeger et al. 1996).

0.4 mm

Figure 8. Gradual slippage at contacts—Mindlin-type response—Annular fretting at a contact subjected to small-displacement cyclic shear from Johnson (1961).

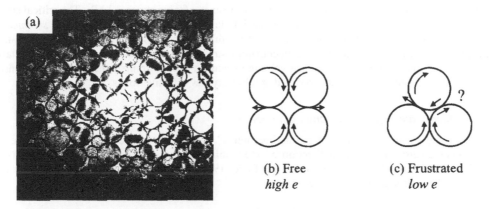

Figure 9. Particle-level mechanisms during normal and shear loading: (a) Formation of granular columns. (b) Free rotation. (c) Rotational frustration (cause dilation or contact slippage).

The formation of columnar structures that resist most of the applied boundary stress is a salient feature in granular materials (Fig. 9a—see Drescher & De Josselin De Jong 1972, Oda et al. 1985). The particles that make the columns are primarily loaded in the direction of the applied principal stress. Particles that are not part of the main columns play a secondary yet very important role of preventing the buckling of the main chains; hence, the main forces acting on these secondary particles are normal to columns (Radjai et al. 1998—see Fig. 9a).

Shear at contacts causes either rotation or slippage. Rotation is possible when the coordination number is low and rotational compatibility is satisfied among all neighboring particles (Fig. 9b). However, high inter-particle coordination at low void ratio leads to rotational arrest or frustration (Fig. 9c). Rotational frustration is overcome by either frictional slippage at contacts, or local *volume dilation* to reduce the number of contacts among particles. The coordination number cannot become too low: the anisotropy in the polar histogram of contacts is limited by the buckling of load carrying columns; following buckling, the local *volume contracts* and the interparticle coordination increases.

The interplay between slippage-vs-rotation and dilation-vs-contraction explains: (1) the lower coordination number during shear when the interparticle friction is higher (angularity and roughness—Fig. 4—see Thornton 2000), (2) the higher frictional resistance in well graded sediments as they tend to be denser and with higher coordination number (Fig. 5a), and (3) the continuous rebuilding of granular columns during shear to maintain a low coordination number.

The formation of particle columns highlights the development of marked fabric anisotropy within the sediment. In fact, the mobilized friction angle ϕ_{mob} is directly related to the evolution of the internal anisotropy in contacts a_c, normal force a_n and shear contact forces a_t (Rothenburg & Bathurst 1989)

$$\sin\phi_{\text{mob}} = \frac{a_c + a_n + a_t}{2} \qquad (3)$$

Within this particle-level perspective, the macroscale friction angle at failure reflects the maximum internal anisotropy a sediment may sustain.

The two competing volume change tendencies of *dilation* and *contraction* reach statistical balance at large strains, and the granular mass shears at constant volume. This is the "critical state". Both loose and dense sediments evolve towards critical-state constant-volume shear at large strains. The void ratio at critical state depends on the effective mean confining stress p' as

$$e_{cs} = e_{1\text{kPa}} - \lambda \log\left(\frac{p'}{\text{kPa}}\right) \quad \text{where } p' = \frac{\sigma_1' + \sigma_2' + \sigma_3'}{3} \qquad (4)$$

where $e_{1\text{kPa}}$ is the critical state void ratio when the mean confining stress is 1 kPa. The critical state line defines two regions in agreement with particle-level mechanisms. (1) *Contractive sediments* start above the critical state line and contract during drained shear; these are either high void ratio sediments or sediments subjected to high confining stress. And (2) *Dilative sediments* which start below the critical state and experience volume expansion during drained shear; these are dense sediments and sediments that are confined at relatively low effective stress.

4.3   *Macroscale frictional response: $\phi_p, \psi, \phi_{cv}, \phi_r$*

The work to shear a granular material in direction 'x' ($\tau\delta x$) is consumed in frictional slippage at contacts to overcome friction in direction 'x' and in dilation against the normal confining stress σ' in direction 'y', that is $\tau\delta x = \sigma'\mu\,\delta x + \sigma'\delta y$ (Taylor 1948). Therefore, the measured peak angle of internal shear strength $\tan\phi_p = \tau/\sigma'$ is

$$\tan\phi_p = \mu + \frac{\delta y}{\delta x} = \tan\phi_{cv} + \tan\psi \qquad (5)$$

where ϕ_{cv} is the *constant volume angle* of shear strength at critical state, and ψ is the *dilatancy angle*. The constant volume friction angle requires a minimum strain level in the order of $\gamma \sim 100\%$ to attain particle rearrangement into a new fabric that is compatible with constant volume shear, independently of the initial fabric. Bolton (1986) simplified Rowe equations for stress-dilatancy under plane strain conditions to obtain an approximate expression for the peak friction that satisfies experimental data gathered for sands

$$\phi_p = \phi_{cv} + 0.8\psi \qquad (6)$$

Particle alignment takes place at large strains when eccentric particles are involved (the particle length ratio may be as low as $\sim$1.1 to bring about the effects of particle eccentricity—Rothumberg & Bathurst 1993). The friction angle decreases as particle alignment takes place. The *residual friction angle ϕ_r* is attained when strains exceed $\gamma > 100\%$ and particles become aligned with respect to the failure plane.

4.4   *Failure envelope*

The constant volume friction angle and the residual friction angle do not change with the normal effective stress, therefore, a linear Coulomb-type model applies in both cases

$$\tau = \sigma'\tan\phi \text{ applies to constant volume and residual conditions} \qquad (7)$$

This is not the case for the peak friction angle, due to the stress-dependent dilatancy ψ. Therefore, the envelope for peak strength is steep at low stress and gradually approaches the constant volume shear (i.e., no dilation) at large confining stress. Then, the peak strength envelope can be piece-wise fitted with a straight line of the form

$$\tau_p = c^* + \sigma' \tan \phi^* \text{ piece-wise fit to peak strength} \tag{8}$$

where c* and ϕ^* are stress dependent. This expression conveys the wrong sense of cohesive strength c* when none may be present in the granular material. A better approach is to adopt a curved envelope (e.g., jointed rock mass—Hoek & Brown 1980)

$$\tau_p = \alpha \sigma'^\beta \text{ global fit to peak strength} \tag{9}$$

The effect of the intermediate stress σ_2 on shear strength, and alternative failure envelopes are discussed later in this manuscript.

5 EMPIRICAL AND NUMERICAL RESULTS

5.1 *Constant volume friction angle*

Experimental and discrete element numerical studies have been conducted to evaluate the relationship between the interparticle friction coefficient μ and the constant volume friction angle ϕ_{cv}. The group of studies summarized in Figure 10 lacks a clear trend, however, some observations can be extracted by focusing on individual investigations. The increase in inter-particle friction produces a pronounced increase in macroscopic friction ϕ_{cv} for low μ-values, but a minor change at high inter-particle friction; this macroscale response reflects the transition from a dominant particle sliding motion at low μ, to dominant rolling motion at high μ (Skinner 1969—Clearer trends shown for ϕ_p in Section 5.3). In terms of tangent values, tanϕ_{cv} increases by only 7% for a change in interparticle friction from $\mu = 0.2$ to $\mu = 0.5$ (Kruyt & Rothenberg 2006). Furthermore, numerical results show that constant volume friction angle does not exceed $\phi_{cv} \approx 40°$ even when the interparticle friction approaches infinite, $\mu \rightarrow \infty$ (see Yimsiri 2001, Rothenburg & Kruyt 2004).

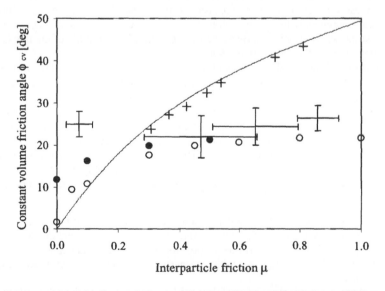

Figure 10. Constant volume friction angle ϕ_{cv} as a function of interparticle friction coefficient μ. Symbols: (—) theoretical solution from Horne (1969), (⊢—⊺—⊣) experiments from Skinner (1969), (○) DEM 3D from Thornton (2000), (•) DEM 2D from Kruyt and Rothenberg (2006), (+) experiments from Rowe (1969).

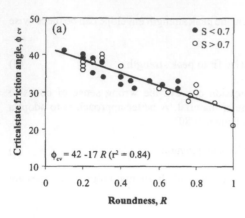
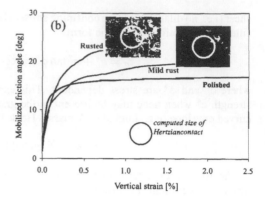

Figure 11.   Constant volume friction angle and particle shape. (a) Particle roundness—Natural and crushed sands; S indicates sphericity (Cho et al. 2006), (b) Particle roughness (Santamarina & Cascante 1998).

The constant volume friction angle ϕ_{cv} is more than just mineral friction: particle eccentricity, angularity and roughness hinder particle rotation, promote looser packing and affect the evolution of anisotropy. This is confirmed by experimental data gathered with natural and crushed sands (ϕ_{cv} increases with particle angularity—Fig. 11a—see also Sukumaran & Ashmawy 2001), and triaxial compression tests run on specimens made of steel ball bearings with different degrees of rust (ϕ_{cv} increases with surface roughness—Fig. 11b). Once again, roughness may be of electrical nature. For example, when the molecular weight of a steric stabilizer increases, the layer around mono-dispersed polystyrene particles conveys higher "electrical roughness", and the viscosity of suspensions increases (Castle et al. 1996).

5.2   Dilatancy angle

The rate of volume change defines the dilatancy angle ψ. The dilatancy angle can be inferred from strain measurements in triaxial compression testing as $\tan \psi = \Delta \varepsilon_{vol}/\Delta \varepsilon_{axial}$. The maximum rate of dilation coincides with the peak strength in uncemented dense soils. This is not the case in cemented soils, where dilation starts as the peak strength is overcome; often, this is the case in highly OC clays as well (Terzaghi, et al. 1996).

Dilatancy is determined by the distance between the initial $e_0 - p_0'$ state of the sediment and the critical state line. In *clays*, the $e_0 - p_0'$ state is defined by the stress history: "normally consolidated" clays are contractive, while "heavily overconsolidated" clays are dilative.

The initial void ratio e_0 in *sands* depends on the depositional method and ensuing packing density; in general, most natural sands are dilative at shallow depth. Let's define the "state parameter" $E = e_0 - e_{cs}$ as the distance between the initial void ratio e_0 and the void ratio at critical stress e_{cs} at the initial mean effective stress (Equation 4). Then, the dilatancy angle is inversely related to the state parameter as conformed by experimental data summarized in Figure 12.

The following additional observations and special cases are relevant to the understanding of dilatancy:

- Particle geometry (eccentricity, angularity and roughness) and grain size distribution affect dilatancy in two ways: (1) they affect initial fabric formation: the higher μ the higher the values of e_{min} and e_{max}, and (2) they determine the position of the critical state line. For the same initial packing density, dilatancy increases with interparticle friction μ (Kruyt & Rothenburg 2006, Guo & Su 2007).
- A dense packing of identical spheres exhibits a dilatancy angle as high as $\psi = 30°$. Such a high angle of dilatancy is also observed in dense sands at very low confinement (microgravity study by Sture et al. 1998), and in locked sands ($\psi \sim 31°$—Dusseault & Morgenstern 1979).

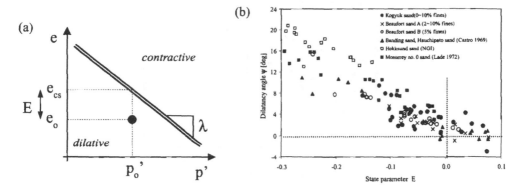

Figure 12. Dilatancy in *sands* as a function of the initial $e_0 - p'_0$ state. (a) Definition of the state parameter $E = e_0 - e_{cs}$. (b) Dilatancy angle ψ as a function of the state parameter E (Been & Jefferies 1985).

- Lightly cemented soils shear into blocks. This blocky granular material exhibits high dilation during shear (even when the uncemented medium would be contractive at the void ratio).
- Dilatancy is hindered by high confinement, and grains may crush rather than override (discussed later in the text).
- Shear localization limits the dilatancy a soil may exhibit at the macroscale.

5.3 *Peak friction angle*

The peak friction angle is a combination of dilatancy and constant volume shear (Equations 5 and 6). Therefore, the peak shear strength can be estimated from data presented in the previous sections. Nevertheless, corroborating information is presented herein, most of it based on DEM simulations that permit testing exactly the same fabric but with different interparticle friction coefficients (Note: it is easier to study peak strength than large-strain constant volume shear in DEM simulations). Results compiled in Figure 13 confirm the early rise in peak friction angle at low interparticle friction coefficient μ, and the asymptotic trend towards strength saturation at large μ when dominant rotational motion reduces the impact of interparticle friction.

5.4 *Residual friction angle—platy particles (clayey sediments and micaceous sands)*

Platy particles align at large strains (in excess of 100%) and cause a decrease in the shear strength of the medium, which is reflected in a low residual friction angle ϕ_r. The residual shear strength depends on particle shape, mineralogy and hardness, grain size (indirectly related to shape), and grain size distribution. Eccentric particles do not affect the residual strength when their mass fraction is lower than ~10%, and they fully determine ϕ_r when they exceed ~25% (Koerner 1970, Lupini et al. 1981, Skempton 1964, 1985). These observations apply to micaceous soils as well (Lee et al. 2007).

The following clay characteristics are correlated: thin atomically smooth small particles, low skeletal force on each particle, high relevance of electrical interparticle forces, and high plasticity index. Then, the friction angle can be correlated with the plastic index IP (in this context, IP becomes an indirect measure of geometric characteristics—Fig. 14):

$$\phi_r = 32 \, e^{\frac{-IP}{250}} \tag{10}$$

Soils that experienced high biological activity during formation deviate from this trend. Diatoms and foraminifers increase the porosity of the soil (i.e., its ability to retain water and the measured

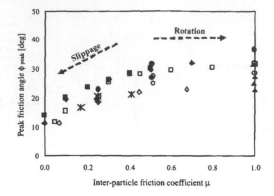

Figure 13. Peak friction angle ϕ_{peak} as a function of interparticle friction coefficient μ. Symbols: *Experiments*: ∗ (Suiker & Fleck 2004). *2D-DEM simulations*: ■ (Kruyt & Rothenberg 2006). *3D-DEM simulations*: □ (Thornton 2000); ◊ (Suiker & Fleck 2004); the following four groups of 3D-DEM simulations are by Yimsiri (2001): Drained TC △, Undrained TC ▲, Drained TE ◯, Undrained TE ●. Finally, the symbol ◆ represents 3D-DEM simulations by the authors.

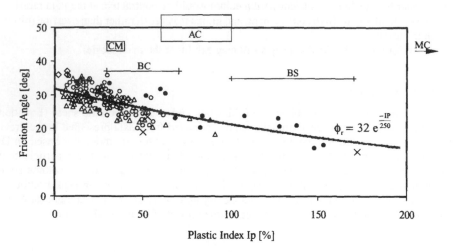

Figure 14. Residual Friction Angle vs Plastic Index—Data digitized from Terzaghi et al. (1996). Blocks show anomalous soil response data discussed in the text (Sections 5.4 and 6.7).

plastic index), contribute roughness and promote interlocking. Therefore, diatomaceous soils exhibit high IP and high ϕ values. Examples of these soils include (Fig. 14):

- Ariake Clay "AC" (Japan. Ohtsubo et al. 1995, Tanaka et al. 2001). Marine, smectite, clay fraction 50%, diatoms: PI = 60-to-100, ϕ_r = 46°-to-57°.
- Bangkok Clay "BC" (Tanaka et al. 2001). Normally consolidated marine clay, smectite, clay fraction 50%, pellets: PI = 30–70, ϕ_r = 37°.
- Bogota Soil "BS" (Moya & Rodriguez, 1987). Volcanic, lacustrian. Kaolinite, montmorillonite and diatoms: PI = 100-to-170, ϕ_r = 35°.
- Cooper Marl "CM" (Charleston, USA. Camp et al. 2002). Marine, soft, very fine grained (≤0.002 mm) impure carbonate deposit with fossils (foraminifers): PI = 30–40, ϕ_r = 43–46.
- Mexico City Soils "MC" (Diaz-Rodriguez et al. 1992, Diaz-Rodriguez et al. 1998). Volcanic, lacustrine. Montmorillonite and illite, clay fraction 20–55%. Silica polymorphs (e.g., biogenic opal, cristobalite). Microfossils (diatoms and ostracods): PI = 400-to-500, ϕ_r = 43°.

Additional insight is gained from data gathered at very high effective stress levels, typically above $\sigma' > 100$ MPa. The coefficient of friction measured under dry conditions ranges from $\mu_{dry} = 0.2$ (graphite) to $\mu_{dry} = 0.8$ (e.g., montomorillonite, gibbsite and kaolinite), and friction correlates with interlayer bond strength, i.e., shear consists of creating new cleavage surfaces and breaking the interlayer bonds in the minerals (when $\mu_{dry} < 0.8$; if $\mu_{dry} = 0.8$, shear involves abrasion, wear, fracture and rolling). The addition of water causes μ_{wet} to decrease (e.g., $\mu_{wet} \sim 0.3$ in montmorillonite); in this case μ_{wet} reflects the shear resistance along structured adsorbed water in thin films (Moore & Lockner 2004). The value of μ_{wet} increases as the effective stress increases and adsorbed layers become thinner. In addition, the value of μ_{wet} increases with increasing temperature (in the hundreds of degrees Celsius), unless other effects arise such as mineral dissolution (Moore & Lockner 2007).

6   OTHER EFFECTS ON FRICTION

Other material properties and boundary conditions that affect the friction angle in sediments are reviewed in this section.

6.1   *Grain crushing*

Grain crushing may replace contact shear and dilation as an alternative, lower-energy deformation mechanism at high confining stress (relative to the particle strength). Therefore, the emergence of crushing is accompanied by volume contraction and results in a lower peak friction angle (experimental results in Fig. 15. DEM numerical results with crushable particles show similar trends—Cheng et al. 2004). The decrease in friction angle reaches asymptotic conditions: crushing produces angular particles and an increase in coordination number as smaller particles fill voids and add contacts among pre-existing particles. The large-strain critical state friction angle shows limited sensitivity to crushing (Vesic & Clough 1968, Coop & Atkinson 1993).

6.2   *Inherent fabric anisotropy*

Sedimentation in a gravity field produces inherent fabric anisotropy because eccentric particles tend to align transverse to gravity. Idealized cases in Figure 16 suggest that dilation and peak strength are affected by inherent fabric anisotropy, so that measured parameters depend on the direction of the specimen α with respect to the deposition direction.

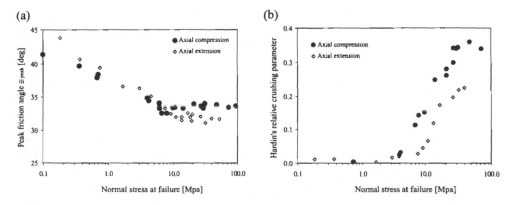

Figure 15.   Grain crushing and friction angle. (a) Peak friction angle, (b) Hardin's relative breakage parameter (Yamamuro & Lade 1996).

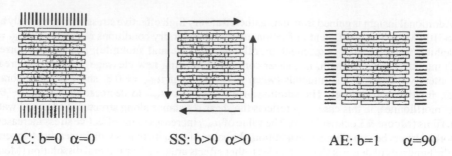

AC: b=0 α=0   SS: b>0 α>0   AE: b=1 α=90

Figure 16. Friction angle, inherent anisotropy and stress path. Axial compression AC. Simple shear SS. Axial extension AE. Parameters: $b = (\sigma_2 - \sigma_3)/(\sigma_1 - \sigma_3)$; α: angle between the particle normal and σ_1.

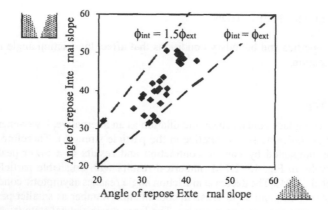

Figure 17. Friction anisotropy (natural and crushed sands). The angle of repose measured on an internal flow cone is significantly larger than the angle of repose measured on an external cone.

Consolidation under zero lateral strain K_0-conditions produces inherent fabric anisotropy as well. The minimum peak strength is measured when the shear plane aligns with the particle orientation in both sands and clays (Ladd 1977, Jamiolkowski et al. 1985, Tatsuoka et al. 1986, Vaid & Sayao 1995, Kurukulasuriya et al. 1999, Lade & Kirkgard 2000, Guo 2008).

6.3 *Stress induced anisotropy—intermediate stress*

The angle of repose is a simple measurement of the critical state friction angle of sands (Cho et al. 2006). However, different values are obtained when the angle of repose is measured using a standard cone geometry, ϕ_{ext}, a planar sliding surface, or an inside flow cone ϕ_{int} (i.e., by removing a central plug beneath a filled cylinder). Data in Figure 17 show that the internal angle of repose ϕ_{int} is significantly greater than the external angle ϕ_{ext}. A possible explanation follows: (1) the internal flow cone experiences a gradual reduction in cross section, sliding particles come closer together, and the interparticle coordination in the annular direction increases, while (2) flow on the external slope experiences a gradual increase in cross section, sliding particles move away from each other, and the coordination number decreases.

Further insight is obtained by exploring discrete element simulations. The evolution of the polar distribution of contacts, average normal contact force $N (\theta)$ and average shear contact force $T (\theta)$ in direction θ during axial compression AC ($b = 0$) and axial extension AE ($b = 1$) tests are shown in Figure 18 (3D micro-mechanical simulations—Chantawarangul 1993; see also Rothenburg & Bathurst 1989, Thornton 2000). The following observations can be made:

- Contact normals during anisotropic loading become preferentially oriented in the direction of the main principal stress σ_1, in agreement with observations made above (see also Oda 1972).

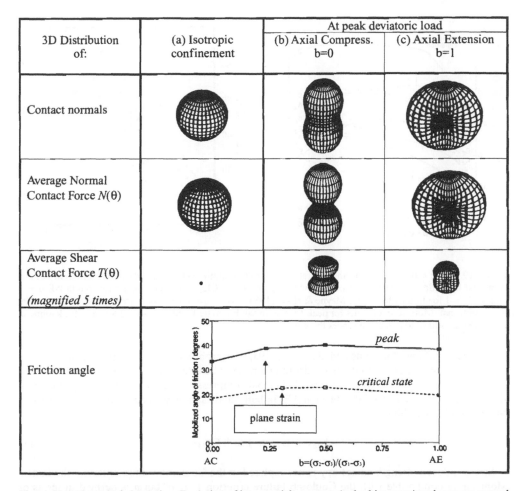

3D Distribution of:	(a) Isotropic confinement	At peak deviatoric load	
		(b) Axial Compress. b=0	(c) Axial Extension b=1
Contact normals			
Average Normal Contact Force $N(\theta)$			
Average Shear Contact Force $T(\theta)$ *(magnified 5 times)*			
Friction angle			

Figure 18. Numerical simulation: Evolution of inter-particle contacts (polar histogram) and average normal and shear contact forces (polar plots) during axial compression and axial extension loading. Variation in friction angle with the intermediate stress σ_2. Figure compiled from Chantawarangul (1993).

- The main reduction in inter-particle contacts takes place in the direction of the minor principal stress: σ_2 and σ_3 directions in AC, and σ_3 direction in AE. This situation allows for more degrees of freedom for particle rotation and for chain buckling in AC (even when the total coordination number at failure is about the same in both cases).
- These volume-statistics of the particle-scale response provide insight into the observed effective peak friction angle (macroscale—numerical results presented in the lower frame of Fig. 18): higher friction angle is mobilized in AE than in AC. Furthermore, the lack of particle displacement in the direction of plane strain hinders rearrangement and causes an even higher peak friction angle in plane strain loading. The critical state friction angle obtained in numerical simulations follows a similar trend, but with less pronounced differences.
- Results by Chantawarangul (1993—not presented here) also show that early volume contraction before the peak strength is more pronounced in AE than in AC tests; this observation is relevant to the interpretation of undrained strength.

Experimental and numerical results presented in Figures 17 and 18 show that the shear strength of a soil reflects restrictions to particle motion established at the level of contacts (sliding resistance, frustration, and crush-resistance), but conditioned by the boundaries.

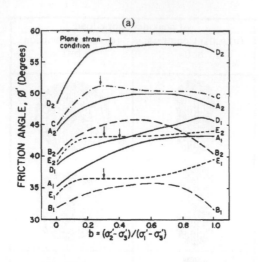
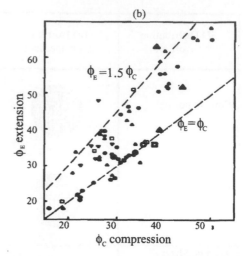

Figure 19.   The effect of the intermediate stress on friction angle. (a) Sands—data from different authors compiled by Ladd et al. (1977)—Compare with Figure 18. (b) Clays: friction angle measured in AE b = 0 and AC b = 1 loading paths—from Mayne & Holtz (1985—Most specimens are normally consolidated under K_0 conditions). More recent results for peak friction in sands can be found in Lade (2006) and for kaoline in Lin & Penumadu (2005) and references therein.

Experimental data for sands and clays (Fig. 19a, b) corroborate angle of repose measurements in Figure 17 and DEM simulations results in Figure 18. The intermediate stress is captured in the coefficient b = $(\sigma_2 - \sigma_3)/(\sigma_1 - \sigma_3)$: b = 0 in axial compression AC, b ~ 0.3–0.4 in plane-strain shear DSS, and b = 1 in axial extension AE. In both sands and clays (based on data compiled by Mayne and Holtz, 1985):

$$\phi_{DSS} > \phi_{AE} \approx 1 \text{ to } 1.5\phi_{AC} \tag{11}$$

Stress-induced strength anisotropy is not captured in the Coulomb failure criterion: the largest Mohr circle compatible with the Coulomb failure criterion $\tau = \sigma' \tan\phi$ is defined in terms of $\sigma_1 - \sigma_3$, so there is no effect of σ_2.

$$\sin\phi = \left.\frac{\sigma_1 - \sigma_3}{\sigma_1 + \sigma_3}\right|_{\text{at failure}} \tag{12}$$

Other failure envelopes have been proposed to take into consideration the intermediate stress. These are summarized in Table 2 and plotted in Figure 20.

6.4   *Temperature*

The increase in temperature—above frozen ground conditions—can cause densification (phenomenon is known as thermosmotic consolidation—Campanella & Mitchell 1968), increase in pore fluid pressure with the consequent decrease in effective stress and shear strength, and even mineral melting. Within the range of common geomechanical applications (T < 100°C), the effect of temperature on the frictional resistance of soils can be summarized as follows

- coarse grained sediments: they experience limited or no effect.
- fine grained sediments: (1) no effect on the constant volume friction angle (Fig. 21—left pane), and (2) increase in peak friction angle with temperature (Fig. 21—right pane) probably due to densification by thermal consolidation (Cekerevac & Laloui 2004, Abuel-Naga et al. 2007—contradictory results are reported in Hueckel & Baldi 1990 for Pontida clay).

Table 2.   Failure Criteria.

Failure criterion	Reference
$\sqrt{J_{2D}}\left(\cos(\theta) + \dfrac{1}{\sqrt{3}}\sin(\phi)\sin(\theta)\right) - \dfrac{1}{3}J_1\sin(\phi) = 0$ where $\sin(3\theta) = -\dfrac{3\sqrt{3}}{2}\dfrac{J_{3D}}{J_{2D}^{3/2}}$	Mohr-Coulomb
$I_1 I_2 - \eta I_3 = 0$ where $\eta = \dfrac{9 - \sin^2\phi}{1 - \sin^2\phi}$	Matsuoka & Nakai (1974)
$I_1^3 - \eta I_3 = 0$ where $\eta = \dfrac{(3 - \sin\phi)^3}{(1 - \sin\phi)\cos^2\phi}$	Lade & Duncan (1975)

Note: stress invariants are $I_1 = \sigma_1 + \sigma_2 + \sigma_3$, $I_3 = \sigma_1\sigma_2\sigma_3$, $J_{2D} = \dfrac{1}{2}\sigma'_{ij}\sigma'_{ij}$, $J_{3D} = \dfrac{1}{3}\sigma'_{ij}\sigma'_{jk}\sigma'_{ki}$.

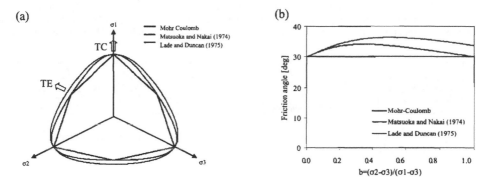

Figure 20.   Failure envelopes. (a) In the $\sigma 1$-$\sigma 2$-$\sigma 3$ space. (b) Friction angle anisotropy predicted for the different failure criteria—values computed using the definition in Equation 12 (Note: a friction angle of $\phi = 30°$ is assigned to triaxial compression $b = 0$).

Note that even normally consolidated clays exhibit a peak friction angle at high temperature, even though they are expected to be contractive at room temperature (Fig. 21—right pane).

6.5   *Large strain segregation*

Particle alignment at large strains ($\gamma > 100\%$) is followed by granular segregation as the strain increases even further ($\gamma \gg 100\%$). Segregation is determined by particle mobility, and leads to the spatial grouping of alike particles according to mass density, eccentricity, roughness, and size. In all cases, the segregated medium exhibits lower shear resistance than the homogenously mixed medium. For example, segregation by size results in two zones of low coordination number where coarse particles ride over a surface of low "effective roughness" made by the smaller particles (Fig. 22a); likewise, shape-based segregation creates a smooth surface of platy particles (Fig. 22b). Therefore, the post-segregation friction angle ϕ_{ps} is lower or equal to the residual and the constant volume friction angles: $\phi_{ps} \le \phi_{res} \le \phi_{cv}$.

6.6   *Strain rate*

Strain rate effects in friction reflect the time scales of internal processes involved in shear: friction changes with strain rate $\dot{\varepsilon}$ when the characteristic time for shear is shorter than the time required for the completion of internal processes. Let's estimate the characteristic time for shear t_d as the time required to cause a shear displacement equal to the particle diameter, thus $t_d = 1/\dot{\varepsilon}$.

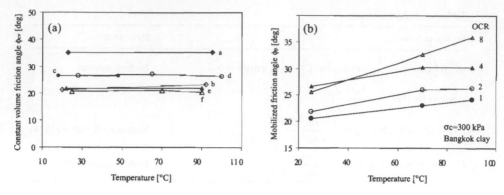

Figure 21.   Effect of temperature on friction in clayey soils. *Left pane*: Constant volume friction References: (a) Pontida clay from Hueckel & Pellegrini (1989) and Hueckel & Baldi (1990); (b) illite from Graham et al. (2001); (c) Tody clay from Burghignoli et al. (2000); (d) Pontida clay from Hueckel & Bakli (1990); (e) kaoline from Cekerevac & Laloui (2004); (f) Bangkok clay from Abuel-Naga et al. (2007). *Right pane*: Peak friction as a function of temperature and overconsolidation (OCR values shown for each line—Abuel-Naga et al. 2007).

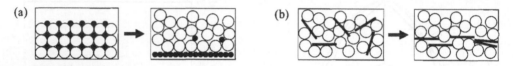

Figure 22.   Segregation at very large strains ($\gamma \gg 100\%$). Examples: (a) by size (see Savage & Lun 1988) and (b) by shape (see Lee et al. 2007).

Inertial effects arise at high strain rates and lead to collisions between grains. The characteristic time for inertial effects t_i can be estimated as the time required for a particle of mass m to travel a distance equal to its diameter d when accelerated due to the skeletal force $F = \sigma'd^2$ (viscous forces are disregarded in this analysis). The travel distance is $d = at_i^2/2$ and the acceleration $a = F/m = \sigma'd^2/m$; then $t_i = \sqrt{(2\,m/\sigma'd)}$. The ratio between the time scale for shear and inertial effects defines the dimensionless inertial ratio

$$I = \frac{t_i}{t_d} = \dot{\varepsilon}\sqrt{\frac{2\,m}{\sigma'd}} \sim \mathrm{eq}\dot{\varepsilon}d\sqrt{\frac{\rho}{\sigma'}} \tag{13}$$

where ρ is the mass density of the mineral that makes the grains. We can distinguish three rate-dependent shear regimes: the quasi-static regime $I \rightarrow 0$, the transition regime $10^{-2} < I < 0.2$, and the collisional regime $I > 0.2$ (GDR MiDi 2004; da Cruz et al. 2005).

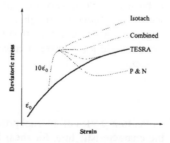

Figure 23.   Strain rate effects. A step increase in strain rate is assumed (continuous line: response at a constant strain rate $\dot{\varepsilon}_0$; dashed lines: responses when the strain rate is suddenly increased to $10\,\dot{\varepsilon}_0$) [Tatsuoka et al. 2008a].

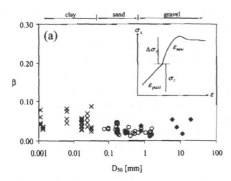
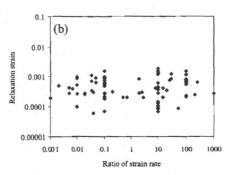

Figure 24.   Strain rate effects—General trends (a) amplitude of the effect in axial compression AC tests (data from Di Benedetto et al. 2005, Enomoto et al. 2009). (b) transition duration (Data compiled from Matsushita et al. 1999, Tatsuoka et al. 2002, Komoto et al. 2003, Li et al. 2003, Nawir et al. 2003, AnhDan et al. 2006, Kiyota & Tatsuoka 2006, Tatsuoka et al. 2008b, Kawabe et al. 2008).

6.6.1   *Quasi-static, low strain-rate regime (I → 0):*

The strain-rate dependent frictional response in sediments sheared at low strain rate has character-istics that are similar to those observed when two large surfaces are subjected to shear (Sections 2.2 and 2.3). Sketches in Figure 23 show the transients observed for a sudden increase in strain rate from $\dot{\varepsilon}_0$-to- $10\ \dot{\varepsilon}_0$. The macroscale sediment response can be categorized as follows (Vaid & Campanella 1977, Tatsuoka et al. 2002, Di Benedetto et al. 2005, Tatsuoka et al. 2008b):

- Isotach response: strength and stiffness increase with strain rate; this is observed in low plasticity clays, either NC or OC, and soft rocks (Šuklje 1957, Hayano et al. 2001, Leroueil 2006, Sorensen et al. 2007).
- Positive and Negative (P&N): the transient is a positive increase in shear strength, but the response eventually converges to a lower stress-strain trend; this response is observed in triaxial compression of poorly graded, stiff and relatively round particles and in drained direct shear tests on poorly-graded, stiff and relatively round particles (Duttine et al. 2008).

There are two other intermediate behaviors (Fig. 23):

- Combined: it runs between Isotach and the original, unaffected stress-strain trend.
- Viscous evanescence (also known as TESRA for transient effects of strain rate and strain accel-eration—Di Benedetto et al. 2002): the same stress-strain behavior is observed at all strain rates however a transient is measured when the strain rate changes (either positive or negative). This is a common response in sands; it is also observed in cemented clays at large strains (Sorensen et al. 2007).

The normalized magnitude of the relaxation amplitude is proportional to the change in strain rate; for axial compression AC loading (Fig. 24a),

$$\frac{\Delta\sigma_z}{\sigma_z} = \beta \, \log \frac{\dot{\varepsilon}_{\text{new}}}{\dot{\varepsilon}_{\text{past}}} \tag{14}$$

The β-factor is $\beta = 0.02$-to-0.05 in sands and gravels; it increases with particle size when particles are round, but the effect of particle size vanishes when angular particles are involved (Duttine et al. 2008). The factor ranges between $\beta = 0.02$-to-0.08 in fine-grained clayey sediments, where higher values correspond to higher degrees of saturation (Di Benedetto et al. 2005).

The duration of the transient, herein called "relaxation strain", does not seem to depend on the ratio between the previous and current strain rates (Fig. 24b). Furthermore, we found no correlation with stress level, particle size or shape in the compiled data. Apparently, the relaxation strain is between $\varepsilon = 10^{-4}$-to-10^{-3} (Note: as a reference, the strain at peak strength is typically between

$\varepsilon = 0.05$ and 0.2). The relatively minor effect of grain size on the relaxation strain suggests that shear takes place along planes rather than distributed throughout the soil mass (even in the absence of shear bands). In support of this hypothesis, consider a homogenously distributed relaxation strain of $\varepsilon = 10^{-4}$ in a $d = 100$ nm size clays: the resulting subatomic particle-to-particle displacement $\delta = \varepsilon d = 10^{-4} \times 10^{-7}$ m $= 0.1$Å contradicts relaxation distances observed in mineral-to-mineral shear ($\delta \sim> 10$ µm—discussion in Section 2.2), and threshold strain levels expected for these materials (Section 4.1 and references therein). In fact, individual particles move together in the form of "wedges of correlated displacement" that displace relative to other wedges along inter-wedge planes where the deformation localizes; eventually, the displacement becomes kinematically restricted, wedges break and new inter-wedge planes form (Drescher & De Josselin De Jong 1972). Domains made of fine particles form aggregations that can move in wedges similarly to coarse grains; this observation can facilitate the interpretation of global similarities between fine and coarse grained soil response.

Following the discussion in Section 2, *one is tempted to conclude* that transients reflect the combination between (1) the current sediment state that is a consequence of the previous strain rate, and (2) the new strain rate that has been suddenly imposed. In fact, the isotach and combined responses appear to be upscaled manifestations of hydrodynamic friction (Fig. 3), while both P&N and TESRA seem to reflect contact level transients observed in dry friction (Fig. 2). *However, other particle-scale mechanisms can contribute to the observed transients in strength.* In particular, we seek to identify processes with long time scales that can interfere with the long time $t_d = 1/\dot{\varepsilon}$ when shear takes place at low strain rate $\dot{\varepsilon}$. Processes that involve successive interactions between multiple grains have longer time scales than the inertial time t_i for a single particle (e.g., a single domino falls much faster than the complete chain):

- Fine-grained soils: contact creep; diffusion of local gradients (chemical, electrical, mechanical or thermal); fabric structuring-destructuring cycles; other particle-level processes involved in soil thixotropy (van Olphen 1977; Díaz-Rodríguez & Santamarina 1999).
- Coarse-grained soils: creep or slippage at contacts "diffuses" in a domino-type propagation throughout the granular skeleton; contact friction coupling with slippage-generated noise (details in Section 7.2).
- Coarse and fine: force redistribution and alteration of load-carrying chains (Kuhn & Mitchell 1993, Rothenburg 1993; Cascante & Santamarina 1996; Hartley & Behringer 2003).

These meso-scale processes affect regions that are much larger than the grain size, tend to follow an exponential evolution in time, and are slow enough to affect the shear response in the quasi-static strain rate regime.

6.6.2 *Transitional regime* $(10^{-2} < I < 0.2)$ *and high strain-rate collisional regime* $(0.2 < I)$

Equation 13 predicts that inertial effects and collisions will arise when shearing large particles at high strain rates in a medium subjected to low effective stress conditions.

There are two immediate consequences of interparticle collisions: (1) the volume dilates and (2) the friction coefficient *increases* due to the energy loss in shear and collision (i.e., restitution coefficient). Studies with discs in the transition regime $10^{-2} \leq I \leq 0.2$ show approximately linear trends for the *increase in porosity* $n = n_0 + aI$, and the *increase in friction coefficient* $\mu = \mu_o + bI$ (GDR MiDi 2004, da Cruz et al. 2005). Both dilation and strengthening saturate as the shear rate approaches the high-strain rate collisional regime $I > 0.2$. Even in this regime, the relationship between global friction angle and the internal anisotropy in the granular medium remains (Equation 3—Rothenburg & Bathurst 1989—discussion applicable to the high strain regime in da Cruz et al. 2005).

6.7 *Pore fluid*

The increase in salinity causes a reduction in double layer repulsion. When salinity exceeds ~ 0.1 molar, van der Waals attraction prevails and the soil fabric tends to adopt parallel aggregation (See Fig. 5b). Fabric effects prevail at low effective stress. As the effective stress increases, and/or as particles become aligned at large strains, the primary effect of salinity is to allow for thinner water films and less effective boundary lubrication. Therefore, an increase in friction with increasing ionic concentration should be expected in most cases.

Data on pore fluid chemical effects on the shear strength of clayey soils are inconclusive in part due to differences in specimen preparation; studies with single-mineral soils show that (Warkentin & Yong 1960, Kenney 1967, Mesri & Olson 1970, Olson 1974, Sridharan & Venkatappa Rao 1979, Moore 1991, Di Maio & Fenelli 1994, Wang & Siu 2006): (1) at a given effective confinement, friction increases when permittivity decreases, ionic concentration increases or the valance of the prevailing ion increases. (Note: the void ratio and fabric are not the same for soils with different fluids); and (2) flocculated clays exhibit higher shear strength than dispersed clays at the same void ratio. In general, a linear relation can be identified between soil strength and the number of bonds (Mitchell 1993, and references therein). Fluids that can alter the mineral surface or hinder hydration can also cause pronounced changes in shear strength (bentonite-water "B" has a $PI = 416$ and $\phi_r = 7°$ versus organo-bentonite "OB" $PI = 7$-to-14 and $\phi_r = 34°$-to-$37°$—Fig. 14; from Soule and Burns, 2001; see also data in Sridharan 2001).

6.8 *Internal interparticle forces of adhesion*

The "effective normal force" at interparticle contacts is a combination of skeletally transmitted forces and contact-level capillary and electrical forces (attraction and repulsion). Therefore, the acting normal contact force may differ from the value imposed by the applied effective stress when fine-grained soils ($d < 10\,\mu m$) or even coarse soils with a significant amount of fines (say %fines >10% by mass) are involved.

The consequences of interparticle forces of adhesion on the computed friction angle diminish—but do not cancel—when friction is defined as the slope of the strength envelope in the $\tau\sigma'$ space, $\tan\phi = \Delta\tau/\Delta\sigma'$, rather than from the origin. The residual effect interparticle forces of adhesion have on ϕ reflects differences in force-displacement among the various particle-level forces; for example, water menisci at interparticle contacts may break and not reform during shear (applies to the pendular regime at low water content).

7 BOUNDARY EFFECTS – MEASUREMENT

The measurement of friction is intimately related to boundary conditions. For example, the angle of repose cannot be measured for smooth spherical particles when a polished base is used because particles roll away; however, the same spherical particles will readily form a pile on a corrugated base (Fig. 25—data in Kalman et al. 1993, Matuttis et al. 2000, Chik & Vallejo 2005, Li et al. 2005).

This experiment highlights that the measurement of soil friction is inherently a boundary-value problem. In this section we explore boundary conditions and interfacial friction, the effects of vibration, the emergence of localization and close with a brief review of measurement difficulties. (Note: stress path/anisotropy was addressed in Section 6.3.)

(a)　　　　　　　　　　　　　　　　　(b)

Figure 25.　Boundary conditions and friction. Smooth spherical particles do not form a granular pile on a polished flat surface (a), but readily form the pile on a corrugated surface (b).

7.1   *Interfacial friction: sediments against surfaces*

Scales in intergranular friction related to grain size and shape identified in Figure 4 are relevant to interfacial friction between sediments and surfaces as well: the sediment slides along the interface when the surface is planar and polished, however, shear takes place within the soil mass when the surface is rough and undulations approach the scale of the particle size (Similar to Fig. 25). Specifically, experimental results show that the interfacial peak ϕ_p^* and constant volume ϕ_{cv}^* friction angles equal the corresponding internal friction angles of the soil mass ϕ_p and ϕ_{cv} when the ratio between the average surface roughness R_a and the mean particle size D_{50} exceeds $R_a/D_{50} \geq 0.01$ (Fig. 26a). Note that there is an implicit correlation between higher roughness amplitude and longer roughness wavelength. Similar results can be found in Subba Rao et al. 1998, Frost & DeJong 2005, Dietz & Lings 2006).

The relative grain-surface hardness affects the mobilized friction. When particles are harder than the surface, plowing adds energy loss to sliding and the angle of interfacial friction increases as the normal stress increases (Fig. 26b).

7.2   *Vibration—noise—cyclic loading*

Friction and vibration are interrelated: friction causes vibration, and vibration (or noise) affects friction. In fact, while stick-slip is a clear cause of vibration, vibration can be used to reduce stick-slip (Popp & Rudolph 2003). Salient implications on the vibration-friction coupling are identified next (references include the work by Fridman & Levesque 1959, Eaves et al. 1975, Budanov et al. 1980, Serdyuk & Mikityanskii 1986, Tworzydlo & Becker 1991, Skare & Stahl 1992, Adams 1996, Thomsen 1999, Bengisu & Akay 1999, Bucher & Wertheim 2001, Littmann et al. 2001).

Vibration either normal or parallel to the inclined sliding direction reduces the effective friction angle (Fig. 27). The higher the slope angle, the lower the acceleration required to trigger slippage. Furthermore, the acceleration required to cause slippage increases with frequency. The duration of the incursion into instability is very short at high frequencies and low vibration amplitudes. Results in Figure 27 show that the displacement in each cycle must exceed a minimum "displacement threshold" to cause block sliding. This length scale can range from the size of asperities in rough surfaces to the angstrom scale for atomically smooth surfaces. A single "threshold displacement" of 0.1 μm per cycle adequately satisfies all the data in Figure 27.

Nonlinear dynamic coupling effects and stochastic resonance may develop in frictional geomaterials. For example, if the granite slider that sits on the substrate is excited with a periodic driving signal in the presence of a background "noise" signal, the noise level required to cause slippage decreases as the amplitude of the periodic signal approaches the static limit; in fact, the peak output signal-to-noise ratio increases, inducing stochastic resonance.

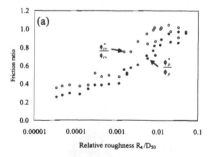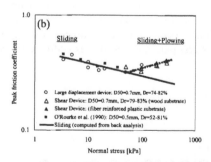

Figure 26.   Interfacial friction. (a) Normalized friction versus relative roughness R_a/D_{50} for sand-metal interaction (data: Subba Rao et al. 1998); relative roughness is defined in terms of the average surface roughness R_a (average absolute value of surface elevation) and the mean particle size D_{50}. (b) Plowing effects on peak interfacial friction coefficient (Dove & Frost 1999—Geomembrane-Ottawa sand).

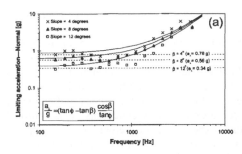 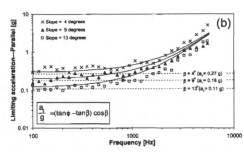

Figure 27.   Base block acceleration required to bring a frictional block to the verge of slippage for different slope angles and excitation frequencies. Sinusoidal vibration (a) *normal* to the sliding surface. (b) *parallel* to the dip vector. Dashed lines correspond to the limit equilibrium prediction—equations shown. Solid lines indicate the acceleration that generates a constant relative displacement of 0.1 μm. (details in Claria & Santamarina 2008).

We do not observe the classical signature of stochastic resonance when we test a granular sand specimen, nonlinear energy coupling effects readily appear: if the sand is simultaneously excited with two sinusoidal signals of different frequencies, the output response at the frequency of the primary driving signal increases as the amplitude of the secondary "noise" signal increases (Wang & Santamarina 2002). In fact, coupling increases as the driving signal brings the specimen to its nonlinear regime.

Repetitive cyclic loading has other manifestations in frictional granular media, including the evolution of volumetric and shear strains towards stable states (terminal density and shakedown), or the evolution towards continuous shear deformation or ratcheting (Johnson 1985, Alonso-Marroquin & Herrmann 2004, Werkmeister et al. 2004, Wichtmann et al. 2004, Narsilio & Santamarina 2008).

7.3   *Strain localization*

Positive feedback mechanisms lead to bifurcation when the returned energy exceeds losses. Strain localization along shear bands is a bifurcation-type response during the shear of granular media (Rudnicki & Rice 1975, Vardoulakis 1979). The development of localization depends on inherent material characteristics, drainage during shear, stress and boundary conditions.

The following sediment characteristics and drainage conditions favor localization: dilative sediments during drained shear, dilative sediments during undrained shear (if the pore fluid cavitates), contractive sediments during undrained shear, sediments that reach very low residual strength (e.g., when platy particles exceed ~15%), initially cemented sediments (cement breakage), unsaturated soils under high suction (menisci failure), non-homogeneous specimen subjected to either drained and undrained shear, grain crushing or void collapse (experimental evidence in Cho & Santamarina 2003).

In the absence of perfectly homogeneous initial conditions (i.e., all real specimens), strain localization results in specimen-size dependent stress-strain-volume response due to progressive failure: the measured global response will reflect a lower peak strength and lower dilation than the local material response. This volume-averaging effect is exacerbated in soft yet brittle materials and large specimens.

The orientation of shear bands is determined by the peak strength friction angle. Therefore, the analysis of the constant volume and the residual friction angles must take into consideration the orientation of the shear plane that formed at peak.

7.4   *Measurement*

It follows from the previous discussions that the measurement of friction is a boundary value problem that gains additional complexity due to inherent and stress-induced anisotropy, interfacial friction against cell walls, drainage conditions (local and global), particle-level forces (that are not determined at the boundaries), localization, spatial variability within the sediment, strain-rate dependency and vibrations.

Great effort has been devoted to developing experimental methods that attempt to create pure boundary conditions (e.g., lubricated polished platens) so that the specimen can be considered as "a point" (lubricated boundaries are discussed in Lade & Duncan 1973, Chu et al. 1996, Frost & Yang 2003). Inevitably, all these efforts have been hindered by inherent physical limitations.

More recent developments explore the intensive use of information technology and numerical modeling to perform information-rich tests with complex boundary conditions. This approach explicitly recognizes that each point in the specimen experiences a different stress history: the specimen is discretized into finite elements, boundary conditions are properly represented, and the constitutive parameters of a robust constitutive model are identified by fitting simulation results to the extensive dataset of measurements gathered around the whole specimen (Bayoumi 2006).

8   CONCLUSIONS

The friction angle is not an inherent property of a soil mass. It depends on grain characteristics (topography, hardness, strength), sediment properties (grain size distribution, packing density, inherent anisotropy), effective confining stress, stress path (stress-induced anisotropy), strain rate, drainage (local and global), temperature and vibration (reduces frictional resistance).

Particle size is the *inherent length scale* in granular materials. All other particle-level characteristics such as eccentricity, angularity, roughness and double layer thickness must be related to particle size to determine their effect on the macroscale friction angle of a soil. In addition, particle size determines the role that particle-level electrical and capillary forces may exert relative to the boundary-determined skeletal forces.

The region between 10-and-50 μm defines a drastic transition in shape, grain formation, mineralogy, governing forces, fabric formation and the role of adsorbed layers relative to surface roughness. Friction in fine-grained media exhibits similar macroscale characteristics to those observed in coarse-grained sediments, even though particle interaction is dominated by electrical effects in fine-grained media.

Upscaling from grain-to-grain interactions to the whole mass takes place through various competing mechanisms that include contact slippage, grain rotation and rotational frustration, dilation against confinement, grain crushing, formation-and-buckling of granular columns. From a particle level perspective, friction is a measure of the medium ability to sustain anisotropy.

The macroscale frictional behavior of the soil mass is a Coulomb-type response, but the Coulomb model must be modified to accommodate stress-dependent dilatancy, as well as inherent and stress-induced anisotropy.

Strain rate effects on friction reflect the *time scales* of internal processes involved in the shear of granular materials: friction changes with strain rate when the characteristic time for shear is shorter than the time required for the completion of internal processes. The relevant time scales during quasi-static shear are related to contact-level mechanics (friction between dry minerals or hydrodynamic effects) and the diffusion of either contact or pore scale processes through the granular medium; available data suggest that strain rate effects are dominated by localized sliding between particle wedges rather than by distributed sliding throughout the soil mass. The high-strain rate regime takes place when large particles subjected to low confinement are sheared at high strain rates; in this case, inertial effects arise leading to interparticle collisions and the increase in friction.

A minimum strain is required to cause contact-level slippage; the threshold strain depends on grain size and effective confinement. As the strain level increases, a sediment reaches the peak friction angle (in dilative soils with respect to critical state), the constant volume friction angle (determined by angularity and roughness), the residual friction angle (if eccentric particles exceed ~10%), and the post-segregation friction angle.

Dilatancy depends on the current packing density and state of stress in relation to the critical state line. Cementation increases the dilative tendency of a soil mass.

The measurement of soil friction is a boundary value problem and is inherently hindered by incomplete knowledge of boundary conditions. Information-intensive measurement methods may help overcome current limitations. Progressive failure leading to strain localization affects the interpretation of boundary measurements.

ACKNOWLEDGMENTS

Support was provided by the National Science Foundation, the Georgia Mining Industry, and The Goizueta Foundation. We are grateful to reviewers and colleagues for insightful comments and suggestions.

REFERENCES

Abuel-Naga, H.M., Bergado, D.T. & Lim, B.F. 2007. Effect of temperature on shear strength and yielding behavior of soft Bangkok clay. *Soils and Foundations.* 47(3): 423–436.

Adams, G.G. 1996. Self-excited oscillations in sliding with a constant friction coefficient—A simple model. *Transactions of the ASME. Journal of Tribology.* 118(4): 819–23.

Alonso-Marroquin, F. & Herrmann, H.J. 2004. Ratcheting of granular materials. *Physical Review Letters.* 92(5): 054301–1.

AnhDan, L., Tatsuoka, F. & Koseki, J. 2006. Viscous effects on the stress-strain behavior of gravelly soil in drained triaxial compression. *Geotechnical Testing Journal.* 29(4): 330–340.

Archard, J.F. 1957. Elastic deformation and the laws of friction. *Proceedings of the Royal Society of London, Series A (Mathematical and Physical Sciences).* 243: 190–205.

Bayoumi, A.M. 2006. New laboratory test procedure for the enhanced calibration of constitutive mode, Georgia Institute of Technology.

Been, K. & Jefferies, M.G. 1985. State parameter for sands. *Geotechnique.* 35(2): 99–112.

Bengisu, M.T. & Akay, A. 1999. Stick-slip oscillations: Dynamics of friction and surface roughness. *Journal of the Acoustical Society of America.* 105(1): 194–205.

Bolton, M.D. 1986. Strength and dilatancy of sands. *Geotechnique.* 36(1): 65–78.

Bowden, F.P. & Tabor, D. 1950. *The friction and lubrication of solids.* New York: Oxford University Press.

Bucher, I. & Wertheim, O. 2001. Reducing friction forces by means of applied vibration. *Proceedings of the ASME Design Engineering Technical Conference* 6 C. American Society of Mechanical Engineers, Pittsburgh, PA, United states, 2995–3000.

Budanov, V., Kudinov, V.A. & Tolstoi, D.M. 1980. Interaction of friction and vibration. *Sov. J. Frict. Wear.* 1: 79–89.

Burghignoli, A., Desideri, A. & Miliziano, S. 2000. A laboratory study on the thermomechanical behaviour of clayey soils. *Canadian Geotechnical Journal.* 37(4): 764–780.

Camp, W.M., Mayne, P.W. & Brown, D.A. 2002. Drilled Shaft Axial Design Values: Predicted Versus Measured Response in a Calcareous Clay. In: *ASCE's 2002 Deep Foundation Congress.*

Campanella, R.G. & Mitchell, J.K. 1968. Influence of temperature variations on soil behavior. *American Society of Civil Engineers Proceedings, Journal of the Soil Mechanics and Foundations Division.* 94(3): 709–734.

Cascante, G. & Santamarina, J.C. 1996. Interparticle contact behavior and wave propagation. *Journal of Geotechnical Engineering.* 122(10): 831–839.

Castle, J., Farid, A. & Woodcock, L.V. 1996. *The effect of surface friction on the rheology of hard-sphere colloids.* Berlin: Springer.

Cekerevac, C. & Laloui, L. 2004. Experimental study of thermal effects on the mechanical behaviour of a clay. *International Journal for Numerical and Analytical Methods in Geomechanics.* 28(3): 209–228.

Chantawarangul, K. 1993. Numerical Simulation of Three-Dimensional Granular Assemblies, University of Waterloo.

Cheng, Y.P., Bolton, M.D. & Nakata, Y. 2004. Crushing and plastic deformation of soils simulated using DEM. *Geotechnique.* 54(2): 131–141.

Chik, Z. & Vallejo, L. 2005. Characterization of the angle of repose of binary granular materials. *Canadian Geotechnical Journal.* 42(2): 683–692.

Cho, G.C., Dodds, J. & Santamarina, J.C. 2006. Particle shape effects on packing density, stiffness, and strength: Natural and crushed sands. *Journal of Geotechnical and Geoenvironmental Engineering.* 132(5): 591–602.

Cho, G.C. & Santamarina, J.C. 2003. The Omnipresence of Localization in Geomaterials. In: *Proc. 3rd Int. Sym. on Deformation Characteristics of Geomaterials, IS Lyon 03*, Lyon, 465–473.

Chu, J., Lo, S.C.R. & Lee, I.K. 1996. Strain softening and shear band formation of sand in multi-axial testing. *Geotechnique.* 46(1): 63–82.

Claria, J.J. & Santamarina, J.C. 2008. Interfacial Friction and Vibration. In: *Proc. Symposium on Characterization and Behavior of Interfaces* (edited by Frost, J.D.). Atlanta, USA: Balkema.

Coop, M.R. & Atkinson, J.H. 1993. The mechanics of cemented carbonate sands. *Geotechnique.* 43(1): 53–67.

Cubrinovski, M. & Ishihara, K. 2002. Maximum and minimum void ratio characteristics of sands. *Soils and Foundations.* 42 (6): 65–78.

da Cruz, F., Emam, S., Prochnow, M., Roux, J.-N. & Chevoir, F. 2005. Rheophysics of dense granular materials: Discrete simulation of plane shear flows. *Physical Review E—Statistical, Nonlinear, and Soft Matter Physics.* 72(2): 1–17.

Dantu, P. 1968. Statistical study of intergranular forces in powdered medium. *Geotechnique.* 18(1): 50–55.

Di Benedetto, H., Tatsuoka, F. & Ishihara, M. 2002. Time-dependent shear deformation characteristics of sand and their constitutive modelling. *Soils and Foundations.* 42(2): 1–22.

Di Benedetto, H., Tatsuoka, F., Lo Presti, D., Sauzéat, C. & Geoffroy, H. 2005. Time effects on the behaviour of geomaterials. In: *Deformation Characteristics of Geomaterials, Recent Investigations and Prospects* (edited by Di Benedetto, H., Doanh, T., Geoffroy, H. & Sauzéat, C.). London: Taylor & Francis Group.

Di Maio, C. & Fenelli, G.B. 1994. Residual strength of kaolin and bentonite: The influence of their constituent pore fluid. *Geotechnique.* 44(2): 217–226.

Diaz-Rodriguez, J.A., Leroueil, S. & Aleman, J.D. 1992. Yielding of Mexico City clay and other natural clays. *Journal of geotechnical engineering.* 118(7): 981–995.

Diaz-Rodriguez, J.A., Lozano-Santa Cruz, R., Davila-Alcocer, V.M., Vallejo, E. & Giron, P. 1998. Physical, chemical, and mineralogical properties of Mexico City sediments: A geotechnical perspective. *Canadian Geotechnical Journal.* 35(4): 600–610.

Díaz-Rodríguez, J.A. & Santamarina, J.C. 1999. Thixotropy: The Case of Mexico City Soils. In: *XI Panamerican Conf. on Soil Mech. and Geotech. Eng.* 1, Iguazu Falls, Brazil, 441–448.

Dieterich, J.H. 1979. Modeling of rock friction. I. Experimental results and constitutive equations. *J. Geophys. Res.* 84: 2161–8.

Dieterich, J.H. & Kilgore, B.D. 1994. Direct observation of frictional contacts: New insights for sliding memory effects. *Pure and Applied Geophysics.* 143(1–3): 283–302.

Dietz, M.S. & Lings, M.L. 2006. Postpeak strength of interfaces in a stress-dilatancy framework. *Journal of Geotechnical and Geoenvironmental Engineering.* 132(11): 1474–1484.

Dobry, R., Ladd, R.S., Yokel, F.Y., Chung, R.M. & Powell, D. 1982. Prediction of pore water pressure buildup and liquefaction of sands during earthquake by the cyclic strain method. In: *Building Science Series 138.* US Department of Commerce, National Bureau of Standards, 150.

Dove, J.E. & Frost, J.D. 1999. Peak friction behavior of smooth geomembrane-particle interfaces. *Journal of Geotechnical and Geoenvironmental Engineering.* 125(7): 544–555.

Drescher, A. & De Josselin De Jong, G. 1972. Photoelastic verification of a mechanical model for the flow of a granular material. *Journal of the Mechanics and Physics of Solids.* 20(5): 337–51.

Dusseault, M.B. & Morgenstern, N.R. 1979. Locked Sands. *Engineering Geology.* 12: 117–131.

Duttine, A., Tatsuoka, F., Kongkitkul, W. & Hirakawa, D. 2008. Viscous behaviour of unbound granular materials in direct shear. *Soils and Foundations.* 48(3): 297–318.

Eaves, A.E., Smith, A.W., Waterhouse, W.J. & Sansome, D.H. 1975. Review of the application of ultrasonic vibrations to deforming metals. *Ultrasonics.* 13(4): 162–70.

Enomoto, T., Kawabe, S., Tatsuoka, F., Di Benedetto, H., Hayashi, T. & Duttine, A. 2009. Effects of particle characteristics on the viscous properties of granular materials in shear. *Soils and Foundations.* 49(1): 25–49.

Fraser, H.J. 1935. Experimental study of the porosity and permeability of clastic sediments. *Journal of Geology.* 13(8–1): 910–1010.

Fridman, H.D. & Levesque, P. 1959. Reduction of static friction by sonic vibrations. *Journal of Applied Physics.* 30(10): 1572–1575.

Frost, J.D. & DeJong, J.T. 2005. In situ assessment of role surface roughness on interface response. *Journal of Geotechnical and Geoenvironmental Engineering.* 131(4): 498–511.

Frost, J.D. & Yang, C.-T. 2003. Effect of end platens on microstructure evolution in dilatant specimens. *Soils and Foundations.* 43(4): 1–11.

GDR MiDi. 2004. On dense granular flows. *European Physical Journal E.* 14(4): 341–365.

Gherbi, M., Gourves, R. & Oudjehane, F. 1993. Distribution of the contact forces inside a granular material. In: *Powders and Grains 93, Prod. 2nd International Conference on Micromechanics of Granular Media.* Elsevier, Brimingham, 167–171.

Graham, J., Tanaka, N., Crilly, T. & Alfaro, M. 2001. Modified Cam-Clay modelling of temperature effects in clays. *Canadian Geotechnical Journal.* 38(3): 608–621.

Greenwood, J.A. & Williamson, J.B.P. 1966. Contact of nominally flat surfaces *Proceedings of the Royal Society of London. Series A, Mathematical and Physical Sciences.* 295(1442): 300–319.

Guo, P. 2008. Modified direct shear test for anisotropic strength of sand. *Journal of Geotechnical and Geoenvironmental Engineering.* 134(9): 1311–1318.

Guo, P. & Su, X. 2007. Shear strength, interparticle locking, and dilatancy of granular materials. *Canadian Geotechnical Journal.* 44(5): 579–591.

Hartley, R.R. & Behringer, R.P. 2003. Logarithmic rate dependence of force networks in sheared granular materials. *Nature.* 421(6926): 928–31.

Hayano, K., Matsumoto, M., Tatsuoka, F. & Koseki, J. 2001. Evaluation of time-dependent deformation properties of sedimentary soft rock and their constitutive modeling. *Soils and Foundations.* 41(2): 21–38.

Hoek, E. & Brown, E.T. 1980. *Underground Excavations in Rock.* Institution of Mining and Metallurgy, London.

Horne, M.R. 1969. The behaviour of an assembly of rotund, rigid, cohesionless particles. III. *Proceedings of the Royal Society of London, Series A (Mathematical and Physical Sciences).* 310(1500): 21–34.

Hueckel, T. & Baldi, G. 1990. Thermoplasticity of saturated clays—Experimental constitutive study. *Journal of Geotechnical Engineering.* 116(12): 1778–1796.

Hueckel, T. & Pellegrini, R. 1989. Modeling of thermal failure of saturated clays. In: *Numerical Models in Geomechanics* (edited by Pietruszczak, S. & Pande, G.N.). New York: Elsevier. 81–90.

Israelachvili, J.N., McGuiggan, P.M. & Homola, A.M. 1988. Dynamic properties of molecularly thin liquid films. *Science.* 240(4849): 189–91.

Jaeger, H.M., Nagel, S.R. & Behringer, R.P. 1996. The physics of granular materials. *Physics Today.* 49(4): 32–8.

Jamiolkowski, M., Ladd, C.C., Germaine, J.T. & Lancellotta, R. 1985. New Developments in Field and Laboratory Testing of Soils. In: *Proc. 11th ICSMFE* 1, San Francisco, 67–153.

Johnson, K.L. 1961. Energy dissipation at spherical surfaces in contact transmitting oscillating forces. *Journal of Mechanical Engineering Science.* 3(4): 362–368.

Johnson, K.L. 1985. *Contact mechanics.* Cambridge: Cambridge University Press.

Kalman, H., Goder, D., Rivkin, M. & Ben-Dor, G. 1993. Effect of the particle-surface friction coefficient on the angle of repose. *Bulk Solids Handling.* 13(1): 123–128.

Kawabe, S., Maeda, Y., Tobisu, Y., Hara, D., Tatsuoka, F. & Ohta, A. 2008. Viscous properties of round granular materials in drained triaxial compression. In: *Fourth International Symposium on Deformation Characteristics of Geomaterials* (edited by Burns, S.E., Mayne, P.W. & Santamarina, J.C.) 1. ISO press, Atlanta, USA, 503–510.

Kenney, T.C. 1967. The influence of meneralogical composition on the residual strength of natrual soils. In: *Proceedings of the geotechnical conference on shear strength properties of natural soils and rocks.* 1: 123–129.

Kiyota, T. & Tatsuoka, F. 2006. Viscous property of loose sand in triaxial compression, extension and cyclic loading. *Soils and Foundations.* 46(5): 665–684.

Koerner, R.M. 1970. Limiting density behavior of quartz powders. *Powder Technology.* 3(4): 208–12.

Kogut, L. & Etsion, I. 2004. A static friction model for elastic-plastic contacting rough surfaces. *Journal of Tribology.* 126(1): 34–40.

Komoto, N., Tatsuoka, F. & Nishi, T. 2003. Viscous stress-strain properties of undisturbed Pleistocene clay and its constitutive modelling. In: *Proc. 3rd Int. Sym. on Deformation Characteristics of Geomaterials, IS Lyon 03.* Balkema, 579–587.

Kragelsky, I.V. & Alisin, V.V. 2001. *Tribology—lubrication, friction, and wear.* London: Professional Engineering Publishing Ltd.

Kruyt, N.P. & Rothenburg, L. 2006. Shear strength, dilatancy, energy and dissipation in quasi-static deformation of granular materials. *Journal of Statistical Mechanics: Theory and Experiment.* 7: 1–13.

Kuhn, M.R. & Mitchell, J.K. 1993. New oerspectives on soil creep. *ASCE J. Geotechnical Engineering.* 119(3): 507–524.

Kurukulasuriya, L.C., Oda, M. & Kazama, H. 1999. Anisotropy of Undrained Shear Strength of an Over-Consolidated Soil by Triaxial and Plane Strain Tests. *Soils and Foundations.* 39(1): 21–29.

Ladd, C., Foott, R., Ishihara, K., Schlosser, F. & Poulos, HG. 1977. Stress-deformation and strength characteristics: State-of-the-art report. In: *Proceedings of 9th International Conference on Soil Mechanics, Foundation Engineering* 2: 421–494.

Lade, P.V. 2006. Assessment of test data for selection of 3-D failure criterion for sand. *International Journal for Numerical and Analytical Methods in Geomechanics.* 30(4): 307–333.

Lade, P.V. & Duncan, J.M. 1973. Cubical triaxial tests on cohesionless soil. *American Society of Civil Engineers, Journal of the Soil Mechanics and Foundations Division.* 99(10): 793–781.

Lade, P.V. & Duncan, J.M. 1975. Elastoplastic stress-strain theory for cohesionless soil. *Journal of the Geotechnical Engineering Division.* 101(10): 1037–1053.

Lade, P.V. & Kirkgard, M.M. 2000. Effects of Stress Rotation and Changes of b-Values on Cross-Anisotropic Behavior of Natural, ko-Consolidated Soft Clay. *Soils and Foundations.* 40(6): 93–105.

Lee, J.-S., Guimaraes, M. & Santamarina, J.C. 2007. Micaceous sands: Microscale mechanisms and macroscale response. *Journal of Geotechnical and Geoenvironmental Engineering.* 133(9): 1136–1143.

Leroueil, S. 2006. The isotache approach. Where are we 50 years after its development by Professor Suklje? In: *Proc. 13th Danube Eur. Conf. on Geotechnical Engineering* 1, Ljubljana, 55–58.

Li, J.Z., Tatsuoka, F., Nishi, T. & Komoto, N. 2003. Viscous stress-strain behaviour of clay under unloaded conditions. In: *Proc. 3rd Int. Sym. on Deformation Characteristics of Geomaterials, IS Lyon 03.* Balkema, 617–625.

Li, Y., Xu, Y. & Thornton, C. 2005. A comparison of discrete element simulations and experiments for 'sandpiles' composed of spherical particles. *Powder Technology.* 160(3): 219–228.

Lin, H. & Penumadu, D. 2005. Experimental investigation on principal stress rotation in Kaolin clay. *Journal of Geotechnical and Geoenvironmental Engineering.* 131(5): 633–642.

Littmann, W., Storck, H. & Wallaschek, J. 2001. Sliding friction in the presence of ultrasonic oscillations: Superposition of longitudinal oscillations. *Archive of Applied Mechanics.* 71(8): 549–554.

Lupini, J.F., Skinner, A.E. & Vaughan, P.R. 1981. Drained residual strength of cohesive soils. *Geotechnique.* 31 (2): 181–213.

Matsui, T., Ito, T. & Mitchell, J.K. 1980. Microscopic study of mechanisms in soils. *American Society of Civil Engineers, Journal of the Geotechnical Engineering Division.* 106(2): 137–152.

Matsuoka, H. & Nakai, T. 1974. Stress-deformation and strength characteristics under three different principal stresses. In: *Proc., Jpn. Soc. Civ. Eng.* 232: 59–70.

Matsushita, M., Tatsuoka, F., Koseki, J., Cazacliu, B., Di Benedetto, H. & Yasin, S.J.M. 1999. Time effects on the pre-peak deformation properties of sands. In: *Proc. Second Int. Conf. on Pre-Failure Deformation Characteristics of Geomaterials, IS Torino '99* (edited by al., J. e.) 1. Balkema, 681–689.

Matuttis, H.G., Luding, S. & Herrmann, H.J. 2000. Discrete element simulations of dense packings and heaps made of spherical and non-spherical particles. *Powder Technology.* 109 (1–3): 278–292.

Mayne, P.W. & Holtz, R.D. 1985. Effect of Principal Stress Rotation on Clay Strength. In: *Proceedings, 11th International Conference on Soil Mechanics and Foundation Engineering* 2, San Francisco, 579–582.

Mesri, G. & Olson, R.E. 1970. Shear strength of montmorillonite. *Geotechnique.* 20(3): 261–270.

Mitchell, J.K. 1993. *Fundamentals of Soil Behavior.* New York: John Wiley & Sons, Inc.

Miura, K., K., M., Furukawa, M. & Toki, S. 1998. Mechanical characteristics of sands with different primary properties. *Soils and Foundations.* 38: 159–172.

Moore, D.E. & Lockner, D.A. 2004. Crystallographic controls on the frictional behavior of dry and water-saturated sheet structure minerals. *Journal of Geophysical Research.* 109(B3): 16.

Moore, D.E. & Lockner, D.A. 2007. Comparative Deformation Behavior of Minerals in Serpentinized Ultramafic Rock: Application to the Slab-Mantle Interface in Subduction Zones. *International Geology Review.* 49(5): 401–415.

Moore, R. 1991. Chemical and mineralogical controls upon the residual strength of pure and natural clays. *Geotechnique.* 41(1): 35–47.

Moya, J. & Rodriguez, J. 1987. El subsuelo de Bogota y los problemas de cimentaciones. In: *Proc. 8th Panamerican Conf. On Soil Mech. and Found. Engrg.*, Universidad Nacional de Colombia, 197–264.

Narsilio, G.A. & Santamarina, J.C. 2008. Terminal densities. *Geotechnique.* 58(8): 669–674.

Nawir, H., Tatsuoka, F. & Kuwano, R. 2003. Experimental evaluation of the viscous properties of sand in shear. *Soils and Foundations.* 43(6): 13–31.

Oda, M. 1972. The mechanism of fabric changes during compressional deformation of sand. *Soils and Foundations.* 12(2): 1–18.

Oda, M., Nemat-Nasser, S. & Konishi, J. 1985. Stress-induced anisotropy in granular masses. *Soils and Foundations.* 25 (3): 85–97.

Ohtsubo, M., Egashira, K. & Kashima, K. 1995. Depositional and post-depositional geochemistry, and its correlation with the geotechnical properties of marine clays in Ariake Bay, Japan. *Geotechnique.* 45(3): 509–509.

Okawara, M. & Mitachi, T. 2003. Basic research on mechanism of the residual strength of clay. In: *The Third International Symposium on Deformation Characteristics of Geomaterials*, (edited by Benedetto, H.D.), IS-Lyon 2003.

Olson, R.E. 1974. Shearing strength of kaolinite, illite, and montmorillonite. *Journal of soil mechanics and foundations division, ASCE.* 100(11): 1215–1229.

Palomino, A.M. & Santamarina, J.C. 2005. Fabric map for kaolinite: Effects of pH and ionic concentration on behavior. *Clays and Clay Minerals.* 53(3): 211–223.

Persson, B.N.J. 1998. *Sliding friction: physical principles and applications.* Berlin: Springer.

Popp, K. & Rudolph, M. 2003. Prevention of stick-slip vibrations by passive normal force control. *PAMM.* 2(1): 68–69.

Rabinowicz, E. 1995. *Friction and wear of materials.* New York: Wiley.

Radjai, F., Wolf, D.E., Jean, M. & Moreau, J.-J. 1998. Bimodal character of stress transmission in granular packings. *Physical Review Letters.* 80(1): 61.

Rothenburg, L. 1993. Effects of particle shape and creep in contacts on micromechanical behavior of simulated sintered granular media. *Mechanics of Granular Materials and Powder Systems.* 37: 133–142.

Rothenburg, L. & Bathurst, R.J. 1989. Analytical study of induced anisotropy in idealized granular materials. *Geotechnique.* 39(4): 601–614.

Rothenburg, L. & Bathurst, R.J. 1993. Influence of particle eccentricity on micromechanical behavior of granular materials. *Mechanics of Materials.* 16(1–2): 141–152.

Rothenburg, L. & Kruyt, N.P. 2004. Critical state and evolution of coordination number in simulated granular materials. *International Journal of Solids and Structures.* 41(21): 5763–74.

Rowe, P.W. 1969. The relation between the shear strength of sands in triaxial compression, plane strain and direct shear. *Gotechnique.* 19(1): 75–86.

Rudnicki, J.W. & Rice, J.R. 1975. Conditions for the localization of deformation in pressure-sensitive dilatant materials. *Journal of the Mechanics and Physics of Solids.* 23(6): 371–394.

Ruina, A. 1983. Slip instability and state variable friction laws. *Journal of Geophysical Research.* 88(B12): 10359–10370.

Santamarina, C. & Cascante, G. 1998. Effect of surface roughness on wave propagation parameters. *Geotechnique.* 48(1): 129–136.

Santamarina, J.C. 2002. Soil Behavior at the Microscale: Particle Forces, Soil Behavior and Soft Ground Construction—The Ladd Symposium. *ASCE Special Publications* MIT, Boston, 25–56.

Savage, S.B. & Lun, C.K.K. 1988. Particle size segregation in inclined chute flow of dry cohesionless granular solids. *Journal of Fluid Mechanics.* 189: 311–25.

Serdyuk, L.M. & Mikityanskii, V.V. 1986. Reliability of clamping devices of machine tool fixtures with machining system vibrations. *Soviet Journal of Friction and Wear.* 7(2): 58–66.

Shimobe, S. & Moroto, N. 1995. A new classification chart for sand liquefaction. In: *Earthquake Geotechnical Engineering* (edited by Ishihara, K.). Balkema, 315–320.

Skare, T. & Stahl, J.E. 1992. Static and dynamic friction processes under the influence of external vibrations. *Wear.* 154(1): 177–92.

Skempton, A.W. 1964. Long-term stability of clay slopes. *Geotechnique.* 14(2): 77–101.

Skempton, A.W. 1985. Residual strength of clays in landslides, folded strata and the laboratory. *Geotechnique.* 35(1): 3–18.

Skinner, A.E. 1969. A note on the influence of interparticle friction on the shearing strength of a random assembly of spherical particles. *Geotechnique.* 1: 150–177.

Sleep, N.H. 1997. Application of a unified rate and state friction theory to the mechanics of fault zones with strain localization. *Journal of Geophysical Research.* 102(B2): 2875–95.

Sorensen, K.K., Baudet, B.A. & Simpson, B. 2007. Influence of structure on the time-dependent behaviour of a stiff sedimentary clay. *Geotechnique.* 57(9): 783–7.

Sridharan, A. 2001. Engineering Behaviour of Clays. In: *Influence of Mineralogy, in Chemo-Mechanical Coupling in Clays* (edited by Di Maio, C., Hueckel, T. & Loret, B.), Maratea, Italy, 3–28.

Sridharan, A. & Venkatappa Rao, G. 1979. Shear strength behaviour of saturated clays and the role of the effective stress concept. *Geotechnique.* 29(2): 177–193.

Sture, S., Costes, N.C., Batiste, S.N., Lankton, M.R., AlShibli, K.A., Jeremic, B., Swanson, R.A. & Frank, M. 1998. Mechanics of granular materials at low effective stresses. *Journal of Aerospace Engineering.* 11(3): 67–72.

Subba Rao, K.S., Allam, M.M. & Robinson, R.G. 1998. Interfacial friction between sands and solid surfaces. *Proceedings of the Institution of Civil Engineers, Geotechnical Engineering.* 131 (2): 75–82.

Suiker, A.S.J. & Fleck, N.A. 2004. Frictional collapse of granular assemblies. *Journal of Applied Mechanics, Transactions ASME.* 71(3): 350–358.

Šuklje, L. 1957. The analysis of the consolidation process by the isotache method. In: *Proc. 4th Int. Conf. on Soil Mech. and Found. Engng.* 1, London, 200–206.

Sukumaran, B. & Ashmawy, A.K. 2001. Quantitative characterisation of the geometry of discrete particles. *Geotechnique.* 51(7): 619–627.

Tabor, D. 1992. Friction as a dissipative process. In: *Fundamentals of Friction: Macroscopic and Microscopic Processes* (edited by Singer, I.L. & M., P.H.). Academic Publishers.

Tanaka, H., Locat, J., Shibuya, S., Thiam Soon, T. & Shiwakoti, D.R. 2001. Characterization of Singapore, Bangkok, and Ariake clays. *Canadian Geotechnical Journal.* 38(2): 378–400.

Tatsuoka, F., Di Benedetto, H., Enomoto, T., Kawabe, S. & Kongkitkul, W. 2008a. Various viscosity types of geomaterials in shear and their mathematical expression. *Soils and Foundations.* 48(1): 41–60.

Tatsuoka, F., Di Benedetto, H., Kongkitkul, W., Kongsukprasert, L., Nishi, T. & Sano, Y. 2008b. Modelling of ageing effects on the elasto-viscoplastic behaviour of geomaterial. *Soils and Foundations.* 48(2): 155–174.

Tatsuoka, F., Ishihara, M., Di Benedetto, H. & Kuwano, R. 2002. Time-dependent shear deformation characteristics of geomaterials and their simulation. *Soils and Foundations.* 42(2): 103–129.

Tatsuoka, F., Ochi, K., Fujii, S. & Okamoto, M. 1986. Cyclic undrained biaxial and torsional shear strength of sands for different sample preparation methods. *Soils and Foundations.* 26(3): 23–41.

Taylor, D.W. 1948. *Fundamentals of soil mechanics.* New York: John Wiley & Sons.

Taylor, P.M. & Pollet, D.M. 2000. A novel technique to measure stick-slip in fabric. *International Journal of Clothing Science and Technology.* 12(2): 124–133.

Terzaghi, K., Peck, R.B. & Mesri, G. 1996. *Soil mechanics in engineering practices.* New York: Wiley.

Thomsen, J.J. 1999. Using fast vibrations to quench friction-induced oscillations. *Journal of Sound and Vibration.* 228(5): 1079–1102.

Thornton, C. 2000. Numerical simulations of deviatoric shear deformation of granular media. *Geotechnique.* 50(1): 43–53.

Tworzydlo, W.W. & Becker, E. 1991. Influence of forced vibrations on the static coefficient of friction. Numerical modeling. *Wear.* 143(1): 175–196.

Vaid, Y.P. & Campanella, G. 1977. Time-dependent behaviour of undisturbed clay. *ASCE Journal of the Geotechnical Engineering Division.* 103(7): 693–709.

Vaid, Y.P. & Sayao, A. 1995. Proportional Loading Behavior in Sand Under Multiaxial Stresses. *Soils and Foundations.* 35 (3): 23–29.

van Olphen, H. 1977. *An introduction to clay colloid chemistry: for clay technologists, geologists, and soil scientists.* New York: Wiley.

Vardoulakis, I. 1979. Bifurcation analysis of the triaxial test on sand samples. *Acta Mechanica.* 32(1–3): 35–54.

Vesic, A.S. & Clough, G.W. 1968. Behavior of granular materials under high stresses. *American Society of Civil Engineers Proceedings, Journal of the Soil Mechanics and Foundations Division.* 94(SM3): 661–688.

Vijaywargiya, R. & Green, I. 2007. A finite element study of the deformations, forces, stress formations, and energy losses in sliding cylindrical contacts. *International Journal of Non-Linear Mechanics.* 42(7): 914–27.

Wang, Y.-H. & Santamarina, J.C. 2002. Dynamic coupling effects in frictional geomaterials—Stochastic resonance. *Journal of Geotechnical and Geoenvironmental Engineering.* 128(11): 952–962.

Wang, Y.H. & Siu, W.K. 2006. Structure characteristics and mechanical properties of kaolinite soils. I. Surface charges and structural characterizations. *Canadian Geotechnical Journal.* 43(6): 587–600.

Warkentin, B.P. & Yong, R.N. 1960. Shear Strength of Montmorillonite and Kaolinite Related to Interparticle Forces. *Clays and Clay Minerals.* 9: 210–218.

Werkmeister, S., Dawson, A.R. & Wellner, F. 2004. Pavement design model for unbound granular materials. *Journal of Transportation Engineering.* 130(5): 665–674.

Wichtmann, T., Niemunis, A. & Triantafyllidis, T. 2004. Strain accumulation in sand due to drained uni-axial cyclic loading. In: *Int. Conf. on Cyclic Behaviour of Soils and Liquefaction Phenomena.* Balkema, Rotterdam, 233–46.

Yamamuro, J.A. & Lade, P.V. 1996. Drained sand behavior in axisymmetric tests at high pressures. *Journal of Geotechnical Engineering.* 122(2): 109–119.

Yimsiri, S. 2001. Pre-failure deformation characteristics of soils: anisotropy and soil fabric, University of Cambridge.

Youd, T.L. 1973. Factors controlling maximum and minimum densities of sands, Evaluation of Relative Density and Its Role in Geotechnical Projects Involving Cohesionless Soils. In: *ASTM STP 523,* 98–112.

What controls the effective friction of shearing granular media?

Makedonska Nataliya & Liran Goren
Department of Environmental Sciences and Energy Research,
Weizmann Institute of Science, Rehovot, Israel

David Sparks
Department of Geology and Geophysics, Texas A&M University, College Station, Texas, USA

Einat Aharonov
Institute of Earth Science, Hebrew University, Jerusalem, Israel

ABSTRACT: In order to better understand the physics of shear on geological faults and the physics of earthquakes we simulate the behavior of granular matter undergoing shear. This work investigates in depth (by using numerical DEM simulation and theoretical calculations) the apparent friction, μ_a, of a two-dimensional granular layer composed of mono-sized particles confined between two rough walls. The behavior of uniform sized grains is an end member case that is easy to predict theoretically. The study is then extended to include the effect of heterogeneity in grain properties, implemented by introducing a certain percentage of 'defect' grains: a fraction of grains is chosen randomly, and one of their physical properties is set differently from the rest of the aggregate. We demonstrate that the apparent, large-scale, friction of the whole shearing granular system stems from two physical effects: 1. 'surface friction' between rubbing surfaces of neighboring grains, and 2. 'geometrical friction'—i.e. the shear stress required to force grains to climb past each other, dilate the system, and allow finite shear strain to occur. We calculate the exact contribution by each of these two components of the apparent friction, and find that 'geometrical friction' has a much larger contribution than the surface friction. The affect of the grain geometry on strain weakening and other frictional characteristics is also studied.

1 INTRODUCTION

Natural faults commonly contain significant accumulations of rock fragments, product of the wear process, termed 'fault gouge'. Fault gouge is actually a confined layer of granular matter that is formed during sliding. It affects profoundly the friction on geological faults and hence the mechanics of earthquakes and faulting (e.g. Byerlee et al. 1978, Engelder 1974, Marone et al., 1990, Chester et al., 1993, Marone 1998, Scholz 2002, Aharanov & Sparks 2004). Many experimental and theoretical studies aimed to identify the mechanical conditions and constitutive properties that affect the transition from stable to unstable sliding. It was found that fault gouge, its dilatancy, and its shear displacement play a pivotal role in controlling sliding stability (see review by Marone, 1998). However, experimental observations do not yet provide a clear mapping of the conditions under which stable or unstable behavior is expected; many questions on mechanisms of fault gouge shearing remain poorly understood. A fruitful direction for understanding mechanics of earthquakes and fault gouge is numerical simulations of granular media, where one of the major questions is the relation between the properties of a single grain and the behavior of systems with many grains (Mora & Place 1998, Mora & Place 1999, Morgan et al. 1999, Morgan 1999). Here we use such numerical modeling to address the micromechanical origin of the measured macroscopic strength of granular layers.

Forces are transmitted within granular materials by elastic and frictional forces at grain contacts, which are controlled by the material properties of these contacts. The macroscopic material properties of the aggregate material will depend on some average of these contact properties, but also on their spatial and temporal distributions, which change as the system deforms. We define a macroscopic, or apparent, friction, μ_a, of a layer filled with grains as the ratio of the shear stress,

τ, needed to maintain a constant rate of shear of the upper wall, to the stress applied normal to the layer boundaries, σ_n.

The 1st part of this study focuses on highly ordered systems: two-dimensional grains with circular cross-section and uniform size and properties are confined between two rough surfaces. This simple geometry by itself provides a wealth of shear motion behaviors as function of grain organization, ability to roll and surface friction. This highly ordered system may be envisioned as the basis from which natural systems (with a large natural variability) diverge. Highly ordered systems are relatively easy to explain analytically and provide useful insights that may be used for treating more heterogeneous systems. Thus we suggest that this work may serve as a point of departure from which to study natural systems and their frictional properties. After performing basic studies on the uniform system, the effect of heterogeneity in grain properties, such as rolling or surface friction, is studied by looking at a series of simple binary systems. These binary systems have two populations of particles, one of 'regular' grains and one of 'defects'. Two types of defects are considered: 'non-rolling' defects are grains that are unable to roll, like particles with a very angular shape; and 'frictionless' defects—grains with frictionless surfaces, which may be simulated as low friction grains such as clay, talc etc.

We combine here analytical calculations and modeling results of apparent friction, exhibited by systems under shear. The modeling is used to validate the analytical calculations but it also shows that the actual behavior is not always as simple as expected even for the simplified cases studied here. Lastly we discuss the important conclusion that geometrical friction, the frustration of the compacted system in its attempt to reorganize, is the most important component in determining the value of the apparent friction and its strain weakening properties.

2   SYSTEM OF UNIFORM GRANULAR MEDIA

2.1   *The numerical model*

We consider the highly ordered system of uniform granular media in 2D. Circular grains are placed initially in a perfect hexagonal packing which has porosity of 9.3%, and a coordination number of six. Two parallel rigid walls are made from half grains that are glued to each other. The side boundaries are periodic, so that the confining box geometry resembles an annulus. The bottom wall is fixed, and a layer-normal force σ_n is applied to each grain of the top wall. The grain positions are set such that the distribution of forces between all grains in a system will be uniform in the initial packing (Fig. 1).

The system is modeled using a 2D Discrete Element Granular Dynamics code, following Cundall & Strack (1979). Grains interact at their contact points through linear elastic shear and normal forces and viscous damping forces. The distance between the centers of two grains, i and j, is described by the vector $\mathbf{r}_{ij}$. When the $\mathbf{r}_{ij}$ is less than the sum of the radii, R_i and R_j (i.e., the grains overlap), an interaction force is exerted on each grain at the point of contact. The grains are noncohesive, so this force vanishes when the overlap reaches zero. The normal component consists of a linear elastic repulsive force and a damping force dependent on the relative grain velocities, $\dot{\mathbf{r}}_{ij}$:

$$F_{ij}^n(t) = [k_n(R_i + R_j - r_{ij}) - \gamma m(\dot{\mathbf{r}}_{ij} \cdot \hat{\mathbf{n}})]\hat{\mathbf{n}} \tag{1}$$

Here $\hat{\mathbf{n}} = (\mathbf{r}_{ij} \cdot \hat{\mathbf{x}}, \mathbf{r}_{ij} \cdot \hat{\mathbf{y}})/r_{ij}$ is the unit vector parallel to the contact; k_n is the normal spring stiffness; m is grain mass, and γ is a damping coefficient. Shear forces on contacts are determined using the friction law detailed in Cundall & Strack (1979).

$$F_{ij}^s(t) = -[\min(k_s \Delta_s, \mu(\mathbf{F} \cdot \hat{\mathbf{n}}))]\hat{\mathbf{s}}, \tag{2}$$

where $\hat{\mathbf{s}} = (\mathbf{r}_{ij} \cdot \hat{\mathbf{x}}, -\mathbf{r}_{ij} \cdot \hat{\mathbf{y}})/r_{ij}$ is the unit vector tangent to the contact, k_s is the contact shear stiffness, and Δ_s is the shear displacement since the formation of the contact. As seen from Eq. 2

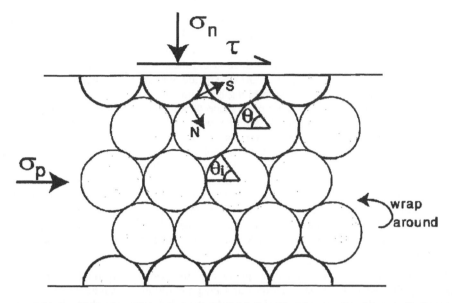

Figure 1.   Initial configuration of 2*D* system: mono-sized circular grains are placed in a perfect hexagonal packing. Two parallel rigid walls are made from half grains that are glued to each other. The side boundaries are periodic. Layer-normal stress σ_n is applied to each grain of the top wall. The layer-parallel stress σ_p is dictated by a prescribed overlap between grains within a row. σ_p is set such that the distribution of forces between all grains in the system will be uniform in the initial packing. The externally applied forces, τ and σ_n, are transmitted to individual contacts, where they are resolved into shear and normal force components, S and N. θ is a contact angle between wall grains and interior grains, θ_i is a contact angle between grains in interior rows. Initially the contact angle between the grains is 60°.

Table 1.   The main dimensionless parameters, which are used in our Granular Dynamics Model.

Parameter	Value
Applied layer-normal force, σ_n	10^{-4}
Applied shear velocity, v	10^{-4}
Normal spring stiffness, k_n	1
Shear stiffness, k_s	0.5
Surface friction coefficient, μ	0.5
Damping coefficient, γ	0.8
Layer-parallel contact force between grains in an interior row, σ_p	$5.77 \cdot 10^{-5}$

shear forces are first elastic and then obey a Mohr-Coulomb frictional sliding criteria, proportional to the normal force and a fixed, uniform grain surface friction coefficient, μ. The parameters used in the model are presented in Table 1 in a non-dimensional form. Distances are scaled by a characteristic grain diameter, d, stresses by a single grain Young's modulus, E, and masses by the mass of a grain of diameter d, m. Force is scaled by Ed^2, the spring constants by Ed, and velocity by $d\sqrt{Ed/m}$. A fuller description of the modeling technique is given in Aharonov & Sparks (1999), Aharonov & Sparks (2002), Aharonov & Sparks (2004). Gravitational forces are not included in our simulations.

2.2   Modeling uniform grains

In this class of simulations we shear a system uniform grains that are initially in perfect hexagonal packing. The hexagonal packing causes all of the grain contacts to fall into one of three populations, with orientations (direction normal to the grain of surfaces at the contact point) at 0°, 60°, and 120°

measured from the horizontal. Initially all contact forces carry identical forces. During shear, the orientations and forces change, and some of the contacts are broken. The numerical results show that deformation is generally localized along one or two boundaries between rows of grains, and is periodic in time. Each period consists of dilation and shear, followed by compaction back to the initial hexagonal packing. The behavior in the next period of motion is similar to the previous ones because the relaxed state of hexagonal packing holds no memory of previous motion. However, as some vibrations always exist in the model, succeeding periods may evolve slightly differently. As expected, the behavior across periods is similar when the system has fewer degrees of freedom (a small number of grains), and is less similar as the number of grains increases.

Shear motion is triggered instantaneously by applying to the top wall a constant layer parallel velocity v. When the shear motion is applied to the top wall, the force τ, needed to keep the upper wall moving in a layer-parallel direction at a constant velocity v, is measured. The value of $\mu_a = \tau/\sigma_n$, which varies with the displacement of the top wall, is interpreted as the apparent friction of the system. Theoretical predictions of apparent friction μ_a, described below, are presented for a single period of motion, which is started from hexagonal packing, and are compared to the simulations of the system performed by DEM code. The following sets of 'gouge shear' simulation are performed:

i. A system with a single interior row of grains confined between two rough walls. The grains are only allowed to slide past each other, and any grain rolling is prohibited.
ii. The same system with a single interior granular layer, but grains are allowed to both roll and slide.
iii. Systems with multiple interior layers confined by two rough walls, where interior grains are allowed to both slide and roll during shear.

2.2.1   *Theoretical model of a single layer of grains undergoing shear*

We investigate the basic physical process of a shearing thin layer of fault gouge between two rough surfaces by simulating first a single interior row of circular grains sheared between two rough walls. For simplicity we assume the walls are made of glued half grains. Generally the interior particles can accommodate shear by either rolling or by sliding on their contacts with each other or with the wall grains. However, as a 1st step we assume that rolling of interior grains is prohibited. In this case shear occurs only by frictional sliding on the wall contacts. In order to find the conditions for sliding we translate the externally applied macroscopic forces, τ and σ_n, into the forces transmitted to individual contacts. These are resolved into shear and normal forces, S and N, on each contact:

$$S = \tau \sin\theta - \sigma_n \cos\theta$$
$$N = \tau \cos\theta + \sigma_n \sin\theta.$$
(3)

Here θ is the contact angle between grains, and it evolves during motion.

Sliding of the top wall is initiated when the shear force on all contacts surpasses the frictional resistance to sliding, i.e. $S = \mu N$, where μ is the surface friction coefficient. This criterion is most difficult to meet in contacts oriented initially in the 60° direction, i.e contacts for which the layer-parallel force τ applied to the top wall pushes the grains forcefully into their neighbors. As motion proceeds, the contact angle θ (defined in Figure 1) evolves: as the grains start climbing on each other, θ progressively increases, as shown in Figure 2a,b and c. The apparent friction, exhibited by this system that shears by pure sliding, can be calculated from Eq. 3 and the sliding criteria $S = \mu N$:

$$\mu_a = \frac{\tau}{\sigma_n}(\mu, \theta) = \frac{\cot\theta + \mu}{1 - \mu \cot\theta}$$
(4)

Eq. 4 dictates that as deformation continues, and θ increases, the apparent friction μ_a decreases; thus the system is predicted to exhibit slip-weakening behavior as θ evolves from 60° to 120°, at which point the system returns to a new hexagonal packing. The motion from one hexagonal

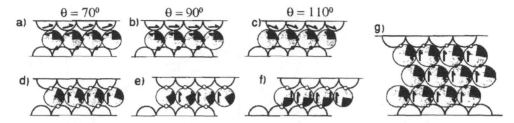

Figure 2. (a), (b) and (c): a cartoon of the evolving granular configuration in a system with one interior row undergoing shear, at three progressive values of the evolving contact angle between the wall grains and the grains in the interior row: $\theta = 70°$, $\theta = 90°$, $\theta = 110°$, respectively. In this set we artificially prohibit grain rolling, and frictional sliding of the top wall accommodates motion. (d), (e) and (f): 'bearing' motion of one interior row at three progressive values of the evolving contact angle: $\theta = 70°$, $\theta = 90°$, $\theta = 110°$, respectively. (g) Grain configuration in a system with multiple interior rows when the contact angle $\theta \sim 70°$. Initially dilation occurs along all the interior rows while they undergo 'beam' rotation. The behavior of each contact is depicted by the symbols drawn on the contact: arrows symbolize frictional sliding; white circles symbolize rolling motion, and black circles symbolize a fixed contact with no relative motion.

packing to a new one is termed one period. During one period of sliding motion the top horizontal wall displacement is equal d, one grain diameter.

When interior grains are allowed to roll, they roll along the wall contacts and slide on the interior contacts, as seen in Figure 2(d–f). In this case shear occurs in a manner similar to 'ball-bearing' motion, where all interior grains rotate in the direction of motion. The layer parallel force required to maintain constant shear rate is determined by the torque balance for a rolling grain. Frictional forces on the contacts between the rolling grains may oppose rolling motion, as seen in Figure 2(d–f), and a detailed calculation (presented in a manuscript to be submitted elsewhere) leads to an apparent friction given by

$$\mu_a = \frac{\tau}{\sigma_n}(\mu, \theta) = \cot\theta + \frac{\mu}{\sin\theta} \cdot \frac{\sigma_p}{\sigma_n}. \tag{5}$$

The second term of Eq. 5 depends on the friction coefficient between grains and the ratio of the force σ_p to the normal force σ_n. As in the case of pure sliding (Eq. 4) μ_a of rolling decreases with strain. For the case of $\sigma_p/\sigma_n = 1/\sqrt{3}$, which is the case where all the contacts initially have equal force on them, μ_a in rolling (Eq. 5) is always smaller than μ_a in sliding (Eq. 4), so that rolling is always preferred over sliding during the whole cycle of motion. In all of our numerical models, the initial system is set up such that $\sigma_p = \sigma_n/\sqrt{3}$, so that all contact forces are equal in the initial hexagonal packing (Table 1). Under these conditions, the apparent friction (Eq. 5) for the numerical simulations is expected to be fitted by

$$\mu_a = \frac{\tau}{\sigma_n}(\mu, \theta) = \frac{\mu}{2\sin\theta\sin 60°} + \cot\theta \tag{6}$$

2.2.2 *Comparing numerical simulations of a single granular layer to the theoretical model*
Figure 3a shows the apparent friction μ_a as function of the layer-parallel wall displacement scaled by d, in the numerical model. The dashed line presents results from simulations where grains are prohibited from rolling. The solid line plots results from simulations where grain rolling is also allowed. The first set is perfectly fitted by the theoretical predictions for pure sliding motion (Eq. 4, Fig. 2a,b,c), while the second set is perfectly fitted by predictions for 'bearing' rolling (Eq. 6, Fig. 2d,e,f). The eqs were plotted using wall displacement, x', instead of contact angle:

$$x' = \cos 60° - \cos\theta \tag{7}$$

Eq. 7 applies for sliding motion. In the case of 'bearing' rolling Eq. 7 should be multiplied by a factor of 2.

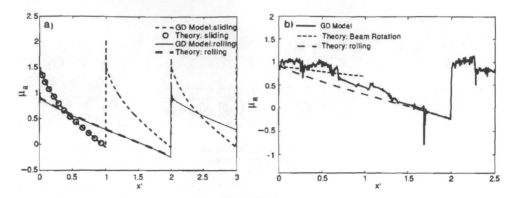

Figure 3.   Apparent friction as function of scaled top wall displacement, x', measured from Granular Dynamics simulations: (a) Apparent friction of a system with one interior row of grains. The dashed line shows results from simulations where grains are prohibited from rolling. The solid line plots results from simulations where grain rolling is also allowed. The first set is perfectly fitted by the theoretical predictions for pure sliding motion (Eq. 4, Fig. 2a,b,c), while the second set is perfectly fitted by predictions for 'bearing' rolling (Eq. 6, Fig. 2d,e,f). (b) Shows apparent friction of a system with multiple interior rows. At the beginning of the period the numerical results agree with prediction for 'beam' rotation (Eq. 8, Fig. 2g), then apparent friction decreases in steps until it converges with the prediction for 'bearing' rolling (Eq. 6).

As theory predicts, the apparent friction in simulations decreases with continued motion from the onset of each period. For the initially isotropic single layer, the apparent friction in the 'bearing' motion is always less than in the pure sliding case for all values of θ, and rolling is always the preferred mechanism, as is well predicted by theory. The apparent friction is negative at the end of the period, where the wall is dynamically moving and so it actually accelerates and then provides a negative force.

2.2.3   *Shearing system with multiple interior rows*
After performing simulations of single layers of grains more complex systems, with several grain layers, are simulated. Simulations were run with 5, 10 and 25 interior rows and show similar behavior in all of those systems. Here we present simulation results of a system with 10 interior rows as a representative system with multiple interior layers between walls. In those systems the basic form of deformation under shear remains the single-row 'bearing' motion mechanism described above due to an observed rapid localization: two random adjacent rows of grains dilate and translate to a new hexagonal packing, followed in the next period of motion by the same localized process in some other row within the system. However, because of the extra degrees of freedom in larger systems, other behaviors may arise. The numerical models commonly show short phases of distributed shear, and also counter-rotation (opposite to the sense of shear) of some grains.

The question of localization merits further discussion: since the rows of grains are identical and hexagonally packed, there is no preference for a particular row to shear, and one might expect that shear would be distributed across all rows. The numerical runs indeed show a phase of distributed shear at the beginning of periods of motion, in which the entire system dilates until random (in our case numerical) fluctuations cause localization along some row. This distributed shear tends to occur only for contact angles near 60°, i.e. for small deviations from hexagonal packing. However, in systems with many rows, this can translate into large displacements of the top wall.

From observations of the numerical model, we develop a conceptual model that approximates the deformation during the distributed shear phase. This model allows us to assess the conditions under which such deformation is favored over localized shear.

During distributed shear motion, sets of grains with contacts near 60° translate as a unit. Those contacts are neither sliding nor rolling, but have no motion (Fig. 2g). These sets of grains rotate clockwise like parallel rigid beams. The grains at the top and bottom of a 'beam' roll on their contacts with the walls. Because all grains are rotating clockwise, there is frictional sliding between adjacent grains in the same row, as in bearing-like rolling. The apparent friction for beam rotation may be calculated from the torque balance on grains that are near a wall:

$$\mu_a = \frac{\tau}{\sigma_n}(\mu, \theta) = \frac{2\mu\sigma_n}{\sigma_p(\sin\theta + \sin\theta_i)} + \frac{\cos\theta + \cos\theta_i}{\sin\theta + \sin\theta_i}, \tag{8}$$

where $\theta_i = 2\theta - 60°$ is now defined as the angle of the contact between interior rows, and θ is still the contact angle on wall grains. θ increases with strain only half as fast as the contact angle θ_i.

It is easily seen that for a reasonable value of μ (say 0.5), for any value of θ the apparent friction predicted by Eq. 8 is always greater than that for the single row rolling mechanism (Eq. 6). Therefore, localized rolling of one interior row should always be the preferred mechanism over distributed shear. Distributed shear by beam-rotation is expected to be only a metastable mechanism; minute deviations from uniformity will cause the shear to localize. In the GD numerical simulations we observe that localization generally occurs spontaneously after a few degrees of contact rotation.

Figure 3b compares the theoretical predictions of one row rolling, according to Eq. 6, and beam rotation, according to Eq. 8 to a numerical simulation run with multiple interior rows. During the first $\sim 0.65d$ of top wall displacement (until $\theta \sim 62°$), the system deforms by distributed shear. The apparent friction in these larger systems is much noisier than in shear of a single layer due to the greater degrees of freedom of motion of the grains. However, friction is consistent with the predictions from the beam-rotation model (Eq. 8). When θ exceeds $\sim 62°$ shear motion localizes onto a single row, which begins 'bearing' rolling, while all other rows remain fixed. The apparent friction undergoes a small drop immediately, and then begins to evolve along a curve parallel to the prediction for a single rolling row. The friction is larger than the value predicted by Eq. 8 due to the distributed dilation that accumulated during the beam rotation phase. The small dilations of rows outside of the rolling row are unstable and these rows collapse (backward) into their original hexagonal packing as shear continues. There is a major collapse event at $x' \sim 1.3d$. At $x' \sim 1.7d$, the system has returned to its original hexagonal packing, except for the rolling row. For the rest of the period, until $x' = 2$, the numerical results agree with the prediction of Eq. 6.

3   INTRODUCING DEFECTS INTO UNIFORM GRANULAR SYSTEM

The above discussed theoretical predictions and simulation results of highly ordered media can be used now for an investigation of the effect of disorder. The effect of disorder is studied by modeling simple binary systems, which have two populations of particles, one of 'regular' grains and one of 'defects': grains that are randomly chosen to have different value of particular physical property. Varying the concentrations of both populations and measuring the apparent friction exhibited by systems under shear the effect of disorder is studied. Such a progressive introduction of disorder allows isolation of the effects of heterogeneity and a systematic approach to the 'real world', which is highly disordered.

3.1   *Non-rolling defects*

As shown above, rolling has a large affect on the apparent friction of a system of circular grains. When angular grains are present, rolling may be inhibited. To study this effect more systematically, 'non-rolling defects' are introduced into the system: randomly chosen grains that are not allowed to rotate. These grains can move past others only by frictional sliding on their contacts.

Simulation results show that adding a small number of non-rolling defects has little effect on the apparent shear behavior. This can be seen in Fig. 4a, comparing concentration of defects, $C_d = 0$ with $C_d = 0.1$. Up to $C_d = 0.3$ (30% of grains are not allowed to roll, 70% of grains are allowed

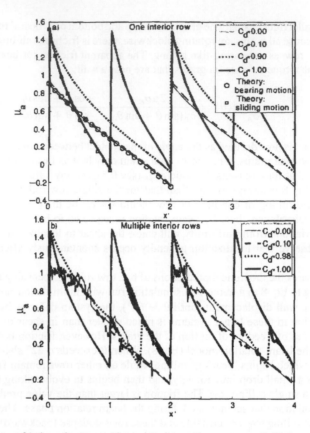

Figure 4. (a) Apparent friction of system with one interior row versus top wall displacement, x', for concentration of non-rolling defects: $C_d = 0.0$ (fitted by theoretical prediction for rolling motion by Eq. 6), $C_d = 0.1$, $C_d = 0.9$, $C_d = 1.00$ (fitted by theoretical prediction for sliding motion by Eq. 4). (b) Apparent friction of system with multiple interior rows versus top wall displacement, x', for concentration of non-rolling defects: $C_d = 0.0$, $C_d = 0.1$, $C_d = 0.98$, $C_d = 1.0$.

to roll), deformation of systems with a single interior row continues to occur through bearing-like rolling. Systems with multiple rows undergo distributed shear at the beginning of period and then localizes into one row rolling, as seen in Fig. 4b. However, we find that while a small fraction of non-rolling grains has little effect on an aggregate composed of a majority of 'rolling' grains, a small number of grains that are able to roll can have a large effect on a system with a majority of non-rolling grains.

Consider the case when all grains are not able to roll, $C_d = 1.0$. Shear motion occurs then by pure sliding, since no other type of motion is allowed in this system. Dilation occurs between two rows and the top wall displacement is d per one period of motion. The apparent friction obtained in the GD simulation is perfectly fitted by theoretical prediction for sliding motion, expressed by Eq. 4 (Fig. 4, $C_d = 1.0$). In system with multiple interior rows the most probable scenario is that the top wall slides over the top interior row, while all the rest of the interior rows show no motion.

Introducing even a small number of rolling grains into this non-rolling system will change the shear motion mechanism. For $C_d = 0.9$ in a system with a single interior row the motion is similar to the 'bearing' motion, even though most of the grains are not rolling: dilation occurs between both walls and interior row of grains, and the top wall moves $2d$ in one period of motion. The apparent friction gradually decreases as more rolling grains are added (Fig. 4a).

In systems with multiple rows, the simulations show that rolling grains may stop rotating partway through the period (Fig. 5). When this happens, motion continues by pure sliding between two

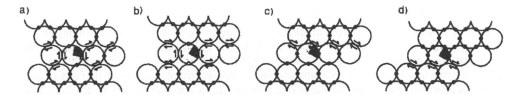

Figure 5. Motion of system with mostly non-rolling grains (mono-gray circles) and one rolling grain (depicted as a colored circle) at four stages of motion period. The behavior of each contact is shown by arrows when the frictional sliding occurs on that contact, a white circle—rolling motion, or a black circle when a contact is fixed and no relative motion occurs.

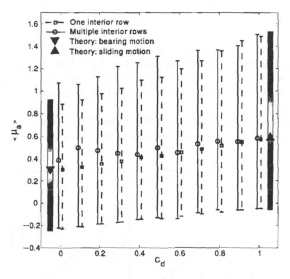

Figure 6. Average value of apparent friction, exhibited by a system under shear, as a function of the concentration of non-rolling defects. Symbols show the mean value of μ_a, error bars show maximum and minimum values, respectively.

rows. Eq. 5 predicts this behavior under specific conditions: when either the σ_p or the inter-granular friction are large enough, then at a critical contact angle sliding will require a lower shear force than rolling (and will thus be preferred over rolling). This transition to sliding is visible in the apparent friction curve as a slope break (at $x' = 1$ in Fig. 4b, $C_d = 0.98$): at this point two of three dilating rows have reached a new hexagonal packing, and sliding resumes between the two remaining rows; as a result a 'double collapse' event is observed in the apparent friction during the period. This combined mechanism occurs in about 40% of the periods in the $C_d = 0.9$ case, and becomes less frequent as number of defects decreases, disappearing altogether at $C_d = 0.5$. In systems dominated by rolling grains (i.e. $C_d < 0.5$) only bearing motion occurs, where the row in which dilation occurs, has the minimum number of non-rolling grains.

Figure 6 shows the average values of apparent friction as function of concentration of non-rolling defects. The error bars show mean, maximum and minimum values of measured μ_a compared to the theoretical predictions for pure sliding (Eq. 4) and 'bearing' motion (Eq. 6). Both in systems with a single interior row of grains and in systems with multiple rows, the apparent friction increases linearly with increasing concentration of non-rolling grains. It is important to notice that the theoretical predictions for sliding (Eq. 4) and bearing motion (Eq. 6) constitute the end member situations, and all system behaviors are tuned between these two end member behaviors by their concentration of non-rolling defects.

3.2   *Frictionless defects*

Analyzing the analytical expressions of the above described mechanisms of motion, sliding (Eq. 4), 'bearing' motion (Eq. 6) and 'beam' rotation (Eq. 8), it is easy to see that even when grains have no prescribed surface friction, (i.e. $\mu = 0.0$), systems exhibit nonzero shear resistance. This frictional resistance is causes by geometrical effects—brought about by the need of grains to climb past each other. In order to see how apparent friction, μ_a, is sensitive to surface friction, μ, we perform simulations of uniform media, where randomly chosen grains have no surface friction $\mu = 0.0$, while all the rest of the regular particles have surface friction of $\mu = 0.5$. When a grain with $\mu = 0.5$ is in contact with a frictionless grain then friction on this contact is chosen as zero.

In a system with one interior row the shear motion occurs by bearing rolling for any amount of frictionless defects, except $C_d = 1.0$ (Fig. 7a). In systems with multiple interior rows beam rotation is observed mostly in systems with a small concentration of frictionless particles, $C_d < 0.10$; a further increase of C_d decreases displacement of top wall, which is a consequence of motion by distributed shear and leads to earlier localization. When $0.20 < C_d < 0.50$ the sliding motion occurs in about 30% of the periods, in the case of $0.50 < C_d < 0.80$ sliding occurs in about 70% of motion periods, and it is always the case when $C_d > 0.80$. In systems where all the interior grains are frictionless, $C_d = 1.0$, motion occurs only by pure sliding and is fitted by the theory of Eq. 4 with $\mu = 0.0$ (Fig. 7b).

The results of our Granular Dynamics simulations show that the apparent friction, μ_a, exhibited by the system under shear decreases with increasing concentration of frictionless grains in all

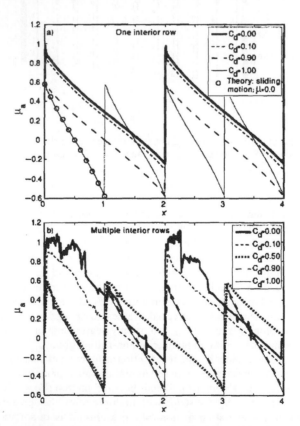

Figure 7.   (a) Apparent friction of system with one interior row and (b) multiple interior rows versus top wall displacement, x', for concentration of frictionless defects: $C_d = 0.0$, $C_d = 0.1$, $C_d = 0.5$, $C_d = 0.9$, $C_d = 1.0$.

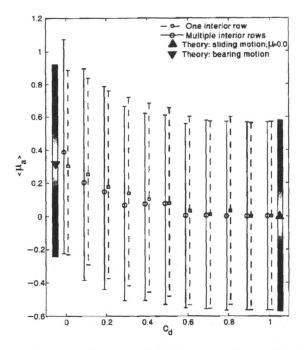

Figure 8. Average value of apparent friction, exhibited by system under shear, as function of frictionless defects concentration. Symbols show the mean value of μ_a, error bars show maximum and minimum values, respectively.

system sizes (one interior row and multiple rows) (Fig. 8). For systems with $C_d > 0.5$ the apparent friction arises mostly from the geometrical effect, and is, therefore, nearly constant with C_d.

4   DISCUSSION: THE IMPORTANCE OF GEOMETRICAL EFFECT

One of the main conclusions of the work presented here is that geometrical controls have a much more important effect on variations in apparent friction of granular material than any material variations expected in the surface friction coefficient of grains. For all the systems studied here, differentiating the apparent friction, μ_a, as given in Eq. (5), Eq. (7) and Eq. (9) with respect to strain x' and with respect to μ, reveals that

$$\left| \frac{\partial \mu_a(\mu, \theta)}{\partial x'} \right| > \left| \frac{\partial \mu_a(\mu, \theta)}{\partial \mu} \right| \tag{9}$$

for any $\mu < 1.7$ (Fig. 9).

Thus, the apparent friction is always more sensitive to the contacts and the positions of the grains ('geometrical friction') than to the surface friction coefficient of individual grains. One can infer from this result that in natural granular systems the strain weakening effect due to grain-climbing and dilatancy (Rowe 1962) is expected to be more important than the resistance to sliding between contacting surfaces that arises from the chemical bonding between them (i.e. their surface friction (Persson 1998)). Thus it is expected that the characteristic distance over which fiction is reduced from its dynamic to its static value will be mostly a result of initial dilation and rearrangements and not of reduction in chemical bonding. A large part of the geometrical friction we calculate is actually the same as the dilatancy affect (Rowe 1962).

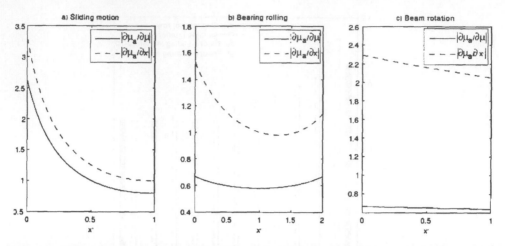

Figure 9.   Differentiation function of apparent friction for sliding motion (a) by Eq. 5, for bearing rolling motion (b) by Eq. 7 and for beam rotation (c) by Eq. 9 with respect to top wall displacement, x', (dashed lines) and with respect to surface friction, μ (solid line).

The value of the top wall velocity, v, (Table 1) used in our simulations, was chosen so that deformation during shear will occur quasi-statically under the applied normal stress (i.e. simulations are in the dense packing granular regime and not in the gas-like collisional granular regime) (Aharonov & Sparks 1999). Varying the top wall velocity, as long as the system is maintained in the dense regime, doesn't influence the apparent friction μ_a. However, applying much higher velocities or reducing the normal stress considerably brings the system to the gas-like deformation regime where the grains collide and jump as molecules in a gas: the map of conditions for transition between the dense and gas-like regimes is shown in Fig. 7 in Aharonov & Sparks (1999). Here we don't show the results of simulations in gas-like regime, but we expect that the shear mechanisms and all other parameters will change, and as a result the apparent friction, μ_a, will also be different in that regime.

On geologic faults, a steadily increasing shear stress is a more realistic loading mechanism than the constant velocity condition. Under such stress loading conditions, strain-weakening systems, in which the frictional resistance decreases with displacement, as seen above, are expected to undergo stick-slip motion. Mair, Frye & Marone (2002) indeed observe stick-slip motion in shear experiments on spherical gouge particles with a small particle size distribution, and no other weakening effects. This behavior we believe arises from the strong strain weakening of apparent friction for circular grains due to geometrical affects. Geometrical effects on shear resistance will be especially important in the transition from stick to slip because of the need to initiate a motion from a tight packing. There is an inherent difference between no motion and motion at velocity v, which does not exist between motion at v and at $2v$. This inherent difference arises due to the need to dilate a tightly packed system in order to produce even the slowest motion, this unique role of initiation of motion means that (i) experiments of continuous sliding of surfaces using stepping of velocity might be inherently different than the situation in earthquakes which start from $v = 0$ and accelerate, and (ii) that the large initial shear resistance measured in all experiments and considered a 'transient' effect which goes away once the surface is 'conditioned', might merit investigation in relation to shear resistance during initiation of earthquakes.

Frye & Marone (2002) suggest that the macroscopic friction of a granular system is a linear combination of the surface friction, μ, and the work done to dilate the system, dy'/dx', where y' is the distance between the shearing walls, scaled by d

$$\mu_a = dy'/dx' + \mu \tag{10}$$

In this way the need of the system to overcome the 'geometric friction' is externally manifested and measured as dilation. This connection between 'geometrical friction', dilation and the theoretical prediction for dilation as a function of grain geometry will be explored in another paper (under preparation). It should be noted that the static friction and its difference from dynamic friction encapsulate this geometrical dilation effect as seen in the saw tooth graphs of μ_a vs x' (Fig. 3a).

REFERENCES

Aharonov E. & Sparks D. 1999. Rigidity phase transition in granular packings. *Physical Review E* 60: 6890–6896.

Aharonov E. & Sparks D. 2002. Shear profiles and localization in simulations of granular materials. *Physical Review E* 65: 051302.

Aharonov E. & Sparks D. 2004. Stick-slip motion in simulated granular layers. *Journal of Geophysical Research—Solid Earth* 109: B09306.

Byerlee, J., Mjachkin, V., Summers, R. & Voevoda, O. 1978. Structures developed in fault gouge during stable sliding and stick-slip. *Tectonophysics* 44:161–171.

Chester, F.M., Evans, J.P. & Biegel, R.L. 1993. Internal structure and weakening mechanisms of the San-Andreas fault. *Journal of Geophysical Research—Solid Earth* 98(B1): 771–786.

Cundall P.A. & Strack O.D. 1979. A discrete numerical model for granular assemblies. *Geotechnique* 29: 47–65.

Engelder, J.T. 1974. Cataclasis and generation of fault gouge. *Geological Society of America Bulletin* 85(10): 1515–1522.

Frye K. & Marone C. 2002. The Effect of Particle Dimensionality on Granular Friction in Laboratory Shear Zones. *Geophysical Research Letters*, 29(19).

Mair, K., Frye, K.M. & Marone, C. 2002. Influence of grain characteristics on the friction of granular shear zones. *Journal Geophysical Research* 107(10): 2219.

Marone, C. 1998. Laboratory-derived friction laws and their application to seismic faulting. *Annual Review of Earth and Planetary Sciences* 26: 643–696.

Marone, C., Raleigh, C.B. & Scholz, C.H. 1990. Frictional behavior and constitutive modeling of simulated fault gouge. *Journal of Geophysical Research—Solid Earth and Planets* 95(B5): 7007–70025.

Mora, P. & Place, D. 1998. Numerical simulation of earthquake faults with gouge: toward a comprehensive explanation for the heat flow paradox. *Journal of Geophysical Research—Solid Earth* 103(B9): 21067–21089.

Mora, P. & Place, D. 1999. The weakness of earthquake faults. *Geophysical Research Letters* 26: 123–126

Morgan, J.K. & Boettcher, M.S. 1999. Numerical simulations of granular shear zones using the distinct element method-1. Shear zone kinematics and the micromechanics of localization. *Journal of Geophysical Research—Solid Earth* 104 (B2): 2703–2719.

Morgan, J.K. 1999. Numerical simulations of granular shear zones using the distinct element method-2. Effects of particle size distribution and interparticle friction on mechanical behavior. *Journal of Geophysical Rresearch—Solid Earth* 104(B2): 2721–2732.

Persson, Bo N.J. 1998. Sliding Friction, Physical Principles and Applications. *Springer-Verlag, Germany.*

Rowe, P.W. 1962. The stress-dilatancy relation for static equilibrium of an assembly of particles in contact. *Proceedings of the Royal Society A* 269: 500–527.

Scholz C.H. 2002. The Mechanics of Earthquakes and Faulting. *Cambridge University Press*: 76–77.

Pore pressure development and liquefaction in saturated sand

S. Frydman, M. Talesnick & A. Mehr
Technion—Israel Institute of Technology

M. Tsesarsky
Ben Gurion University of the Negev

ABSTRACT: Development of pore-water pressure in saturated sands during earthquakes results in a decrease in rigidity and strength of the sand. As a result, structures founded in or on saturated sand profiles may be indanger of damage or even collapse. The extreme state of this pore-water pressure build-up is liquefaction of the sand, resulting in a total loss of shear strength and rigidity. The paper describes an experimental study of the behavior of saturated sand during cyclic loading in a hollow cylinder, torsional shear apparatus developed at the Technion. An energy-based constitutive model for pore-water pressure development is presented, and combined with a modified hyperbolic stress-strain model in a seismic site response code which considers development and dissipation of pore-water pressure. The significance of pore-water development and dissipation is illustrated through examples.

1 INTRODUCTION

Pore-water pressure development in saturated sands is a major cause of damage during earthquakes. The associated decrease in effective confining stress causes a decrease in rigidity and shear strength of the sand mass, resulting in settlement and possibly collapse of structures founded in these soils. Liquefaction is the extreme case of this phenomenon, in which the pore-water pressure actually becomes equal to the confining stress, so that the effective confining stress is reduced to zero. Although attention is often limited to consideration of the danger of full liquefaction, the development of pore-water pressures significantly lower than those required for liquefaction can cause considerable damage to structures. Consequently, there is clearly a need for methods of estimating pore-water pressure development during earthquakes.

Unfortunately very few measurements of pore-water pressure development during earthquakes have been reported in the literature (Ishihara et al., 1981; 1989; Shen et al., 1991; Youd and Holzer 1994). Figures 1 and 2 show two such cases. Fig. 1 shows recordings of ground surface accelerations and pore-water pressure ratio (r_u, equal to the ratio of pore-water pressure to overburden pressure) at 2.9 m depth, at the Wildlife site in California during the Superstition Hills earthquake of 1987. Liquefaction developed during this event. Fig. 2 shows recordings at Owi Island in Japan during the 1980 Mid-Chiba earthquake; in this case, the pore-water pressures developed did not reach liquefaction level. This difference in response is almost certainly due, mainly, to the relative strengths of the earthquakes; the peak ground acceleration at Wildlife was 0.21 g compared to only 0.095 g at Owi Island.

The need for a reasonable approach for prediction of earthquake induced pore-water pressure was the impetus for the work presented in this paper. A review of the literature indicated that a satisfactory predictive model is not presently available. The approach taken in the present research therefore involved the following stages: Review of models presently available for pore-water pressure prediction, performance of a laboratory study for acquisition of data from cyclic shear tests, including development of suitable testing equipment, analysis of test data and choice of suitable constitutive model, and development of a code for seismic site response. These stages are described briefly in this paper.

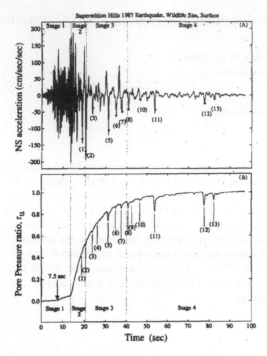

Figure 1. Recorded acceleration and pore-water pressure at Wildlife during Superstition Hills 1987 Earthquake (after Youd & Holzer, 1994).

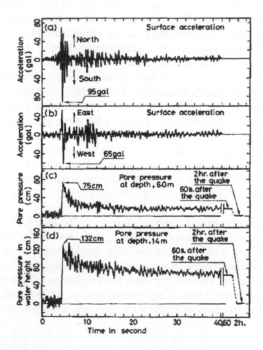

Figure 2.   Recorded accelerations and pore-water pressure at Owi Island during Mid- Chiba 1980 Earthquake (after Ishihara et al., 1981).

2   MODELS FOR PORE-WATER PRESSURE DEVELOPMENT

Two basic approaches are commonly used for prediction of pore-water pressure development during cyclic loading of sands. The first is based on an assumed empirical or elasto-plastic constitutive model for the soil, with the restraint that the volumetric strain, given by the sum of the normal strains, must be zero. Several empirical models (e.g. Martin et al., 1975; Byrne, 1991) or elasto-plastic constitutive models (see Elgamal et al., 2002) have been developed to simulate cyclic mobility and/or flow-liquefaction soil response. One of the major problems associated with the application of elasto-plastic models for prediction of behavior of sand under undrained cyclic loading, is the observation, first pointed out by Finn and his coworkers (Finn et al., 1977; Martin et al., 1975) that: *"In general we have observed in laboratory cyclic simple shear tests that most of the volume changes in dry sands, and the increases in pore-water pressure in undrained saturated sand occur during the unloading portions of the load cycle."* Baker et al. (1981) made a similar observation in undrained cyclic triaxial tests, and concluded that a constitutive formulation must include a non-linear elastic component, so allowing for coupling between spherical and deviatoric tensors. They were able to successfully model one cycle of triaxial loading, but not further cycles, in which they overestimated pore-water pressure development.

The second approach estimates the pore-water pressure generation in undrained conditions empirically. Towhata and Ishihara (1985) conducted a series of cyclic undrained tests on loose sand under various loading schemes (torsional, triaxial and a simultaneous combination of them) and concluded that *"the excess pore pressure development depends solely on the current stress state and the accumulated shear work"*. Other researchers (e.g. Nemat-Nasser & Shokooh, 1979; Mostaghel & Habibaghi, 1979; Davis and Berrill, 1982, 1998, 2001; Berrill and Davis, 1985; Yamazaki et al., 1985; Law et al., 1990; Kim et al., 2007) also related pore-water pressure development to shear work or dissipated energy. The advantages of these energy based models are in their simplicity and their capability to fit any kind of loading. The authors tested some of the above models against triaxial and simple shear tests results obtained by Arulmori et al., (1992), and found reasonably good fit of the tests results with the simple model of Berrill and Davis (1985):

$$R_u = \frac{u}{\sigma'_{v0}} = \alpha \cdot \sqrt{\frac{E}{\sigma'_{v0}}} \le 1 \tag{1}$$

where u = accumulated excess pore-water pressure.
E = deissipated energy.
σ'_{v0} = initial vertical effective stress.

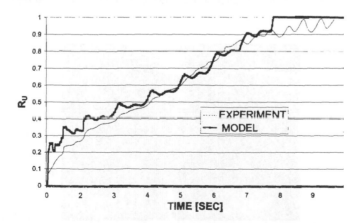

Figure 3.   Normalized pore-water pressure generation versus time; cyclic undrained simple shear (Nevada sand, DR = 40%).

The model was tested for 6 cyclic simple shear tests and 10 cyclic triaxial tests on Nevada sand at a relative density, DR = 40%, under different initial vertical effective stresses and gave good compatibilities using the same calibration parameter ($\alpha = 2.0$). For DR = 60%, 6 cyclic simple shear tests and 14 cyclic triaxial tests showed good compatibility using a calibration parameter $\alpha = 1.2$. Figure 3 shows an example of the compatibility in a cyclic simple shear test on Nevada sand with relative density, DR = 40%.

In order to obtain independent data for selection of a suitable constitutive model to be used for earthquake analysis, an experimental testing program was carried out. After consideration of the various testing methods commonly employed to study shear behavior of soils, it was concluded that the most suitable test procedure for the present purpose is the cyclic torsion of hollow, cylindrical samples.

3   THE HOLLOW CYLINDER EQUIPMENT AND TEST SAND

The hollow cylinder samples were of height 120 mm, outer diameter 70 mm and inner diameter 50 mm. The samples were prepared dry, by raining the sand into an annular space between an inner and outer mold, with inner and outer membranes attached, and then tamped to the desired density using a rod prodder. Following compaction and connection of an annular top cap, a small vacuum was applied to support the sample, and the molds were removed. Transducers were then placed for measurement of sample displacements during later testing. Fig. 4 shows a sample under vacuum prior to attachment of the transducers.

The system adopted in the testing program for measurement of shear and vertical displacements employs local measurements, based on a system developed by Talesnick and Shehadeh (2007). Two light plastic rings are attached around the sample at fixed elevations, and held in place without confining the specimen. Shear and axial strains are related to the original distance between these two rings. Vertical normal strain is obtained from the height change measured between the rings by 3 LVDT's, while shear strain is obtained by measuring the relative rotation of the rings with a proximity gauge (Fig. 5).

Pore-water pressure is measured by a pressure cell at the base of the sample, volumetric strain by a Wykeham Farrance volume change device, and vertical force and torsional moment by a dual-component load cell located above the sample within the test cell. The sample, with all measurement equipment attached, is shown in Fig. 6. The outer, Perspex cell is then attached to the base, the cell and inner bore of the sample are filled with silicon oil, the oil (inner and outer) is pressurized to a low pressure and the vacuum is released.

In order to saturate the sample, CO_2 is flowed through the sample for approximately 2 hours, to displace air in the voids, and then de-aired water is flowed through, displacing the CO_2. The inner

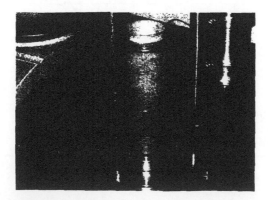

Figure 4.   Hollow cylinder sample after preparation, supported by vacuum.

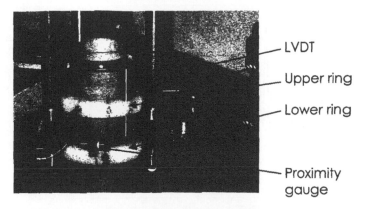

Figure 5.   Sample with transducers attached for axial and torsional displacement measurements.

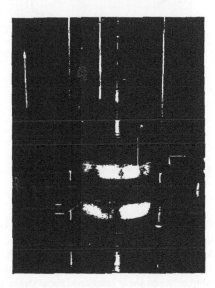

Figure 6.   The measurement set-up.

and outer pressures are then increased to the required confinement pressure, and the sample is allowed to consolidate. The cyclic torsion test is then carried out in the stress control mode applying a sinusoidal wave form, normally under undrained conditions.

The sand tested in the investigation is a dune sand from Caesaria on the Israeli Coastal Plane; it is uniformly graded, with mean particle size of 0.2–0.3 mm. Tests were performed at relative densities, DR, of 40% and 60%.

4   TYPICAL TEST RESULTS

Fig. 7 shows typical results of an isotropically consolidated, undrained, cyclic torsional test performed on a sample at DR = 60%. The confining pressure, p, was constant at 200 kPa; consequently Fig. 7a, which shows the effective confining pressure p′ (=p-u) starting at 200 kPa (point A), and moving left, indicates the development of positive pore-water pressure, u, from cycle to cycle. Fig. 7b shows shear stress—shear strain cycles, indicating the decreasing rigidity of the sand with increasing cycles. Fig. 7c shows the development of pore-water pressure (normalized by confining

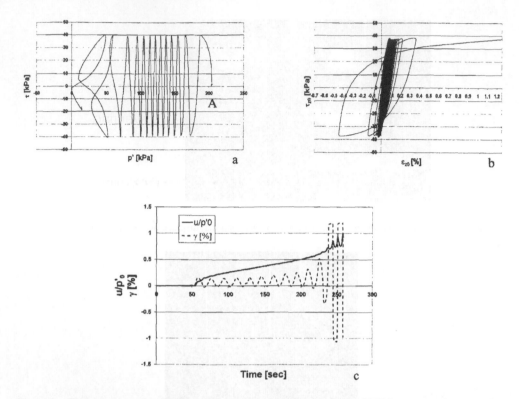

Figure 7.   Typical test results; Sample at DR = 60%, confining pressure = 200 kPa, stress ratio, τ/p = 0.4.

pressure) and shear strain with time during the test. Full liquefaction of the sand occurred after 12 cycles of shearing under a stress ratio, τ/p, of 0.4.

In order to develop a model for prediction of soil response under earthquake shear loading, the following relations are required: (a) A relationship between shear stress and shear strain during virgin loading; (b) A relationship between shear stress and shear strain during subsequent unloading and reloading cycles; and (c) A relationship for development of pore-water pressure during cyclic shearing.

Several empirical stress-strain relations have been developed for virgin shear loading of soils, some of the most popular being the hyperbolic relation (Kondner and Zelasko, 1972), the Ramberg-Osgood relation (Ramberg and Osgood, 1943) and the logarithmic relation (Puzrin and Burland, 1996). Recently, a modified hyperbolic (MKZ) relation has been developed (Matasovic, 1993), appearing to better model soil behavior than the original hyperbolic relation. Fig. 8 shows the shear stress—shear strain curve obtained in the first cycle of the test presented in Fig. 7, together with best predictions made with each of these models. The best correspondence in this test, and overall in the test program, was obtained with the MKZ model.

Fig. 9 shows the development of pore-water pressure measured in the above test compared to that predicted by a number of empirical models. Overall best agreement was obtained with the energy model suggested by Berill and Davis (1985).

Stress-strain behavior during unload and reload cycles following the virgin loading segment are assumed to follow the Masing rules (Pyke, 1979), adapted for the change in effective confining stress due to development of pore-water pressure. Fig. 10 shows measured cyclic shear stress—shear strain curves, together with curves predicted on the basis of the Masing rules modified in accordance with the pore-water pressure developed during the test. Reasonably good agreement is observed in this test, as it was throughout the test program.

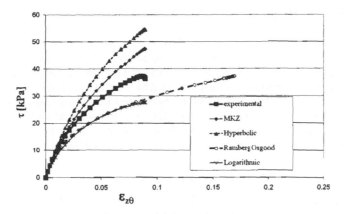

Figure 8.   Comparison between experimental virgin shear stress-strain curve and model predictions.

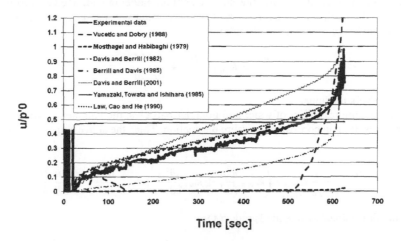

Figure 9.   Comparison between experimental pore-water pressure development and model predictions.

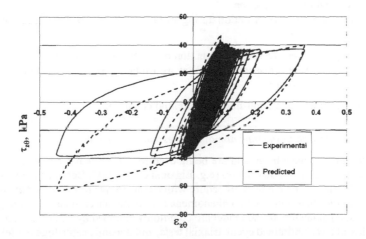

Figure 10.   Comparison between experimental and predicted cyclic shear stress—shear strain curves.

5   SITE RESPONSE ANALYSIS

On the basis of the experimental study described, a code for seismic site response has been prepared. The code considers one-dimensional, vertical propagation of shear waves from bedrock upward through saturated sand layers, and is based on the following assumed components of the material constitutive behavior: Virgin shear loading is modeled by the Modified Hyperpbolic Model (MKZ), pore-water pressure development is modeled by Berrill and Davis' dissipated energy model, and cyclic stress-strain behavior is modeled by the Masing relations modified for pore-water development. The code takes account both of development of pore-water pressure, and also of its dissipation, depending on the permeability of the sand, and on the boundary conditions.

5.1   *One dimensional pore-water pressure dissipation*

The present discussion is limited to a one-dimensional (vertical) model for both propagation of shear waves in the field, and flow of water. Since the shear strain profile is not uniform with depth at any time, different excess pore-water pressures develop at different depths, resulting in vertical flow. Use of an empirical pore-water pressure generation model in undrained conditions, such as an energy based model, enables expression of the pore-water pressure change as a superposition of the energy and volume change contributions:

$$\frac{\partial u}{\partial t} = K_{bulk} \cdot \frac{\partial \varepsilon_v}{\partial t} + \frac{\partial u}{\partial E}\frac{\partial E}{\partial t} \qquad (2)$$

where K_{bulk}—bulk modulus.

Assuming incompressibility of water and granular particles, the volumetric strain, $d\varepsilon_v$, is due to flow only, resulting in pore-water pressure dissipation. Thus, the above equation converts to the following form:

$$\frac{\partial u}{\partial t} = -\frac{1}{\gamma_w}K_{bulk} \cdot \left(k_{perx}\frac{\partial^2 u}{\partial x^2} + k_{pery}\frac{\partial^2 u}{\partial y^2} + k_{perz}\frac{\partial^2 u}{\partial z^2}\right) + \frac{\partial u}{\partial E}\frac{\partial E}{\partial t} \qquad (3)$$

and, for one dimensional flow, eqn (3) reduces to:

$$\frac{\partial u}{\partial t} = -\frac{1}{\gamma_w}K_{bulk} \cdot \left(k_{perz}\frac{\partial^2 u}{\partial z^2}\right) + \frac{\partial u}{\partial E}\frac{\partial E}{\partial t} \qquad (4)$$

where k_{perj} is permeability in direction i.

Thus, the pore-water pressure evolution during an earthquake, according to equation (4), can be resolved into two components:

1. Generation of pore-water pressure as a function of the accumulated energy.
2. Dissipation of pore-water pressure due to flow as a function of the permeability and the bulk modulus of the soil structure.

Eqn (4) is a one dimensional, non-linear differential equation.

5.2   *Non-linear approach in the time domain for response analysis*

Plasticity-based models have been, and are being used for site response analysis including pore-water pressure generation and dissipation (e.g. Elgamal et al., 2002; Bernadie et al., 2006). These models generally use a finite element procedure with a constitutive model aimed at simulating the cyclic mobility and/or flow-liquefaction phenomena commonly observed in laboratory triaxial and simple shear tests. Parameters of the model are commonly obtained by calibration against drained monotonic triaxial tests, undrained cyclic triaxial tests, and dynamic centrifuge model tests. These models are often complex, and evaluation of the model parameters is not simple. Consequently, this presentation offers an alternative approach, aimed at accounting for the effects of pore-water pressure development and dissipation on the basis of the simple, empirical, energy-based model described above, incorporated into a one dimensional wave propagation code.

A procedure to solve nonlinear, one dimensional ground response using step by step integration was proposed by Joyner and Chen (1975) who employed the hyperbolic stress-strain model to describe one dimensional shearing. Shiran (2000) extended Joyner and Chen's code and enabled the use of the R-O model (Ramberg and Osgood, 1943) and the logarithmic model (Puzrin and Burland, 1996) within the same framework. None of these approaches considered pore-water pressure generation or dissipation.

5.3 Site response program

A computer code has been developed for one dimensional site response, incorporating pore-water pressure generation, and dissipation by one dimensional flow. The code was formulated using the finite difference approach, combining the procedure of Joyner (1977) with a numerical, explicit formulation of equation (4), applying the energy based model of Berrill and Davis (1985). The code can use either the hyperbolic, logarithmic (Puzrin and Burland, 1996), or Hooke stress strain model for virgin loading, but the modified hyperbolic model has been found to be preferable, as shown above. Masing rules are adopted for cyclic loading, accounting for pore-water development. The Mohr-Coulomb effective stress failure criterion is assumed, and the maximum shear modulus is updated to account for pore-water pressure generation according to equation (5):

$$G_{max} = k_g \cdot p_a \sqrt{\frac{\sigma'_{v0} - u}{p_a}} \tag{5}$$

where G_{max}—maximum shear modulus.
k_g–calibration parameter.
p_a–atmospheric pressure.

The code uses the input motion and finds the shear stresses and strains at each differential element of the soil at any specific time. The energies computed according to the stresses and strains are then used to obtain the pore-water pressure by the numerical formulation of equation (4) at every element of soil. After obtaining the pore-water pressure at each element of soil, the maximum shear modulus and the failure shear stress are updated and used for update of the virgin stress-strain curve according to the MKZ model, and of the cyclic behavior according to Masing rules. The program then progresses one step in time.

5.4 The numerical solution scheme

Analysis of the one dimensional upward (z) propagation of shear waves through the sand deposit is based on the solution of the wave equation:

$$\frac{\partial \tau}{\partial z} = \rho \frac{\partial^2 x}{\partial t^2} = \rho \frac{\partial \dot{x}}{\partial t} \tag{6}$$

where τ = shear stress, x = horizontal displacement, ρ = mass density and t = time.

The soil column is divided into N elements, each of height Δz_i. The solution then advances in space in small increments of time, Δt, using the following finite difference approximations:

$$\frac{\partial \tau}{\partial z} \approx \frac{\tau_{i+1,t} - \tau_{i,t}}{\Delta z_i}$$
$$\frac{\partial \dot{x}}{\partial t} \approx \frac{\dot{x}_{i,t+\Delta t} - \dot{x}_{i,t}}{\Delta t} \tag{7}$$

Substituting equations (7) into (6) and rearranging:

$$\dot{x}_{i,t+\Delta t} = \dot{x}_{i,t} + \frac{\Delta t}{\rho_i \cdot \Delta z_i}(\tau_{i+1,t} - \tau_{i,t}) \tag{8}$$

Consequently, if the shear stresses and velocities are known at time t, the velocities can be calculated at time $t + \Delta t$. The horizontal displacements are then calculated:

$$x_{i,t+\Delta t} = x_{i,t} + \dot{x}_{i,t} \cdot \Delta t \tag{9}$$

Following update of the displacements over the whole column depth, shear strains are calculated:

$$\gamma_{i,t+\Delta t} = \frac{\partial x_{i,t}}{\partial z} = \frac{x_{i+1,t} - x_{i,t}}{\Delta z_i} \tag{10}$$

Now the constitutive relations are used in order to calculate shear stresses at time $t + \Delta t$, on the basis of the shear strains obtained from eq. (10).

The steps in performance of the analysis can, therefore, be summarized as follows:

1. At time $t = 0$, the displacements, velocities and stresses are known throughout the column.
2. Calculate velocities, at time $t + \Delta t$ from eq. (8).
3. Calculate displacements at time $t + \Delta t$ from eq. (9).
4. Calculate shear strains at time $t + \Delta t$ from eq. (10).
5. Calculate shear stresses at time $t + \Delta t$ from the constitutive model.
6. Advance one step in time and return to step 2.

A similar scheme was used by Matasovic (1993), but his pore-water pressure model was different.

6 EXAMPLE ANALYSIS

Site response analysis was carried out for the simple soil section shown in Fig. 11, as a result of the N-S horizontal rock motion recorded at Gold Hill 4 West during the Parkfield earthquake of 28.09.2004. The peak acceleration recorded in this motion was 0.384 g, and its duration was 20 seconds. The recorded acceleration history is shown in Fig. 12. The major shaking occurred during the first 10 seconds, with only small accelerations evident after that.

The site consists of two sand layers, one 10 m thick and the second 20 m thick, separated by a 2 m normally consolidated clay layer. The lower sand layer is underlain by rock, and water table is at the soil surface. The sand layers have a permeability, $k = 10^{-5}$ m/s, relative density, DR $= 60\%$ (corresponding to a submerged unit weight of 7 kN/m^3), cohesion, $c = 0$, friction angle, $\phi = 35°$ in the upper layer and 37° in the lower layer. Other properties were in accordance with findings of the experimental program; in particular, the pore-water pressure energy coefficient, α, was taken as 8.5.

The present research has not studied the cyclic behavior of clay, although this is planned for the future. The clay layer was included in the site profile in order to provide a comparison between two sand layers, one above and one below a low permeability layer. For the purpose of the present analyses, it was assumed that normally consolidated clay can be basically modeled by constitutive relations similar to those used for the sand. It is known (e.g. Matasovic, 1993) that pore-water pressure development in clays during cyclic loading is smaller than in sands, and so an α value, one-tenth that of the sand value, was assumed (i.e. $\alpha = 0.85$). The small-strain shear modulus, G_{max}, was assumed equal to 20,000 kPa, the submerged unit weight 10 kN/m^3, and the coefficient of permeability, $k = 10^{-9}$ m/s.

Fig. 13 shows development and dissipation of pore-water pressure at three different levels in the profile during the earthquake. The pressure at the base of the top layer reaches 70 kPa, equal to the effective overburden pressure, indicating that liquefaction develops at this depth. On the other hand,

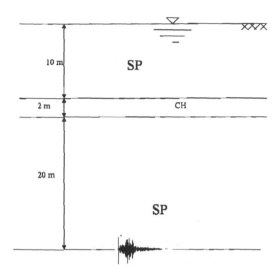

Figure 11.   Soil section for one-dimensional site response analysis.

Input acceleration

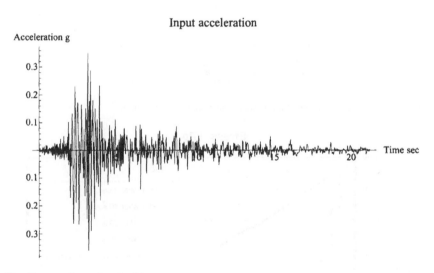

Figure 12.   Input rock acceleration history.

at 1 m depth the maximum pore-water pressure developed is about 3.4 kPa, less than the effective overburden pressure of 7 kPa, and so there is no liquefaction at this depth This is very likely a result of dissipation of pore-water pressure, in parallel to its development, by flow to the soil surface. The maximum pressure that develops at 22 m is less than half that at 10 m, presumeably as a result of the higher rigidity of the sand at this depth. The pore-water pressure begins to decrease towards the end of the earthquake; the rate of decrease is greater at 10 m than at 1 m, as a result of upward flow of water from 10 m towards the surface. As a result of the very low α value assumed for the clay, insignificant pore-water water pressures developed in this layer.

Fig. 14 shows distribution of pore-water pressure with depth at different times following the start of the earthquake, together with the distribution of effective overburden pressure. It is seen, again, that at the base of the upper sand layer at 10 m, the excess pore-water pressure equals the effective overburden pressure, and this results in liquefaction. These high pore-water pressures above the clay layer were found to be a result of severe amplification through the clay layer, resulting in

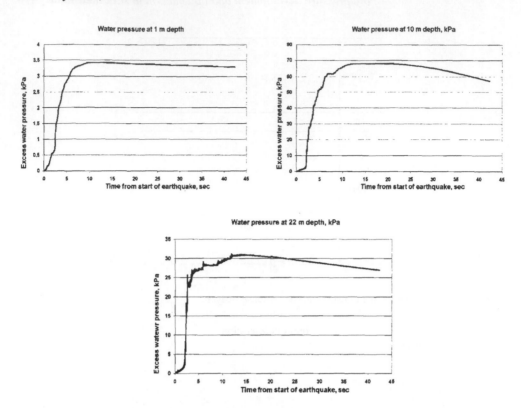

Figure 13.   Development and dissipation of pore-water pressure with time.

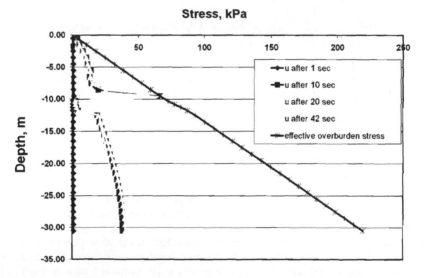

Figure 14.   Distribution of excess pore-water pressure with depth at different times.

high accelerations at base of the upper sand layer. De-amplification then occurred up to the soil surface, which together with drainage at the soil surface resulted in the lower pore-water pressures developed there. Due to the slow rate of dissipation evident in Fig. 13, no significant distinction is seen between the different time curves in Fig. 14.

7 CONCUDING REMARKS

A program of laboratory testing was carried out, in which torsional, hollow cylinder equipment was developed for cyclic loading of sand samples. On the basis of a review of the literature and analysis of the results from this program, a constitutive framework for the behavior of saturated sand under cyclic shear was developed. The central feature of this framework is an energy-based model for pore-water pressure build up during undrained cyclic loading. The constitutive framework formed the basis for preparation of a computer code for analysis of seismic site effect of granular profiles, and results of the analysis of a simple soil profile have been presented. These analyses account for the build up and dissipation of pore-water pressure during and following an earthquake. It is envisaged that this approach will be extended to a study of other soil types such as clays and clayey sands.

REFERENCES

Arulmori, K., Muraleetharan, K.K., Hossain, M.M. and Fruth L.S. 1992. VELACS: Verification of liquefaction analyses by centrifuge studies; laboratory testing program—soil data report. *Res. Rept., The Earth Technology Corp.*

Baker, R., Frydman, S. and Galil, J. 1981. A plasticity model for the load unload behavior of sand. *Proc. Symposium on Limit Equilibrium, Plasticity and Generalized Stress-Strain Applications in Geotechnical Engineering, ASCE*, Editors: R.N. Yong & E.T. Selig, 25–51.

Bernadie, S., Foerster, E. and Modaressi, H. 2006—Non-linear site response simulations in Chang-Hwa region during the 1999 Chi-Chi earthquake, Taiwan. *Soil Dynamics and Earthquake Engineering*, 26, 1038–1048.

Berrill, J.B. and Davis, R.O. 1985. Energy dissipation and seismic liquefaction of sands: revised model. *Soils and foundations*, 25(2), 106–118.

Byrne, P.M. 1991. A cyclic shear volume-coupling and porewater pressure model for sand. *Proc., 2nd Int. Conf. on Recent Advances in Geotechnical Earthquake Engineering & Soil Dynamics*, St. Louis, 47–56.

Davis, R.O. and Berrill J.B. 1982. Energy dissipation and seismic liquefaction is sands. *Earthquake Eng. and Struct. Dyn*, 10, 59–68.

Davis, R.O. and Berrill, J.B. 1998. Rational approximation of shear stress and strain based on downhole acceleration records. *Int. J. Numer. & Analyt. Methods in Geomech.*, 22, 603–619.

Davis, R.O. and Berrill, J.B. 2001. Pore pressure and dissipation energy in earthquakes- field verification. *J. Geotech. and Geoenvir. Engrg. ASCE*, 127(3), 269–274.

Elgamal, A.W., Zeghal, M. and Parra E. 1996. Liquefaction of reclaimed island in Kobe, Japan. *J. Geotech. Eng. ASCE*, 122(1), 39–49.

Finn, W.D.L., Lee, K.W. and Martin, G.R. 1977. An effective stress model for liquefaction. *J. Geotech. Eng. Div., ASCE*, 103(GT6), 517–533.

Ishihara, K., Shimizu, K. and Yamada, Y. 1981. Pore water pressure measured in sand deposits during an earthquake. *Soils and Foundations*, 21(4), 85–100.

Ishihara, K., Muroi, T. and Towhata, K. 1989. In situ pore water pressure and ground motions during the 1987 Chiba-Toho-Oki earthquake. *Soils and Foundations*, 29(4), 75–90.

Joyner, W.B. 1977. NONLI3: A fortran program for calculating nonlinear ground response. *Open File Rpt. 77–761, U.S. Geological Survey*, Menlo Park, California.

Joyner, W.B. and Chen, A.T.F. 1975. Calculation of nonlinear ground response in earthquakes. *Bull. Seis. Soc. America*, 65(5), 1315–1336.

Kim, S.I., Park, K.B., Seo, K.B. and Park, S.Y. 2007. Excess pore pressure generation model for liquefaction and post-liquefaction analysis of sand. *13th Asian Regional Conf. of Soil Mech. & Geotech. Engng.*, Koolkata.

Kondner, R.L. and Zelasko, J.S. 1963. A hyperbolic stress-strain formulation of sands. *Proc. 2nd Pan American Conf. on Soil Mech. and Foundn Eng.*, Brazil, 1, 290–324.

Law, K.T., Cao, Y.L. and He, G.N. 1985. An energy approach for assessing seismic liquefaction potential. *Canadian Geotechnical Journal*, 27(3), 320–329.

Martin, G.R., Finn, L.W.D. and Seed. H.B. 1975. Fundamentals of liquefaction under cyclic loading. *J. Geotech. Eng. Div., ASCE*, 101(GT5), 423–438.

Matasovic, N. 1993. Seismic response of composite horizontally-layered soil deposits. *PhD Thesis*, University of California, Los Angeles.

Mostaghal, M. and Habibagahi, K. 1979. Cyclic liquefaction strength of sands. *Earthquake Engineering and Structural Dynamics*, 7, 213–233.

Nemat-Nasser, S. and Shokooh, A. 1979. A unified approach to densification and liquefaction of cohesionless sand in cyclic shearing. *Canadian Geotechnical Journal*, 16, 659–678.

Puzrin, A.M. and Burland, J.B. 1996. Logarithmic stress-strain function for rocks and soils. *Geotechnique*, 46(1), 157–164.

Pyke, R. 1979. Non linear soil models for irregular cyclic loading. *J. Geotech. Eng. Div., ASCE*, 105(GT5), 715–726.

Ramberg, W. and Osgood, W.R. 1943. Description of stress-strain curves by three parameters. *Technical Note 902, National Advisory Committee for Aeronautics*, Washington, D.C.

Shen, C.K., Wang, Z. and Li, X.S. 1991. Pore pressure response during 1986 Lotung earthquake. *Proc., 2nd Int. Conf on Recent Advances in Geotechnical Earthquake Engineering & Soil Dynamics*, St Louis, 11–15.

Shiran, A. 2000. Kinematic seismic response of piles in sand. *M.Sc Research Thesis.* Technion—I.I.T. (In Hebrew).

Talesnick, M. and Shehadeh, S. 2007. The effect of water content on the mechanical response of a high porosity chalk. *Int. J. for Rock Mech. and Mining Sciences*, 44(4), 584–600.

Towahata, I. and Ishihara, K. 1985. Shear work and pore water pressure in undrained shear. *Soils and Foundations*, 25(5), 73–84.

Yamazaki, F., Towhata, I. and Ishihara, K. 1985. Numerical model for liquefaction problem under multi-directional shearing on horizontal plane. *5th Int. Conf. on Numerical Methods in Geomechanics*, Nagoya, 399–406.

Youd, T.L. and Holzer, T.L. 1994. Piezometer performance at Wildlife liquefaction site, California. *J. Geotech Eng., ASCE*, 120(6), 975–995.

Characterizing localization processes during liquefaction using inverse analyses of instrumentation arrays

Ronnie Kamai & Ross Boulanger

Department of Civil and Environmental Engineering, University of California, Davis, CA, USA

ABSTRACT: This paper presents results from, and an evaluation of, inverse analysis techniques applied to dense instrumentation arrays in a dynamic centrifuge model test for the purpose of better characterizing the localization processes at sand/clay interfaces due to earthquake-induced liquefaction. Three different inverse analysis techniques for defining the dynamic stress-strain responses from an accelerometer array are compared and discussed. Two different inverse analysis techniques for defining pore water flow and volumetric strains during and after shaking from pore-pressure transducer arrays are compared and discussed. Results from the stress-strain and pore-water-flow inverse analyses are combined with data from displacement transducers and high-speed video cameras to describe the processes that affected liquefaction-induced deformations and localizations in the centrifuge model.

1 INTRODUCTION

The in-situ residual strength of soils liquefied by earthquake loading can be greatly affected by seepage-induced void redistribution and possible shear localizations at interfaces with overlying lower-permeability layers. Figure 1 is a schematic illustration of a submerged infinite slope, in which a liquefiable soil that is initially dense-of-critical (i.e. not susceptible to flow deformation at its current state) is confined by lower-permeability layers above and below it. Excess pore water pressures generated by strong earthquake shaking can cause an upward hydraulic gradient through the profile. The excess pore water pressures then dissipate by upward flow, causing a net outflow of water at point C, and hence densification (contraction) of the lower part of the layer, denoted h_c in Figure 1. At the top of the liquefiable layer, point A, the upwards flow is impeded by the lower-permeability barrier, leading to a net inflow of water, and hence to a loosening (dilation) of some upper portion of the layer, denoted h_d in Figure 1. The progressive loosening of the soil at point A can result in a shear localization at the interface, and if the associated strength loss is sufficient, to instability of the slope (Whitman 1985).

Understanding the evolution of the dilating zone in Figure 1 is critical to determining the potential for localization due to void redistribution. The volumetric strains in the dilating zone are directly related to the volume of water that is expelled from the contracting sublayer (V_c) and the thickness of the dilating zone. The volume of water expelled from the contracting sublayer depends on the soil relative density (D_R), slope angle, layer thickness, and ground motion characteristics (e.g. Kulasingam et al. 2004). The thickness of the dilating zone depends on similar factors and it evolves during the process of void redistribution. For example, 1D analyses of post-shaking void redistribution in an infinite slope (Malvick et al. 2006) suggest that the thickness of the dilating zone evolves with changes in the soil's dilatation angle, such that the dilating zone can initially be a significant fraction of the liquefiable layer's thickness (before the soil has loosened significantly), but then decrease to the thickness of a shear band (i.e. several grain diameters thick) when the soil at point A has loosened sufficiently for the dilation angle to become zero.

Numerical modeling of an infinite slope problem by Seid-Karbasi and Byrne (2007) similarly suggests that the thickness of the dilating zone can initially be a significant fraction of the overall liquefiable layer thickness, followed by the formation of a shear band at the top interface. Understanding the capacity of the dilating zone to accommodate the inflow of water from the underlying contracting zone is an important part of being able to distinguish between cases where the process of void redistribution does, or does not, lead to shear localizations or slope instability.

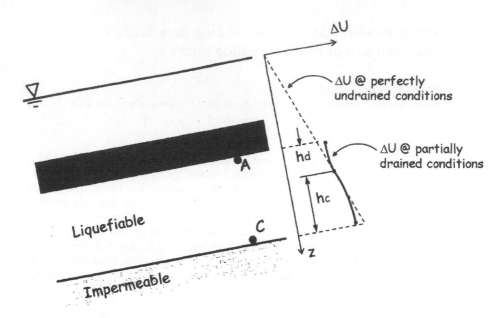

Figure 1.   Void redistribution schematic (Kulasingam et al. 2004).

The mechanics of void redistribution have become better understood in recent years through a combination of physical model studies, laboratory element studies, and theoretical or numerical analyses (e.g. see reviews in Kulasingam et al. 2004; Kokusho 2006; Seid-Karbasi & Byrne 2007). The evolution and timing of localization processes associated with void redistribution have not, however, been well defined by measurements in prior experiments, in large part because of the current inability to directly measure shear and volumetric strains with sufficient resolution as the processes occur. Malvick et al. (2008) attempted to define the evolution of a dilating zone by introducing an inverse analysis method for determining pore water flow and volumetric strain rates from pore pressure transducer arrays; their results for one saturated sand slope indicated that the dilating zone beneath an embedded silt interlayer overlying approximately 10 m of loose sand was initially on the order of 1 m thick prior to the formation of a thin shear band at the silt layer interface.

This paper presents results from, and an evaluation of, inverse analysis techniques applied to dense instrumentation arrays in a dynamic centrifuge model test for the purpose of better characterizing the processes of void redistribution and shear localization at sand/clay interfaces due to earthquake-induced liquefaction. The centrifuge model is described first, followed by the inverse analyses. Results from three different inverse analysis techniques for defining the dynamic stress-strain responses from an accelerometer array are compared and discussed. Results from two different inverse analysis techniques for defining pore water flow and volumetric strains during and after shaking from pore pressure transducer arrays are compared and discussed. The results from the stress-strain and pore-water-flow inverse analyses are combined with data from displacement transducers and high-speed video cameras to describe the processes that affected liquefaction-induced deformations and localizations in the centrifuge model.

2   CENTRIFUGE EXPERIMENT

The dynamic centrifuge model test that is analyzed in this paper was one of three tests performed to evaluate the effectiveness of geosynthetic drains for mitigating liquefaction of sands (Marinucci et al. 2008). A schematic representation of the dynamic centrifuge model is presented in Figure 2. The test was performed in a flexible shear beam container (0.8-m wide, 1.65-m long, 0.5-m deep) on the 9-m radius centrifuge at the Center for Geotechnical Modeling at UC Davis. The test was performed at a centrifugal acceleration of 15 g and the results are presented in equivalent prototype units unless otherwise stated. The scaling factors used to convert from model scale to prototype scale are listed in Table 1.

The soil model was constructed of two symmetrical slopes separated by an open channel between them, as shown in Figure 2. The soil profiles consisted of a 5-m thick (320 mm model scale) liquefiable layer of Nevada sand overlain by a 1-m thick (67 mm model scale) crust of clayey-silt, which consolidated to approximately 0.88-m thickness after spin-up. One side was treated with model-scale plastic drains and the other side was left untreated to evaluate the effectiveness of the drains in liquefaction mitigation. Two dense vertical instrumentation arrays of accelerometers (Acc's) and pore pressure transducers (PPT's) were placed on each side to track wave propagation and water dissipation throughout the profiles. Sand was placed using dry pluviation, calibrated to the desired relative density ($D_R = 40\%$). When sensors had to be placed, pluviation was stopped and excess sand was vacuumed out to reach the appropriate height before sensor placement. The clayey

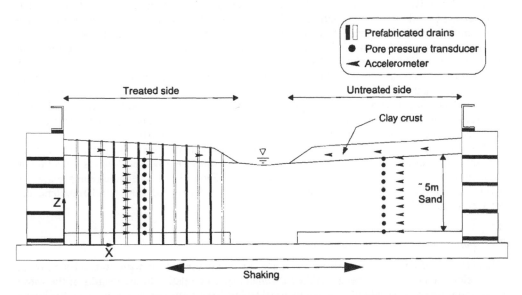

Figure 2. Cross-section of model showing vertical instrumentation arrays (Note that other transducers are omitted from the figure for clarity).

Table 1. Scaling factors for common parameters (Kutter 1994).

Parameter	Ratio of model to prototype[a]
Length	$1/n$
Area	$1/n^2$
Volume	$1/n^3$
Stress	1
Strain	1
Force	$1/n^2$
Velocity	1
Acceleration	n
Frequency	n
Time (dynamic)	$1/n$
Time (Consolidation) [b]	$1/n^2$

[a] n = centrifuge acceleration

[b] Consolidation time scaling based on same pore fluid viscosity in model and prototype.

crust was placed in multiple lifts compacted by gentle tamping. The model was placed under vacuum, flushed with CO_2, and then saturated with de-aired water. The model was subjected to five shaking events separated from each other by sufficient time for full dissipation of any excess pore water pressures. The five successive shaking events consisted of 20 uniform sinusoidal cycles at 2 Hz with single-amplitudes of 0.01, 0.03, 0.07, 0.11 and 0.3 g, respectively. Two of the five shaking events will be discussed later in this paper—one with amplitude of 0.07 g, referred to as "shake 3" and the other with amplitude of 0.11 g, referred to as "shake 4".

3   INVERSE ANALYSES OF ACCELEROMETER ARRAYS TO OBTAIN STRESS-STRAIN RESPONSES

Inverse analyses of accelerometer array data from the field or physical models to define the stress-strain response of the soil has been found useful by other researchers and is being used in ongoing studies. The derivation of shear strain requires integrating the accelerations twice in time to obtain displacements, followed by differentiation of the displacements in space to obtain shear strains. The differentiation of displacement data at discrete points has been performed using different numerical methods. Zeghal and Elgamal (1994) and Zeghal et al. (1995) used piece-wise linear interpolation, whereas Davis and Berrill (1998, 2001) used a cosine chain similarly to conducting a Fourier transform of the displacement profile. Wilson (1998) and Brandenberg et al. (2009) compared a weighted-residual method with several other methods for the differentiation of discrete bending moment data along piles. Most of these past studies were conducted on sparse arrays in soils or on structural features such as piles, and thus the present motivation was to re-evaluate these inverse analysis techniques for use with dense accelerometer arrays in soil profiles.

The inverse analysis techniques compared herein assume a 1D shear-beam response with upward propagation of shear waves. Other forms of waves such as surface waves or P-waves reflecting off the container walls are assumed to be largely negligible. Figure 3 is a schematic representation of a 1D soil column, divided into elements of soil between every two adjacent accelerometers, also referred to as 'nodes'.

3.1   *Importance of instrument spacing*

The accuracy of any inverse analysis technique for interpreting shear wave transmission based on accelerometer records depends on the instrument spacing relative to the lengths of the waves being transmitted. Conceptually, most numerical techniques should perform well when the instrument spacing is less than about one eighth of the shortest wave length ($\lambda_{min}/8$), suggesting a critical dependence on effective shear-wave velocity (V_S) and the highest frequency (f_{max}) that is significantly present in the transmitted motion (Eq. 1).

$$\Delta Z_{max} = \frac{\lambda_{max}}{8} = \frac{V_s}{8 \cdot f_{max}} \tag{1}$$

The effective shear wave velocity of the soil was determined by cross-correlation of individual upwardly-propagating wave pulses during different shaking events. The shear wave velocity was estimated to be approximately 200 m/sec at small shaking levels (i.e., before shake 3). Once stronger shaking was applied and the secant shear modulus started to degrade, the effective shear wave velocities decreased to about 130 m/s before liquefaction during shake 3 (0.07 g) and to approximately 20 m/s after liquefaction during shake 4 (0.11 g). For a dominant shaking frequency of 2 Hz, these velocities suggest wave lengths that range from about 100 m at low shaking levels to only 10 m at strong shaking levels with liquefaction. This suggests an optimal sensor spacing of no larger than 12 m for the smaller shakes, and no larger than 1.25 m for the liquefied profile. The average spacing of functioning transducers for this array was 0.73 m, with the maximum being 1.26 m between depths of 3.5 and 4.7 m. This suggests that the vertical array of accelerometers

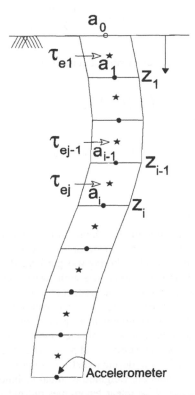

Figure 3.   Schematic of 1D shear-beam model for interpreting the vertical array data.

was dense enough to accurately describe wave transmission even after liquefaction developed in the soil profile.

The results of the inverse analyses may also be affected by any errors in the soil motions that are computed from the accelerometer records, including measurement errors (e.g. noise, drift, calibration factors, instrument bandwidths) and signal processing limitations (e.g. filtering, baseline corrections). The calculation of shear stresses is relatively insensitive to such errors, whereas the calculation of shear strains, which involves both the integration with time and differentiation with space of accelerometer data, is more sensitive to various error sources. The comparison between different numerical techniques for computing shear strains, presented later in this section, illustrates the different degrees of sensitivity to potential measurement errors and signal processing assumptions.

3.2   Calculation of shear stresses

Shear stresses are calculated at the centers of each element (midpoints between accelerometers), as indicated in Figure 3, by summation of the horizontal inertia force of the soil above that point. Accelerations are assumed to vary linearly between accelerometers. The acceleration at node 0 (the surface) is set equal to the acceleration at node 1 (the first accelerometer), which assumes that the soil above the top accelerometer acts as a rigid body. This is a reasonable assumption given that the first accelerometer is located at a depth of only 0.5 m, and the typical wave lengths in the non-liquefied crust are on the order of several meters as discussed previously. The shear stress at the surface is zero, and thus the shear stresses at the first two elements are determined as:

$$\tau_{e,1} = \rho \cdot \frac{z_1}{2} \cdot a_1 \tag{2}$$

$$\tau_{e,2} = \tau_{e,1} + \rho \cdot \frac{z_1}{2} \cdot a_1 + \rho \cdot \frac{z_2 - z_1}{2} \cdot \left(\frac{3 \cdot a_1 + a_2}{4} \right) \qquad (3)$$

and the shear stress at the subsequent elements is determined as:

$$\tau_{e,j} = \tau_{e,j-1} + \rho \cdot \frac{z_{i-1} - z_{i-2}}{2} \cdot \left(\frac{3 \cdot a_{i-1} + a_{i-2}}{4} \right) + \rho \cdot \frac{z_i - z_{i-1}}{2} \cdot \left(\frac{3 \cdot a_{i-1} + a_i}{4} \right) \qquad (4)$$

where ρ is the density of the soil, $\tau_{e,j}$ is the shear stress at element j, and Z_i and a_i are the depth and acceleration at node i.

3.3   Calculation of shear strains

Shear strains are computed as the derivative of the soil displacements with depth. The displacements are obtained by double-integrating the accelerometer records, after filtering out frequencies that are not within the sensor's bandwidth. Filtering was performed in the frequency domain using 7th-order Butterworth high-pass and low-pass filters with corner frequencies of 0.6 Hz and 50 Hz, respectively, based on the signal-to-noise characteristics of the recorded accelerations and knowledge of the instrumentation/data acquisition system's characteristics (e.g. Wilson et al. 1998). The resulting displacements are the transient component only. Five numerical techniques for computing the derivative of the soil displacements between accelerometer depths were evaluated: weighted residual, cubic spline, cosine series, polynomial, and piece-wise linear. The latter four techniques begin by using interpolation functions to define a distribution for the displacements between the discrete measurement points, whereas the weighted residual technique is a direct differentiation scheme. The polynomial and piece-wise linear techniques did not perform as well as the cubic spline or weighted residual techniques, and thus are not presented for purposes of brevity.

A cubic spline is a piecewise third-order polynomial, fit to each element separately, forcing continuity of both first and second derivatives at the nodes. The cubic spline is forced to fit the data perfectly at all measurement points. The form of the cubic spline is,

$$D_i(z) = a_i(z - Z_i)^3 + b_i(z - Z_i)^2 + c_i(z - Z_i) + d_i \qquad (5)$$

where Z_i is the depth for instrument i, and a_i, b_i, c_i and d_i are constants found for each depth interval at each time step. Boundary conditions can be introduced by constraining the values of these constants as appropriate; e.g. zero strain at the ground surface requires $c_0 = 0$ at all times.

A cosine series interpolation function, as suggested by Davis and Berrill (1998), automatically satisfies the free surface boundary conditions of zero shear strain. It is similar in its trigonometric essence to the elastic wave propagation equation, and by that seems to have great potential for representing propagating waves. The displacements are fit with the function:

$$D(z) = a_0 + a_1 \cos(\kappa z) + a_2 \cos(2\kappa z) + a_3 \cos(3\kappa z) + \cdots a_{J-1} \cos((J-1)\kappa z) \qquad (6)$$

where J is the number of instruments in the array, κ is the wave number, and the coefficient a_0 through $a_{(J-1)}$ are determined at each time step for the instantaneous values of computed soil displacements. The value for κ is determined based on instrument spacing in a two-step procedure as follows: first, a trial value for κ, κ_{trial}, is chosen based on the greatest instrument depth being roughly equal to a quarter wavelength (Eq. 7). Then, the Nyquist wave number, κ_N, is determined based on the smallest instrument spacing (Eq. 8). If the value of $J * \kappa_{trial}$ is smaller than κ_N, the trial value for κ_{trial} is accepted; otherwise, κ is reduced to some value smaller than κ_N/J (Eq. 9).

$$\kappa_{\text{trial}} = \frac{2 \cdot \pi}{4 \cdot Z_{\max}} \tag{7}$$

$$\kappa_N = \frac{2 \cdot \pi}{2 \cdot \min[(Z_i - Z_{i-1})]} \tag{8}$$

$$\kappa = \begin{cases} \kappa_{\text{trial}} & \text{IF} \quad J * \kappa_{\text{trial}} > \kappa_N \\ \dfrac{\kappa_N}{J} & \text{otherwise} \end{cases} \tag{9}$$

The weighted residual technique does not use a displacement interpolation function, but rather is a differentiation scheme based on minimizing weighted residuals, as is often used in finite element approximations (Brandenberg et al. 2009; Wilson 1998). According to the Weighted Residual (WR) differentiation scheme, the derivative $g(z)$ to a function $f(z)$, for which only discrete data is available, can be obtained by multiplying $(g\text{-}f')$ by a suitable test function and setting to zero (Eq. 10), such that $g(z)$ is equal to $f'(z)$, "weakly".

$$\int \{g(z) - f'(z)\} \cdot \psi(z) dz = 0 \tag{10}$$

The basis functions $\psi(z)$ can be any arbitrary weighting function. For this study, and following Brandenberg et al. (2009), the weighting functions were chosen to be "finite-element type" piecewise-linear basis functions. The discretized form of $f(z)$ and $g(z)$, using the discrete data obtained at the nodes, results in a linear system of equations (Eq. 11), that can be solved for each time step to obtain $f'(z)$. The number of terms in Equation 11, J, is equal to the number of instruments in the array.

$$\begin{pmatrix} f_2 - f_1 \\ f_3 - f_1 \\ f_4 - f_2 \\ \vdots \\ f_J - f_{J-2} \\ f_J - f_{J-1} \end{pmatrix} = $$

$$\frac{1}{3} \cdot \begin{pmatrix} 2(x_2 - x_1) & (x_2 - x_1) & 0 & 0 & \cdots \\ (x_2 - x_1) & 2(x_3 - x_1) & (x_3 - x_2) & 0 & \cdots \\ 0 & (x_3 - x_2) & 2(x_4 - x_2) & (x_4 - x_3) & \cdots \\ \vdots & \vdots & \vdots & \vdots & \vdots \\ \cdots & 0 & (x_{J-1} - x_{J-2}) & 2(x_J - x_{J-2}) & (x_J - x_{J-1}) \\ \cdots & 0 & 0 & (x_J - x_{J-1}) & 2(x_J - x_{J-1}) \end{pmatrix} \begin{pmatrix} g_1 \\ g_2 \\ g_3 \\ \vdots \\ g_{J-1} \\ g_J \end{pmatrix} \tag{11}$$

3.4 *Comparison of numerical techniques for computing shear strains*

The distributions of displacements and shear strains with depth for the three different numerical techniques at two different times during shake 3 are compared in Figure 4. Figure 4a shows the profile at $t = 9$ sec, before triggering of liquefaction, whereas Figure 4b shows the profile at $t = 12$ sec, after the triggering of liquefaction. The displacements vary more smoothly with depth before liquefaction than after liquefaction, and the shear strains are smaller before liquefaction than after liquefaction. The cubic spline and weighted residual techniques produce reasonably consistent profiles of shear strains throughout the soil, whereas the cosine series produced an erratic response over the lower portions of the array for both strains and displacements.

The poor performance of the cosine series was unexpected, given that the form of the interpolation technique seems physically well founded. Several variations on this form of interpolation function were implemented which covered a range of terms, wave numbers, and combinations of wave numbers. A less erratic response at the lower intervals could be obtained by limiting the largest wave number to a small value (by that keeping the wave lengths of the fitting function long) and

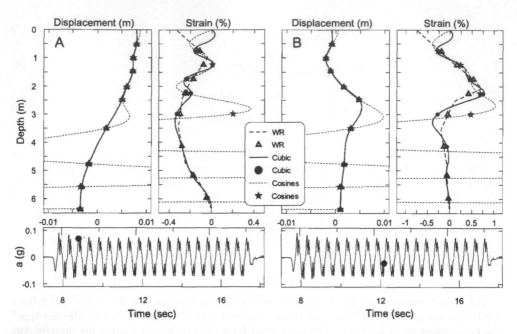

Figure 4.   Profiles of displacements and shear strains for 3 techniques at: (A) $t = 9$ sec, and (B) $t = 12$ sec. (Note that the strain scales are different in (A) and (B)).

reducing the number of terms, but this eliminated the ability of the function to follow sharper variations in soil displacement with depth. Including larger wave numbers (short wave lengths) provided a better fit with soil displacements at shallow depths (with closer instrument spacings), but produced erratic soil displacement profiles between instrument locations at the lower depths.

3.5   *Comparison of stress-strain responses*

A complete representation in time and depth of acceleration time histories alongside with the back-calculated stress-strain response for both the weighted residual and cubic spline techniques is given in Figure 5 for the upper portion of the soil profile during shake 3 (results for the lower depths are similar, but omitted for brevity). Since strain is sampled at the mid-points, there are 6 acceleration time histories, and only 5 corresponding stress-strain plots. The stress-strain responses computed using the weighted residual technique (1st column) are very similar to those obtained using the cubic spline technique (2nd column). Both sets of results show the progressive degradation of shear modulus with each cycle of loading, including a substantial loss of stiffness and increase in shear strains at the time when excess pore water pressures were approaching 100% of the initial vertical effective stresses. This progressive degradation of shear stiffness is further illustrated in the 3rd column of Figure 5 showing a sample of 3 single cycles out of the complete time history; for example at depth 1.75 m, the response is relatively elastic for cycle 1, shows a substantially softer response with a clear strain-hardening (dilation) response at large negative strains for cycle 6, and has very low stiffness in cycle 18 at the end of shaking because liquefaction has been triggered at that depth and the dynamic shear strains are smaller than those that developed in earlier cycles of loading.

The accelerations and stress-strain responses obtained from the next shaking event, where lique-faction has triggered faster and to a greater depth, are shown in Figure 6. The stress-strain responses are not as clearly comparable to those observed in typical laboratory element tests, now that the soil has significantly softened very early in shaking. Nonetheless, both the weighted residual and cubic spline interpolation techniques are consistent in identifying certain general behaviors. For example, consider again the response at a depth of 1.75 m, a strong dilation peak can be seen in

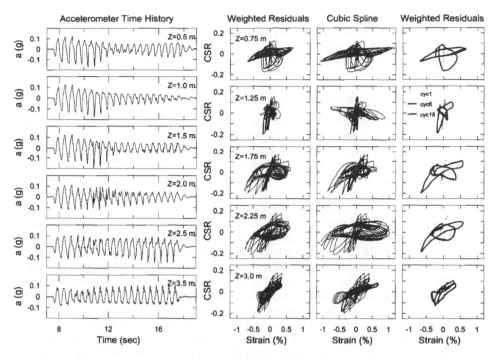

Figure 5. Acceleration time histories for 6 depths, together with stress-strain responses computed using the weighted residual and cubic spline interpolation techniques for shake 3.

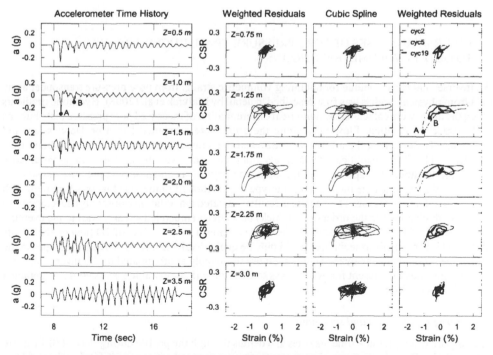

Figure 6. Acceleration time histories for 6 depths, together with stress-strain responses computed using the weighted residual and cubic spline interpolation techniques for shake 4.

cycle 2 (coincides with the transmission of a strong acceleration pulse up through the soil profile), after which the shear modulus degrades rapidly, and by the end of shaking, the soil column is so soft that very little motion is transmitted to the surface. These aspects of behavior are evident in both the acceleration time-histories and the small stress-strain loops at the end of shaking.

3.6   *Summary of evaluation of inverse analysis techniques*

The weighted residual and cubic spline techniques produced the most reasonably consistent results across a range of motions and depths for the dense vertical array from the centrifuge model test described herein. The cosine series technique performed surprisingly poorly, despite the promising similarity to the physical wave propagation equation. The polynomial interpolation technique also performed poorly at certain depths and in different shaking events, regardless of variations in the polynomial degree or imposed boundary conditions. The piecewise-linear interpolation technique performed reasonably well, with a few exceptions, but offers no particular advantage over either the cubic spline or weighted residual techniques. The combined use of the weighted residual and cubic spline techniques provides an independent check on the consistency of results, which can be advantageous for identifying cases where the inverse analysis results are less well behaved.

The quality of the computed stress-strain results depends strongly on the instrument spacing in the array, the quality of accelerometer data, and the degree to which the assumption of 1D shear beam response is appropriate. Limitations in the current analysis results include: (1) calculated strains are transient only and do not include permanent deformations that occur too slowly to be measured within the bandwidth of the accelerometers, (2) the relative influence of the model container on the soil column response becomes more important as the soil liquefies and softens, and (3) the formation of localized cracks, water films, and soil boils when liquefaction occurs introduces additional spatial variations in the soil model that are only represented "on average" by the inverse analyses. Consequently, it is believed that more sophisticated inverse analysis techniques based on a 1D shear beam approximation will not improve the results relative to those presented herein.

4   INVERSE ANALYSES OF PORE PRESSURE TRANSDUCER ARRAYS TO OBTAIN VOLUMETRIC STRAINS

An inverse analysis technique for obtaining volumetric strains from pore pressure transducer array data from physical models was first described in detail by Malvick et al. (2008). Hydraulic gradients along the array were calculated by numerical differentiation of measured pore pressure distributions with respect to position. Flow rates along the array were computed based on Darcy's law and assuming 1D flow along the array. Flow rates were integrated with respect to time to obtain flow quantities. Net flow quantities into intervals along the array were used to determine volumetric strains versus time. They applied their technique to arrays of six and nine pore pressure transducers located within a centrifuge model of a prototype 2 : 1 saturated sand slope that was 9-m high with 7.2 m of sand below the toe. Their analysis results indicated that dilation (loosening) developed in a relatively thick zone (on the order of 1 m thick) beneath an embedded silt arc, prior to the eventual formation of a thin shear localization at the sand/silt interface. The effect of 2D flow on the results of this 1D inverse analysis technique could only be qualitatively discussed. The gentle slopes in the centrifuge model being analyzed herein are more reasonably approximated by 1D inverse analysis models, and thus the present motivation was to determine if comparable results could be obtained under these conditions.

4.1   *Pore pressure data and screening*

Excess pore water pressure time histories at 6 depths through the profile during shake 4 (0.11 g) are represented in Figure 7, in terms of an excess pore water pressure ratio, $r_u = \Delta U/\sigma'_{v0}$, where $\Delta U =$ excess pore water pressure, and $\sigma'_{v0} =$ initial vertical effective stress. The records show that the top

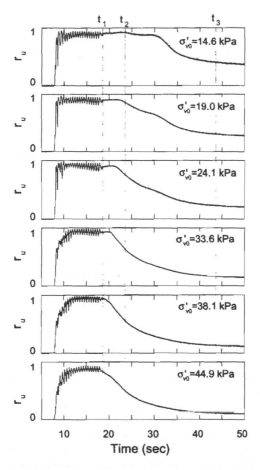

Figure 7. Pore pressures at 6 depths, shake 4 (0.11 g). t_1, t_2 and t_3 are snapshot times for Figure 8.

of the profile has triggered liquefaction (in terms of $r_u = 100\%$) during the first cycle, whereas the bottom of the profile triggered liquefaction after approximately 10 cycles of shaking. Pore water pressures remained elevated at all depths throughout the remainder of shaking. Dissipation started immediately after shaking at the bottom-most sensor, whereas pore pressures remained high for the longest duration at the top sensor. This is due to an upward hydraulic gradient and water flow from the bottom to the top of the array.

The pore pressure transducer data was screened for evidence of measurement errors because any such errors were found to have a strong effect on the volumetric strains obtained from the inverse analyses. Screening criteria and corrections were applied based on the assumption that all sensors should line-up on a consistent hydrostatic reading before shaking. First, the water table height was determined by averaging hydrostatic readings of three sensors at the bottom of the profile. Next, a 'calibration' ratio between the actual sensor reading and the reading it should have had at its recorded depth according to the obtained water table height was calculated. This 'calibration' correction was applied up to a maximum correction of 1% from the original pressure readings, as it was estimated that the cumulative errors from sensor calibration, temperature drifts, and data acquisition system settings should be less than about ±1%. The next correction was a depth correction; sensor locations were measured after they were placed in the soil during model preparation with an accuracy of ±0.1 mm. However, sensor location could have changed during further pluviation, consolidation, or any other treatment of the model. The limit on the depth correction was set to ±4 mm (model scale) as a reasonable shift that might occur during or after sensor placement. A depth correction up to this maximum value was applied so that the final sensor depth would coincide with the appropriate hydrostatic reading, such that all sensor readings are consistent with one identical water table elevation before shaking. Two of the pore pressure transducers were determined to be outliers, in that corrections within the above limits were not sufficient to make them consistent with the water table defined by the other transducers. Accordingly, these two transducers were omitted from the inverse analyses.

4.2   *Inverse analysis technique*

The inverse analyses assuming 1D flow along the array involved the following sequence of steps (after Malvick et al. 2008). Hydraulic gradients (i) along the array are related to the excess pore water pressures (ΔU) as,

$$i = \frac{\partial(\Delta U)}{\partial z} \cdot \frac{1}{\gamma_w} \tag{12}$$

where z is position along the array, and γ_w is the unit weight of water. Volumetric strain rates are calculated based on Darcy's law and assuming 1D flow along the vertical array as,

$$\frac{\partial \varepsilon_v}{\partial t} = \frac{\partial^2(\Delta U)}{\partial z^2} \cdot \frac{k_s}{\gamma_w} \tag{13}$$

where k_s is the vertical hydraulic conductivity of the sand. Numerical double-differentiation of discrete excess pore pressure data with respect to position is sensitive to measurement errors in the data, and thus some form of interpolation or differentiation technique that smoothes the computed responses was required.

Two different numerical techniques for use in the inverse analyses were evaluated. The weighted residual technique (as introduced earlier) achieves smoothing by using a linear combination of basis functions to minimize the weighted residuals of the derivative.

The second technique involved fitting a smooth interpolation function to the pore pressure data, after which the interpolation function can be numerically differentiated at any point in space. The choice of fitting function and imposed boundary conditions can have a strong effect on the analysis results. The fitted function in this study was an exponential of the form (Malvick et al. 2008):

$$\Delta U(z) = a_0 \exp(a_1 z) + a_2 z^2 + a_3 z + a_4 \tag{14}$$

where a_i are constants determined for each time step separately, from the least-squares fit and imposed boundary conditions.

Boundary conditions for the inverse analyses were set such that the hydraulic gradient at the bottom of the box was forced to be zero at any time, since the box is an impermeable boundary for vertical flow. The top boundary condition was set at the interface between the sand and the clay crust, forcing continuity of flow across the interface, such that $i_s k_s = i_c k_c$ where k_s and k_c are the permeability in the sand and the clay, respectively. The hydraulic gradient through the clay was assumed constant at any given time, between the bottom of the clay layer (the interface) and the original water table height within the clay. The permeability ratio (k_s/k_c) between the sand and crust was also assumed constant although cracks and boils through the clay can significantly increase the permeability locally. The actual permeability of the clayey crust is unknown due to the use of a non-homogenous natural soil placed by manual compaction at low stresses. It is believed that the clay permeability while intact and continuous is approximately three orders of magnitude lower than that of the sand (permeability ratio = 1000) but that any cracking, including non-surficial cracks, may significantly increase the crust permeability locally and by that lead to a reduced permeability contrast on average. Hence, three permeability ratios were applied to the data—10, 100 and 100—in order to check the sensitivity of the results to this assumed boundary condition.

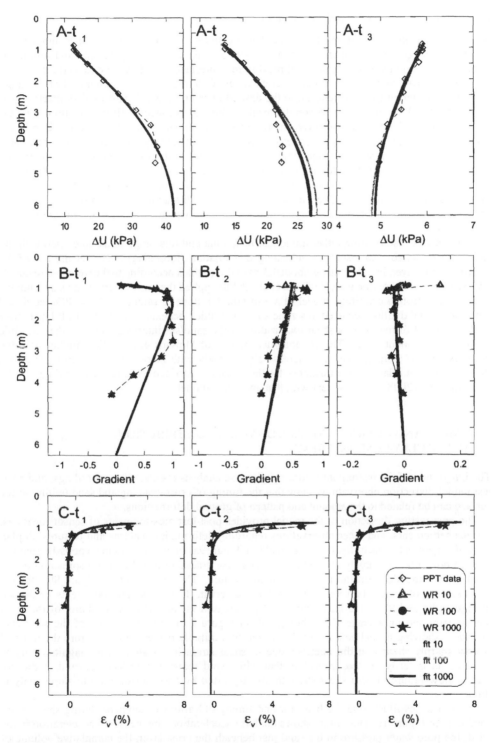

Figure 8. Excess pore pressures (top row), hydraulic gradients (middle row), and volumetric strains (bottom row) at 0 s (left column), 5 s (middle column), and 25 s (right column) after the end of shaking in shake 4 for 3 different premeability ratios and using the exponential interpolation and weight residual differentiation techniques.

4.3   *Analysis of the results and observations*

Results of the inverse analyses for shake 4 using the two different numerical techniques are compared in Figure 8. This figure shows excess pore pressures, hydraulic gradients, and volumetric strains versus depth along the array for three different interface boundary conditions (permeability ratios of 10, 100, and 1000) at three different times (0, 5 and 25 seconds after end of shaking); these three different times correspond to the dashed lines on Figure 7. The applied interface permeability ratio has very little effect on the results using the exponential fitting function. With the weighted residual technique, however, the top two measurement points (at 0.1 and 0.2 m below the interface) display a much greater sensitivity to the permeability ratio, both in the calculated hydraulic gradients as well as in the magnitude of volumetric strains. As a result, the magnitude of computed dilation at the interface is significantly smaller for a permeability ratio of 10 (less than 2%), than with either 100 or 1000 (approximately 6%). For permeability ratios of 100 and 1000, which are more reasonable for this crust material, the amount of dilation as well as the thickness of the dilating zone, are very similar for both the weighted residual and the exponential fit techniques.

The analysis results using either numerical technique and this range of assumed permeability ratios are reasonably consistent in indicating that dilation (loosening) occurs at the top of the confined sand layer, in a zone that is about 0.3-m to 0.6-m thick according to the weighted residuals and exponential fit, respectively. Malvick et al. (2008) applied the same inverse analysis technique to array data from a centrifuge model of a 9-m tall, 2:1 slope of saturated $D_R = 40\%$ sand, and determined that dilation occurred in a zone about 1-m thick beneath an embedded silt layer, prior to the eventual formation of a shear localization at the sand/silt interface and post-shaking slide movements of about 1.4 m. The 1D analytical model of Malvick et al. (2006) predicts that the thickness of the dilating zone will increase significantly with increasing slope angle, and thus the difference in dilating zone thickness for the present centrifuge test and the centrifuge test by Malvick et al. (2008) are consistent with that analytical model.

5   TIMING AND CHARACTERISTICS OF VOID REDISTRIBUTION AT THE CLAY-SAND INTERFACE

The integration of the instrument recordings, inverse analysis results, video recordings, and post-test model dissection observations show that the timing of crust cracking and sand boiling to the surface can be related to the amount and pattern of ground deformations.

The post-testing dissection of the model provides post-test measurements of sensor locations, surface settlements, and other observations regarding deformation patterns and features. A plan view of the ground surface after the end of shaking (Figure 9a) shows a non-homogenous distribution of sand boils that are largely aligned along transverse cracks formed by lateral spreading of the crust toward the slope on the right side of the picture. The photograph in Figure 9b shows a cross section of a sand boil vent through the clay crust. The cracks and boil vents evidenced in Figures 9a and 9b would have dramatically increased the apparent permeability of the crust both locally and on average. The photograph in Figure 9c is a cross section of the sand-clay interface, showing a sand/bentonite column that was initially placed vertical before the test and is now sharply sheared at the interface due to shear strain localization. The magnitude of the shear offset is fairly limited, indicating that while void redistribution had occurred, it did not lead to flow failure or zero shear strength, as suggested by an idealized infinite slope analysis (Figure 1).

Time histories illustrating the behavior and timing of various phenomena during shake 4 are combined in Figure 10. This figure shows the base acceleration, the measured acceleration in the crust, the pore water pressure in the sand just beneath the crust layer, the cumulative volume of water expelled from the sand layer, the lateral crust displacement, the vertical crust settlement, and the times at which water boils were first and last observed on the ground surface (as recorded by the high speed video cameras).

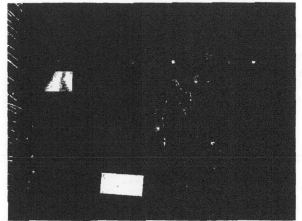

(a) Looking down on the surface, with spreading toward the right.

(b) Excavation in the crust showing a boil vent filled with sand.

(c) Excavation along a colored marker showing an offset at the sand/clay interface.

Figure 9.   Photographs of sand boils and localizations observed during model dissection after testing.

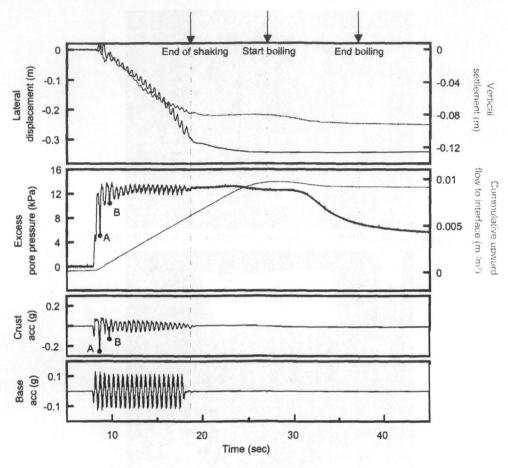

Figure 10.   Time histories of cumulative upward flow to the interface, lateral displacements of the crust, vertical settlements of the crust, excess pore pressure ratio in the sand beneath the crust, base acceleration, and timing of sand and water boils during shake 4.

- Early in shaking ($t = 8$ to 10 s), negative acceleration and shear stress peaks are observed to coincide with transient pore pressure drops (points A and B on Figures 6 and 10). The observed strain-hardening stress-strain responses at these two times are expected for dilative soils during undrained loading to large strains, and are similar to what is observed in undrained cyclic laboratory experiments (e.g. Idriss & Boulanger 2008). The dilative behavior observed at shallower depths is asymmetric, with more dilative behavior appearing in the negative direction, which is the direction of the crust slope and water channel in the centrifuge model. This is consistent with observations from field studies, suggesting that the dilation peaks are asymmetric due to the presence of a free-face which enhances the lateral spreading in one direction (e.g. Zeghal & Elgamal 1994).
- At the end of shaking ($t = 18$ s), pore pressures at the top transducer were at their maximum value, approximately 5 mm^3/mm^2 of water had flown up to the interface, about 76% of the final horizontal displacement had occurred, and about 79% of the final vertical settlement had occurred.
- For the next 7 s after shaking (from $t = 18$ to 25 s), the crust continued to spread laterally another 30 mm, the cumulative flow of water to the interface had increased to about 9 mm^3/mm^2, the pore pressure at the top transducer was just beginning to decrease, and a little heave (as opposed to settlement) was observed by the vertical displacement transducer.

- Two seconds later ($t = 27$ s), the excess pore pressures throughout the sand layer had approximately equalized, such that there was no longer a significant net upward flow of water to the interface (i.e., the computed cumulative flow to the interface had reached about 10 mm^3/mm^2 but was no longer increasing), and water boils first started to appear at the ground surface.
- For the next 10 s ($t = 27$ to 37 s), water boils continued at the ground surface, the ground surface settled approximately 15 mm, pore water pressure at the top transducer progressively decreased, and lateral crust displacements did not change. The surface settlement of about 15 mm during the time of water boiling corresponds well with the calculated volume of cumulative water flow to the interface of about 10 mm^3/mm^2.
- During the first 4 s of water boiling ($t = 27$ to 31 s), the flow rates out of the boils was greatest, the pore pressure at the top transducer remained approximately constant and approximately equal to the total vertical overburden stress, and the majority of the post-shaking ground surface settlement occurred. This aspect of behavior would be consistent with the formation of a water film beneath the crust, which then remained under high pressure until the venting of water through a boil caused the water film to collapse.

These observations suggest that water had accumulated at the sand-clay interface during shaking and for 9 additional seconds after shaking stopped. The accumulated water led to reduced strength, which contributed to shear localization at the sand/clay interface and a slight increase in lateral spreading displacements. When surface cracking enabled water to flow through it to the surface, pore pressure dissipated from the interface, collapsing any possible water films and increasing the effective stresses at the interface and thus providing resistance against any further lateral crust displacements.

The end of water boiling can be related to the time history of excess pore pressure dissipation at the top transducer, as shown in Figure 10, as follows. The initial static water table was about 0.7 m below the ground surface at the location of the pore pressure transducer array. The excess pore water pressure in the sand immediately below the crust would have to be greater than or equal to 7 kPa (for an excess head $\Delta U/\gamma_w \geq 0.7$ m) to maintain water flow up through a crack to the ground surface. Once the excess pore pressure in the sand below the crust dropped below 7 kPa, water would no longer flow up through the cracks, but would instead begin to seep downward back into the sand. This calculation matches with the observation that sand boils ceased when the pore pressure at the upper transducer had decreased to about 7 kPa. Furthermore, the rate of pore pressure dissipation was relatively rapid during the time that water boils were emerging at the ground surface, but then slowed rapidly once the boiling stopped because the boil vents no longer acted as a net sink for outward seepage.

The potential formation and effects of water films beneath the crust during or after shaking can be further qualitatively evaluated from other aspects of the recorded responses. The settlement and flow quantity time histories in Figure 10 suggest that water had accumulated beneath the crust and likely formed water films in at least localized pockets. The quantity of water that had flowed up to the interface (and hence into the dilating zone) during shaking, as computed by the inverse analysis of the pore pressure transducer array, was about 5 mm^3/mm^2. This flow quantity increased to about 10 mm^3/mm^2 over the 9 s after the end of shaking. If the dilating zone at the top of the sand layer was always limited to the eventual localization thickness (e.g. less than a couple mm thick based on the photograph in Figure 9c and considering that a shear band would be expected to be about ten times the median particle size of $D_{50} - 0.14$ mm), then this volume of water would be expected to have produced a water film that averaged 5 to 10 mm thick. If the dilating zone was instead initially about 0.3 m thick, as suggested by the inverse analyses of the pore pressure transducer array using the weighted residuals technique, then 5 to 10 mm^3/mm^2 of water would be expected to have produced an average volumetric dilation of 5% to 10% at the top of the sand layer based on a parabolic distribution of volumetric strains within the dilating zone. This simple calculation gives volumetric strains at the top of the sand layer before the boiling of water to the surface that are similar to those computed by the inverse analyses (e.g. 4 to 7%), with the differences being due to the differences in the distribution of strains within the dilating zone and other flow patterns not accounted for in the 1D flow analysis. A volumetric dilation of 4% in Nevada sand at an initial

$D_R = 40\%$ would reduce its D_R to a final value of about 20%, whereas a volumetric strain of 7% would reduce the final D_R value to almost 0%. At these levels of loosening, the analytical model of Malvick et al. (2006) would suggest that the dilating zone would have collapsed in thickness to the size of the eventual shear localization toward the end of shaking or after shaking, which is consistent with the observation that localization did occur. At the same time, the acceleration time histories for the clay crust, the inverse analysis of stress-strain responses at various depths beneath the crust, and the post-shaking stability of the slope suggest that the sand/clay interface continued to be able to transmit small, but nonetheless significant, shear stresses in both directions on average throughout shaking. The transmission of significant shear stresses would seem inconsistent with the formation of a water film over large portions of the sand/clay interface, but would be consistent with water films being restricted to localized pockets, such that shear stresses could be transmitted through those zones without water films or through the lateral boundaries of the laterally spreading soil (i.e. at the contacts with the container walls).

The mechanisms of void redistribution and shear localization observed in this centrifuge model test complement observations from other physical modeling studies, wherein a broad range of responses have been observed; e.g. cases with and without localizations forming, cases with different magnitudes of deformations during and after shaking. The archiving of these detailed experimental data (Kamai et al. 2008), in combination with data for other model geometries and loading conditions, provide a basis for future research evaluating the ability of nonlinear numerical modeling procedures to differentiate between the conditions under which void redistribution will or will not lead to significant strength loss or shear localization in the field.

6   CONCLUDING REMARKS

The inverse analysis techniques evaluated in this paper provide a means to better characterize the mechanics of void redistribution and shear localization due to earthquake-induced liquefaction in layered soil profiles. The use of these inverse analysis techniques were illustrated by their application to one dynamic centrifuge model test, and the following conclusions were presented.

For inverse analyses of stress-strain responses from accelerometer arrays, the weighted residual and cubic spline interpolation techniques performed equally satisfactorily. The cosine series and polynomial interpolation techniques performed poorly in certain situations. The piecewise-linear interpolation technique performed reasonably well, with a few exceptions, but offers no particular advantage over either the cubic spline or weighted residual techniques. Performing separate analyses with the weighted residual and cubic spline techniques has the advantage that it provides a check on the data quality; i.e. good agreement between analysis results can be expected if the data are of good quality, whereas significant differences in analysis results may indicate that there is a problem in the data quality or the applicability of the underlying assumption of a 1D shear beam response.

For inverse analyses of flow quantities and volumetric strains from pore pressure transducer arrays, both the weighted residual and exponential interpolation techniques performed equally satisfactory under reasonable assumptions of interface permeability ratio. The weighted residual technique was more sensitive to a permeability ratio of 10, which is thought to be too high for the presented model, but produced similar results for permeability ratios of 100 and 1000 as those of the exponential interpolation technique.

The mechanics of void redistribution and shear localization observed in one dynamic centrifuge model test were discussed based on integration of the inverse analysis results, instrument recordings, video recordings, and post-test model dissection observations. The observations indicate that: (1) a significant volume of pore water accumulated beneath the surface clay layer due to upward seepage in the liquefied sand layer during and after shaking, (2) some of the water flowing to the sand/clay interface was accommodated by dilation of the sand in a zone about 0.3 to 0.6 m thick, (3) a thin shear localization developed at the sand/clay interface, but the magnitude of slip at this interface was only about 150 mm, (4) water films likely formed in localized areas beneath the crust after shaking had ended and prior to the formation of sand boils, and (5) sand boils through cracks and vents in the crust layer released water that had accumulated beneath the crust, thereby contributing to both surface settlements and an arresting of lateral movements.

The experimental data described in this paper are publicly archived for use by other researchers. It is hoped that the insights provide by these data, in combination with data from other physical models or case histories, will lead to an improved ability to differentiate between the conditions under which void redistribution will or will not lead to significant strength loss or shear localization in the field.

ACKNOWLEDGEMENTS

The centrifuge test described in this paper was conducted as part of the NEES-Grand Challenge project for Seismic Risk Management for Port Systems, directed by Dr. Glenn Rix. The project was funded by the George E. Brown, Jr., Network for Earthquake Engineering Simulation (NEES) under Award No. CMS-0530478. Seiji Kano, Ellen Rathje, Antonio Marinucci, Rachelle Howell, Carolyn Conlee and Patricia Gallagher collaborated in designing, planning and conducting the test.

The authors appreciate the assistance of Dr. Dan Wilson with signal processing and interpolation algorithms, and the thoughtful comments and suggestions of Dr. Bruce Kutter.

REFERENCES

Brandenberg, S.J., Wilson, D.W. & Rashid, M.M. 2009. A Weighted Residual Numerical Differentiation Algorithm Applied to Experimental Bending Moment Data. *Journal of Geotechnical and Geoenvironmental Engineering* accepted for publication.

Davis, R.O. & Berrill, J.B. 1998. Rational approximation of stress and strain based on downhole acceleration measurements. *International Journal for Numerical and Analytical Methods in Geomechanics* Vol. 22 No. (8): 603–619.

Davis, R.O. & Berrill, J.B. 2001. Pore pressure and dissipated energy in earthquakes—Field verification. *Journal of Geotechnical and Geoenvironmental Engineering* Vol. 127 No. (3): 269–274.

Idriss, I.M. & Boulanger, R.W. 2008. *Soil liquefaction during earthquakes.* Monograph MNO-12, Earthquake Engineering Research Institute, Oakland, CA.

Kamai, R., Kano, S., Conlee, C., Marinucci, A., Boulanger, R.W., Rathje, E., Rix, G. & Howell, R. 2008. *Evaluation of the effectiveness of prefabricated vertical drains for liquefaction remediation—centrifuge data report for SSK01.* (https://central.nees.org/)

Kokusho, T. 2006. Recent Developments in Liquefaction Research Learned from Earthquake Damage. *Journal of Disaster Research* Vol. 1 No. (2): 226–243.

Kulasingam, R., Malvick, E.J., Boulanger, R.W. & Kutter, B.L. 2004. Strength loss and localization of silt interlayers in slopes of liquefied sand. *Journal of Geotechnical and Geoenvironmental Engineering* Vol. 130 No. (11): 1192–1202.

Kutter, B.L. 1994. Recent Advances in Centrifuge Modeling of Seismic Shaking. *3rd International Conference on Recent Advances in Geotechnical Earthquake Engineering and Soil Dynamics.* St. Louis, MO, pp. 927–942.

Malvick, E.J., Kutter, B.L. & Boulanger, R.W. 2008. Postshaking shear strain localization in a centrifuge model of a saturated sand slope. *Journal of Geotechnical and Geoenvironmental Engineering* Vol. 134 No. (2): 164–174.

Malvick, E.J., Kutter, B.L., Boulanger, R.W. & Kulasingam, R. 2006. Shear localization due to liquefaction-induced void redistribution in a layered infinite slope. *Journal of Geotechnical and Geoenvironmental Engineering* Vol. 132 No. (10): 1293–1303.

Marinucci, A., Rathje, E., Kano, S., Kamai, R., Conlee, C., Howell, R., Boulanger, R.W. & Gallagher, P. 2008. Centrifuge Testing of Prefabricated Vertical Drains for Liquefaction Remediation. *Geotechnical Earthquake Engineering and Soil Dynamics IV.* Sacramento, CA: ASCE.

Seid-Karbasi, M. & Byrne, P.M. 2007. Seismic liquefaction, lateral spreading, and flow slides: a numerical investigation into void redistribution. *Canadian Geotechnical Journal* Vol. 44 No. (7): 873–890.

Whitman, R.V. 1985. On liquefaction. *11th International Conference on Soil Mechanics and Foundation Engineering.* San Francisco, CA.: Balkema Rotterdam, pp. 1923–1926.

Wilson, D.W. 1998. Soil-Pile-Superstructure Interaction in Liquefying Sand and Soft Clay. University of California, Davis, 173.

Wilson, D.W., Boulanger, R.W. & Kutter, B.L. 1998. Signal processing for and analyses of dynamic soil-pile interaction experiments. In: T. Kimura, O. Kasusakabe, and J. Takemura, Eds., *Centrifuge*. Tokyo, Japan: Balkema.

Zeghal, M. & Elgamal, A.W. 1994. Analysis of Site Liquefaction Using Earthquake Records. *Journal of Geotechnical Engineering-Asce* Vol. 120 No. (6): 996–1017.

Zeghal, M., Elgamal, A.W., Tang, H.T. & Stepp, J.C. 1995. Lotung Downhole Array. 2. Evaluation of Soil Nonlinear Properties. *Journal of Geotechnical Engineering-Asce* Vol. 121 No. (4): 363–378.

On seismic P- and S-wave velocities in unconsolidated sediments: Accounting for non-uniform contacts and heterogeneous stress fields in the effective media approximation

Ran Bachrach
Department of Geophysics and Planetary Sciences, Tel-Aviv University

Per Avseth
Odin Petroleum AS and Norwegian University of Science and Technology, Tronheim, Norway

ABSTRACT: We show that by treating the contact stiffness as a variable, one can extend the effective medium approximation used to obtain elastic stiffness of a random pack of spherical grains. More specifically, we suggest calibrating the effective media approximation based on contact mechanics by incorporating non-uniform contact models. The simple extension of the theory provides a better fit to many laboratory and field experiments and can provide insight to the micromechanical bonds associated with unconsolidated sediments. The key geophysical observation derived from P and S wave velocities (and potentially anisotropy) can provide insight to loading history and micro-mechanical state of the granular media. We present the application of the theory for experiments including multi-component surface seismic and well log data from a shallow North Sea gas field and deep water Gulf of Mexico oil sands. Potential application may include better understanding of stress changes in the subsurface by monitoring changes in Poisson's ratio.

1   INTRODUCTION

The effective properties of sphere packs have been used as an analog for the behavior of unconsolidated sands for many years. Contact between two spheres has been well characterized for different loading paths, boundary conditions, and grain radius (Johnson, 1985), and the forces acting on two-particle arrangement have been characterized using normal and tangential stiffnesses (Winkler, 1983; Norris and Johnson, 1997; Mavko et al., 1998). The effective media approximation (EMA) associated with granular media attempts to average two-grain contacts into the effective behavior of a pack of aggregates with many contacts. Gassmann (1951) calculated the effective elastic properties of a dense hexagonal pack of spheres and compared the results to that of seismic velocity measurements in unconsolidated sands. Digby (1981) and Walton (1987) developed effective medium averaging techniques to estimate the effective properties of a random sphere pack while considering contact laws for adhesive contacts, rough contacts, and smooth contacts. Winkler (1983) showed that for Hertz-Mindlin contact stiffness, the simple volumetric averaging of Digby can be presented in terms of normal to tangential stiffness ratios. Muhlhaus and Oka (1996) analyzed dispersion and wave propagation in granular media using homogenization of discrete equations of motion. Norris and Johnson (1997) rederived the effective medium approximation of Walton and Digby using energy density functions for different contact models, and showed that, in general, as the contact force between the grains is path dependent (i.e., relates to the history of loading), the EMA will relate to the loading path.

   In the recent years, with the advancement of computer power, the use of granular dynamic models has improved the understanding of the elastic behavior of a random pack of spheres and associated force distributions (Makse et al., 1999, 2004). These studies have shown that, while bulk modulus is well predicted using EMA theory, the shear modulus predictions do not follow conventional EMA predictions as the grains tend to relax from the affine, macroscopic deformation (i.e., each grain translates according to the direction of the macroscopic strain) or rotate. The effect of rotation of grains has also been studied by Pasternak et al. (2006). The assumption of affinity given by the EMA theory is approximately valid for the bulk modulus, but seriously flawed for the shear modulus. Thus, the uniform strain assumption breaks down causing the EMA approximation for granular packs to differ considerably from observed values. Several experimental studies have also demonstrated significant differences between the shear moduli predicted by Hertz-Mindlin effective-medium models and empirical results, including Winkler (1983), Goddard (1990), and Zimmer et al. (2007).

One of the consequences of the inadequacies of using EMA to describe shear behavior is in the prediction of Poisson's ratio and Vp/Vs ratio in unconsolidated sands (e.g., Avseth and Bachrach, 2005; Sava and Hardage, 2006). Manificat and Gueguen (1998) showed that a contact roughness model can explain the higher than expected Poisson's ratio observed in sands. Bachrach et al. (2000) showed that while analyzing P- and S-wave velocities in unconsolidated sands, different contact curvature of grains can be accounted for and will not cause changes in the Vp/Vs ratio. In near-surface sediments, the observed Poisson's ratio can be used to determine fraction of slipping contacts to non-slipping contacts by simple averaging of two representative media: one with and one without tangential contact stiffness.

In this article, we address two dependent problems that often appear when using rock physics models in granular media. This first problem we address is the expansion of the EMA theory of Norris and Johnson to account for variable contact models in granular systems. Specifically, we show that choosing a binary model where tangential stiffness of a contact can be zero, or following Hertz-Mindlin, enables us to derive an EMA for granular pack where not all contacts are the same. The second problem we address deals with the application of the theory to real data, where we show that the calibration process will depend also on additional granular properties such as effective contact ratio, and not only the coordination number. The paper is organized as follows: We first provide a short review of the basic theory associated with EMA of granular media. Next, we introduce the binary contact model that accounts for non-uniform contacts in EMA. We follow with a detailed discussion of the steps associated with model calibration for real data, and finally, we present the examples of how the theory is applied to well log data and rock physics template analysis (Ødegaard and Avseth, 2004; Avseth et al., 2005).

2   THEORETICAL BACKGROUND

In this section we review the basic of EMA for a pack of sphere and highlight the assumptions associated with its derivation. We also provide an exhaustive literature review for the interested party.

2.1   *Contact stiffness*

In this section, we follow closely the derivation of Norris and Johnson (1997). All detailed derivation is given in their paper, and here we only repeat relevant equations associated with non-uniform grain contacts forces.

A single contact between two spheres can be characterized by normal and tangential stiffness, defined as (Winkler, 1983; Mavko et al., 1998)

$$S_n = \frac{\partial F_n}{\partial \delta}, \quad S_\tau = \frac{\partial F_t}{\partial \tau}, \tag{1}$$

where F_n, F_t are the normal and tangential components of the force that is acting on the contact and δ and τ are the normal and tangential displacements, respectively, resulting from such a force. In general, the behavior of a single contact between spheres depends on the loading path and additional parameters such as friction, cement, and adhesions (Norris and Johnson, 1997). The actual normal stiffness between two elastic spheres has been modeled and experimentally verified (Johnson, 1985) using the Hertzian contact model:

$$S_n = \frac{4aG}{1 - \nu},\tag{2}$$

where

$$a = \sqrt{\overline{R}\delta} = \sqrt[3]{3F_n\overline{R}(1 - \nu)/8G}\tag{3}$$

is the radius of contact area between two spheres, G is the shear modulus, and ν is the Poisson's ratio of the sphere material, which is assumed to be isotropic.

The effective contact radius $\overline{R}$ is defined as

$$\overline{R} = 0.5 \left(\frac{1}{R_1} + \frac{1}{R_2} \right)^{-1},\tag{4}$$

where R_1 and R_2 are the radii of the two grains in contact as shown first by Hertz in his 1882 seminal paper "on the contacts of elastic solids" (Love, 1927; Johnson, 1985). Note that $\overline{R}$ is related to the curvature of the actual contact surface.

To derive the tangential stiffness, one needs to assume a specific loading path and boundary conditions. One choice of the tangential model is the perfectly smooth case (Walton, 1987) where the tangential stiffness is zero. Another typical choice associated with Hertzian contact and friction is the Hertz-Mindlin model where the tangential stiffness is given by (Mindlin, 1949):

$$S_t = \frac{8aG}{2 - \nu}.\tag{5}$$

As discussed by Norris and Johnson (1997), the tangential stiffness depends on the boundary conditions and loading path. In practice, for most poorly consolidated sediments, these are often poorly known. This subject will be further discussed.

2.2 Derivation of effective bulk modulus for a dry pack of spherical grains under hydrostatic loading

To illustrate how the effective bulk modulus of a random pack of identical spheres can be obtained, we present the following simple derivation:

Consider the pack of spheres presented in figure (1). The pressure is defined as force/area. If we consider the solid fraction of a spherical volume of radius R with porosity ϕ through which

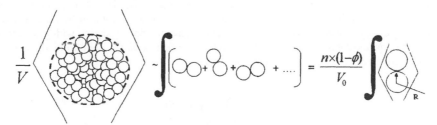

Figure 1. (a). A spherical region with grains is considered for the effective medium averaging. The solid fraction of the spherical volume has porosity ϕ through which the forces are transmitted, and on average n points of contacts per spheres. The EMA is summing over contacts and averaging over sphere of radius R.

the forces are transmitted, and we consider n points of contacts per spheres (also known as the coordination number), we can write the hydrostatic pressure as:

$$\langle P \rangle = \frac{F_n(1 - \phi)n}{4\pi R^2}. \tag{6}$$

This equation is identical to equation 71 in Norris and Johnson, 1997. Note that in equation 6 the normal contact force, which is a local point, is averaged along a spherical radius R, which is associated with the representative volume of the grain. This point will be addressed later.

The effective volumetric increment is defined as:

$$\langle e \rangle = \frac{dV}{V} = \frac{4\pi R^2 dR}{4\pi R^3/3} = \frac{3dR}{R}. \tag{7}$$

Bulk modulus relates the pressure increment to the volume increment, and therefore:

$$K_{eff} = \frac{d \langle P \rangle}{\langle e \rangle} = \frac{(1 - \phi)n}{12\pi R} \frac{dF_n}{dR} = \frac{(1 - \phi)n}{12\pi R} S_n, \tag{8}$$

which is identical to the formulas derived by Norris and Johnson (1997) and Walton (1987).

Note that, in the simple derivation above, equation 6 is the EMA for hydrostatic state of stress. EMA assumes that, as the forces are not interacting with each other, and that on average, the pressure can be represented as the force times the contact number that acts on the solid part of a reference sphere. Equation 7 is the EMA for strain where it is assumed that the volumetric deformation can be related to the normal displacement dR. We also would like to make a distinction here between the effective contact radius $\overline{R}$ and the volumetric averaging radius R (see figure 1). While $\overline{R}$ affects the contact stiffness of a two-grain configuration, the volumetric averaging radius R is the volume over which EMA is being performed. This distinction will be further discussed below.

We also note that the term "random pack" is used in this case because we do not assume a specific packing, and that a pack is characterized by porosity and coordination number independently. For example, an effective media approximation for a hexagonal pack where porosity and coordination numbers are known is given by Gassmann (1951). Also note that the above expression is valid for dry pack only. Fluid effects must be taken into account in real applications (Gassmann, 1951).

2.3   *Effective elastic modulus with normal and tangential stiffness*

While the derivation of effective bulk modulus is very simple, the derivation of effective shear modulus for hydrostatic loading, and the derivation of elastic modulus for non-hydrostatic loading, are more complicated. An elegant derivation of the effective media associated with granular packs is given by Norris and Johnson (1997), who derived the effective elastic moduli by differentiating the strain energy density per unit volume U defined as:

$$U = \frac{1}{V} \sum_{contacts} \int F \cdot du \tag{9}$$

where $F \cdot du = F_n d\delta + F_t \cdot d\tau$ with respect to strain and V is the volume associated with the EMA averaging.

The assumption made by Digby (1981), Walton (1987), and Norris and Johnson (1997) is that all contacts are "the same", and that the volumetric average associated with equation 9 is that

$$U = \frac{1}{V} \sum_{\text{contacts}} \int F \cdot du \approx \frac{n(1-\phi)}{V_0} \left\langle \int F \cdot du \right\rangle \tag{10}$$

(equation 20 in Norris and Johnson), where V_0 is the volume of a single grain. Note that the EMA averaging rule is also used in the simple form of equation 8.

3 EXTENSION OF EMA FOR NON-UNIFORM CONTACTS

3.1 *Motivation*

As was discussed in the introduction, the stress distribution on granular packs is non-uniform. As the tangential contact stiffness associated with two grains is path dependent, it is clear that equation 10 will not capture such heterogeneities. In the next section, we discuss one way to accommodate stress field heterogeneities.

3.2 *Binary scheme*

We now consider a case where not all contacts are the same. A simple way to introduce different contacts is to assume a binary mix where some contacts may behave as smooth contacts with zero tangential stiffness, while other contacts may have finite tangential stiffness as given by equation 5. This assumption tries to account for the heterogeneities in stress chains as have been observed both in laboratory measurements and numerical simulations (Ammi et al., 1987; Mueth et al., 1998, Geng et al., 2001; Makse et al., 2004). The binary model, yet simplified from the true grain pack, can be viewed as a specific probability distribution of non-uniform contacts.

Given the binary model assumption, and because the strain energy is linear, we can rewrite equation 10 as:

$$U = U_{\text{smooth}} + U_{\text{no-slip}} \approx \sum_{\text{smooth}} \int F \cdot du + \sum_{\text{no-slip}} \int F \cdot du$$

$$= \frac{n(1-\phi)}{V_0} \left(f_s \left\langle \int F \cdot du \right\rangle + f_t \left\langle \int F \cdot du \right\rangle \right) \tag{11}$$

where f_s is the fraction of smooth contacts in the system and $f_t = 1 - f_s$ is the fraction of no-slip contacts. What equation 11 says is that the strain energy density for unit volume can be related to more than a specific type of boundary conditions.

The macroscopic stress-strain relations are derived by differentiating equation 11 with respect to the strain (Norris and Johnson, equation 30).

$$\sigma_{ij} = \frac{\partial U}{\partial e_{ij}} = \frac{\partial U_{\text{smooth}}}{\partial e_{ij}} + \frac{\partial U_{\text{no-slip}}}{\partial e_{ij}}. \tag{12}$$

The effective elastic modulus is then given by the combination of smooth and no-slip contacts, which is a simple average of elastic moduli formulas derived by Walton (1987) and Norris and Johnson (1997). Note that similar conclusions can be reached using the forces acting on the contacts as derived by Walton (1987).

For a homogenous system (i.e., a granular pack where all contacts are the same), one can write the stress-strain relations associated with equation 12 as (e.g., Muhlhaus and Oka, 1996; Pasternak et al., 2006):

$$\sigma_{ij} = C^*_{ijkl}(S_n, S_t)e_{kl}, \tag{13}$$

where $C^*_{ijkl}(S_n, S_t)$ is the effective medium approximation of the elastic modulus derived from equation 12 for identical contacts with given normal and tangential stiffnesses. Then, due to the linearity of equation 12, we can write the macroscopic stress-strain relations for non-uniform contacts as:

$$\sigma_{ij} = (f_s C^*_{ijkl}(S_n, S_t = 0) + f_t C^*_{ijkl}(S_n, S_t \neq 0))e_{kl} \tag{14}$$

and the overall effective stiffness is $\langle C^*_{ijkl} \rangle = f_s C^*_{ijkl}(S_n, S_t = 0) + f_t C^*_{ijkl}(S_n, S_t \neq 0)$. We note that the case of $f_t = 0.5$ is similar to the case investigated by Manificat and Gueguen (1998).

It is also interesting to note that equation 14 suggests that the Voigt average, which implies iso-strain (Mavko et al., 1998), is the one appropriate to the binary contact problem. Recall that the macroscopic stress-strain relationships were derived in this case from a general energy density function. Equation 12 assumes that an averaged macroscopic strain can be defined over the volume. Thus, the iso-strain interpretation is consistent with equation 14.

For the specific case of hydrostatic loading, the bulk and shear moduli for Hertz-Mindlin contacts are given by:

$$K_{eff} = \frac{n(1 - \phi)}{12\pi R}S_n, \quad G_{eff} = \frac{n(1 - \phi)}{20\pi R}\left(S_n + \frac{3}{2}S_t\right). \tag{15}$$

The effective shear modulus in terms of volume fraction of no-slip contacts f_t is given by:

$$K_{eff} = \frac{n(1 - \phi)}{12\pi R}S_n, \quad G_{eff} = \frac{n(1 - \phi)}{20\pi R}\left(S_n + f_t\frac{3}{2}S_t\right). \tag{16}$$

Following Bachrach et al. (2000), using equations 2, 3, and 14, we can write the effective Poisson's ratio in terms of f_t as:

$$\nu_{eff} = \frac{S_n - f_t S_t}{4S_n + f_t S_t} = \frac{(2 - \nu) - 2f_t(1 - \nu)}{4(2 - \nu) + 2f_t(1 - \nu)}, \tag{17}$$

which clearly shows that for $f_t = 0$, $\nu_{eff} = 1/4$, as was first derived by Walton (1987). Equation 17 also shows that the effective Poisson's ratio is only a function of the mineral Poisson's ratio and the fraction of no-slip contacts, where an increase in no-slip contacts will reduce Poisson's ratio. In Figure 2, we present the result of applying equation 17 to quartz spheres. We note that a similar result was produced by Bachrach et al. (2000), which was derived using the Hashin-Shtrikman upper and lower bounds.

It is important to bear in mind that the maximum value of $\nu_{eff} = 1/4$ is a *dry* value. The presence of pore fluid in the system should be addressed separately (One approach is to use Gassman's relationships, following Gassman's 1951 example). The theory does not account for surface processes that may occur in the presence of residual water saturation. Also, the theory does not account for cement. We note that contact slip is not likely to occure when small strain seismic waves are passing through the aggregate. For more discussion on smooth contacts and slip during wave propagation see Bachrach and Avseth, 2008.

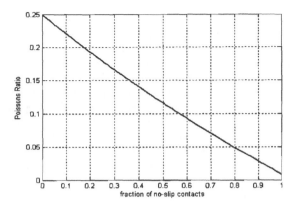

Figure 2.   Effective dry Poisson's ratio as a function of volume fraction of no-slip contacts. (after Bachrach and Avseth, 2008)

3.3   *A note on effective contact curvature, coordination number,*
and model calibration in reservoir settings

From equations 2, 3, 4, 5, 6, and 16, we can write the effective bulk and shear moduli for hydrostatic loading explicitly as:

$$K_{eff} = \left(\frac{(1-\phi)^2 G^2}{18\pi^2(1-\nu)^2} \right)^{1/3} \left(n^2 \frac{\overline{R}}{R} \right)^{1/3} \langle P \rangle^{1/3}$$

$$G_{eff} = \left[\frac{1}{10} \left(\frac{12(1-\phi)^2 G^2}{\pi^2(1-\nu)^2} \right)^{1/3} \left(n^2 \frac{\overline{R}}{R} \right)^{1/3} \langle P \rangle^{1/3} \right]$$

$$+ \left[\frac{3}{10} \left(\frac{12(1-\phi)^2 G^2(1-\nu)}{\pi^2(2-\nu)^3} \right)^{1/3} \left(n^2 \frac{\overline{R}}{R} \right)^{1/3} \langle P \rangle^{1/3} \right] f_t \qquad (18)$$

Here, $\langle P \rangle$ is the effective pressure at grain contacts and is related to the contact forces by equation (6). When predicting bulk and shear moduli for sands at given depth, using this equation, we normally assume mineralogy, porosity, and overburden stress as known parameters. However, as shown in equation 18, calibration parameters are related to *both* coordination number n, and the ratio $\overline{R}/R$, unless the granular aggregate is made out of identical spheres where $\overline{R}/R = 1$. We note that $\overline{R}/R$ can be larger or smaller than 1. We note that the ratio $\overline{R}/R$ captures some aspects of grain angularity and sorting, and can be viewed as a mechanical parameter that characterizes the averaged effective contact radii and general grain size distribution.

Another issue with model calibration is that, in general, the coordination number changes with the porosity of the sediment and with stress. Murphy (1982) established the relationship between coordination number and porosity from theory and observations for porosity ranges between 0.2 and 0.6, where for 20% porosity, $n = 14$ and for 60% porosity, $n = 4.78$. Makse et al. (1999) showed that in their granular dynamic simulation, for low effective stress ranges, the coordination number still can change without a porosity change, then in association with pressure change. Their coordination number/pressure dependency has been formulated as:

$$n(\langle P \rangle) = 6 + \left(\frac{\langle P \rangle}{P_0} \right)^{1/3} \qquad (19)$$

where P_0 is an empirical fitting parameter. It is interesting to note that for random pack at low effective stress with porosity of 36%, Murphy's coordination number is about 9, while Makse et al. predict a coordination number close to 6. In natural materials, stress and compaction are related to each other. In our opinion, it is beneficial to look at the coordination number as a function of both porosity and stress. If we follow Bowers' (1995) definition of "virgin compaction curve", it is possible to relate the coordination number to porosity and stress in a similar fashion to the way stress and porosity are related to each other along the virgin compaction curve. Thus, the choice of relating porosity to the coordination number (Murphy, 1982) is simply a statement that the sedimentary pack is following a specific compaction curve, such as the loading curve (Bowers, 1995). One way to reconcile both Murphy's and Makse's number is to consider different compaction/stress history associated with sediment burial. We also note that when the radius difference is large (i.e., poorly sorted aggregates) small grain will fall in-between the pore space and thus porosity will be reduced while coordination number will increase dramatically. In low porosity material coordination number of 14 has been measured for porosities of 25% (Murphy, 1982).

4   FIELD EXAMPLES: SEISMIC WAVE PROPAGATION IN SOFT UNCONSOLIDATED SEDIMENTS

We present the following examples for application of the Theory. In Figure 3 we present a well log suite from shallow North Sea gas well where unconsolidated gas reservoir is located at depth

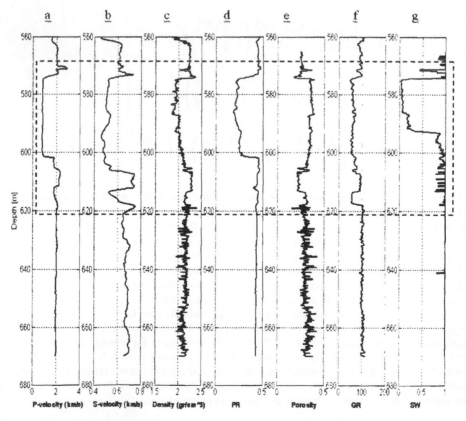

Figure 3.   Well log observations of shallow unconsolidated sands embedded in shales. (a). P-wave velocity. (b). S-wave velocity. (c). Density. (d). Poisson's ratio. (e). Porosity. (f). Gamma ray. (g). Water saturation. (Data from Avseth et al., 2007). Dashed line embraces the zone of interest laminated sands.

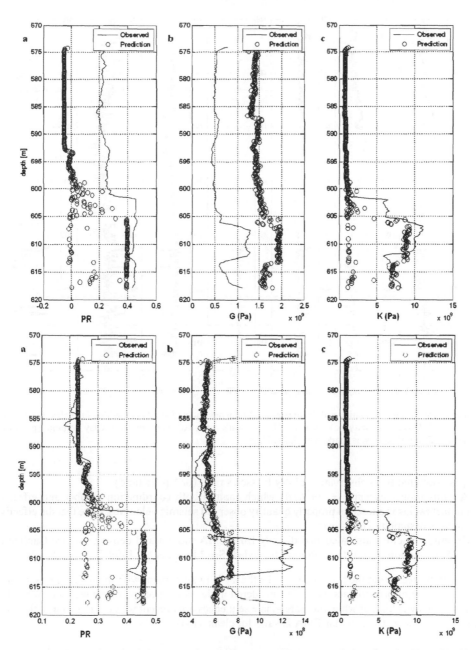

Figure 4. Model vs. well log observation for shallow gas well. Top: (a). Poisson's ratio (PR). (b). Shear modulus (G) and Bulk modulus (K) for model with $f_t = 1$. Bottom. PR, G and K for model $f_t = 0.07$ and obtained a good match for both bulk and shear. Note that the zone where the shear modulus underpredicts the observation is where the depositional environment has changed and a laminated sand-shale sequence is present in the well (see gamma-ray plot in Figure 3).

of about 500–600 m below sea level. In Figure 4 we present the modeling result with $f_t = 1$ and $f_t = 0.07$ which enable us to fit the well log data. In Figure 5 we show another example from deep water Gulf of Mexico well log data where we can fit the data with $f_t = 0.35$. In both examples we followed the following recipe to fit the data:

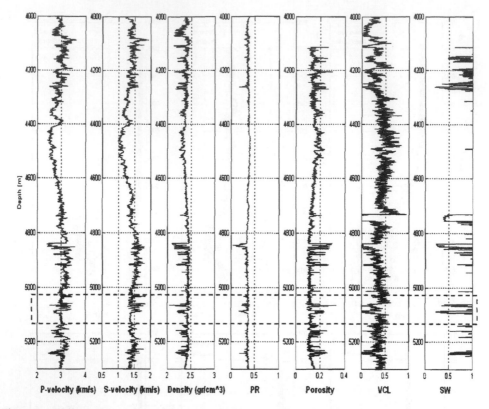

Figure 5.   Well log observations of deepwater GOM unconsolidated sands embedded in shales. (a). P-wave velocity. (b). S-wave velocity. (c). Density. (d). Poisson's ratio. (e). Porosity). (f). Clay volume. (g). Water saturation. Zone of interest is marked by dashed line.

A. f_t is estimated from observation of dry Poisson's ratio (or Vp/Vs ratio)
B. A calibration constant $c = (n^2\overline{R}/R)$ which capture both coordination number and effective contact radius is chosen to properly quantify observed seismic velocities. Note that the effective stress is assumed to be known
C. Seismic velocities are calculated directly from EMA moduli and density. We use Gassman's (1951) equation for fluid substitution if needed (See Mavko et al., 1998 for more details).

As can be seen from the results above, the theory is in agreement with the well log observation provided non uniform contact model is considered in the sandy portion of the well log. When the volume of clay is high or when thin sand-shale sequences are present we obtain scatter in the data. This is expected as the granular EMA is valid for clean sands.

5   SUMMARY AND CONCLUSION

From the data shown here we observe that as depth increases dry Poisson's ratio decreases and is consistent with the idea that larger portion of the grain contacts do not slip. We note that this was observed in laboratory experiment as well as presented in Figure 7 (Zimmer, 2003). It is interesting to note that in surface seismic data in loose sands we observe also high Poisson's ratio (Bachrach et al., 2000).

The EMA approximation is a tool to handle non-uniform stress distribution in the subsurface without resorting to expensive numerical modeling. It has been used to estimate AVA response of

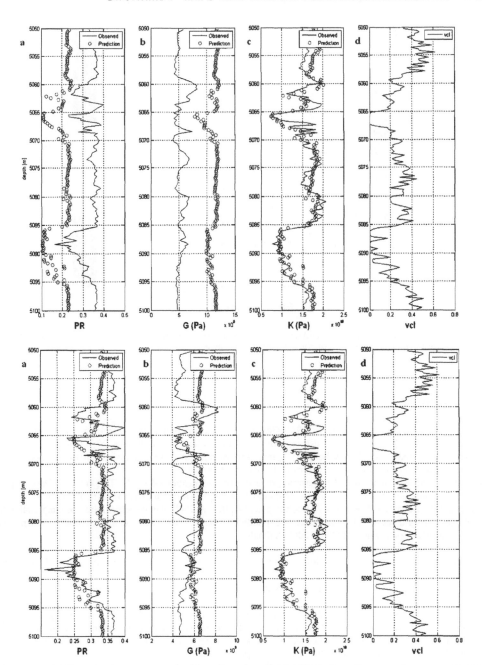

Figure 6.   Model vs. well log observation for deep water Gulf of Mexico well. Top. PR, G, K and volume of clay (VCL) fitted with $f_t = 1$. Bottom: PR, G, K and volume of clay (VCL) fitted with $f_t = 0.35$ and obtained a good match for both bulk and shear.

sands as well as deriving the stress sensitivity of granular sediments for 4D seismic application (Avseth et al., 2009). However, the theory also suggests that the measurement of dynamic Poisson's ratio's can provide us with some insight regarding the state of stress in the subsurface by considering the amount of stiff contact in the sediment.

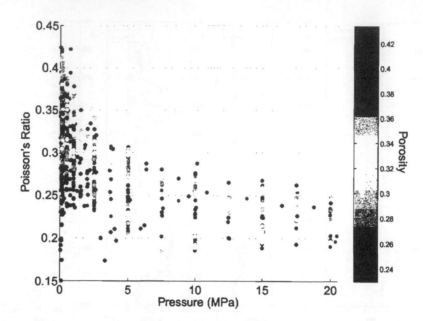

Figure 7.   Poisson's ratio in unconsolidated sands at low pressure (Data after Zimmer, 2003). Unconsolidated measurements shows that at high pressure Poisson's ratio drops.

REFERENCES

Ammi, B., D. Bideau, and J.P. Troadec, 1987, Geometrical structure of disordered packing of regular polygons, comparison with disc packing structure, Journal of Physics D: Applied Physics, 20, 424–428.

Avseth, P., T. Mukerji, and G. Mavko, 2005, Quantitative Seismic Interpretation; Applying Rock Physics Tools to Reduce Interpretation Risk: Cambridge University Press, 416p.

Avseth, P., and R. Bachrach, 2005, Seismic properties of unconsolidated sands: Tangential stiffness, Vp/Vs ratios and diagenesis: 75th Annual International Meeting, SEG, Expanded Abstracts, 1473–1476.

Avseth, P., R. Bachrach, T. Bersaas, A. Drottning, and A. Bruun, 2009, Patchy cementation and implication for stress and fluid sensitivity in sandstones; Expanded Abstract, 11th international congress of the Brazilian Geophysical Society & EXPOGEF 2009, Salvador.

Avseth, P., R. Bachrach, A.-J. van Wijngaarden, T. Fristad, and E. Ødegaard, 2007, Application of contact mechanics to hydrocarbon prediction in shallow sediments: 69th EAGE Conference & Technical Exhibition, Extended Abstracts.

Bachrach, R., J. Dvorkin, and A. Nur, 2000, Seismic velocities and Poisson's ratio of shallow unconsolidated sands: Geophysics, 65, 559–564.

Bachrach R., S. Noeth, N. Banick, M. Sengupta, G. Bunge, B. Flack, R. Utech, C. Sayers, P. Hooyman., L. den Boer, L. Leu., W. Troyer, and J. Moore, 2007, From pore-pressure prediction to reservoir characterization: A combined geomechanics-seismic inversion workflow using trend-kriging techniques in a deepwater basin: Special issue on Geomechanics, The Leading Edge, 26, 590–595.

Bowers, G.L., 1995, Pore pressure estimation from velocity data: Accounting for overpressure mechanisms beside undercompaction: Society of Petroleum Engineers, 27488, 89–95.

Digby, P.J., 1981, The effective elastic moduli of porous granular rocks: Journal of Applied Mechanics, 48, 803–808.

Gassmann, F., 1951, Elastic wave through a packing of spheres: Geophysics, 16, 673–685.

Geng, J., D. Howell, R.P. Behringer, G. Reydellet, L. Vanel, and E. Clement, 2001, Footprints in the sands: The response of granular material to local perturbations: Physical Review Letters, 87, 035506 1–4.

Goddard, J.D., 1990, Nonlinear Elasticity and Pressure-Dependent Wave Speeds in Granular Media: Proceedings of the Royal Society of London. Series A, Mathematical and Physical Sciences, Volume 430, 105–131.

Johnson, K.L., 1985, Contact Mechanics, Cambridge University Press.

Love, A.E.H, 1927, A Treatise on The Mathematical Theory of Elasticity: Cambridge University Press, New York.

Majmaudar, T.S., and R.P. Behringer, 2005, Contact force measurements and stress induced anisotropy in granular materials: Nature, 23, 1079–1082.

Manificat, G., and Y. Guéguen, 1998, What does control Vp/Vs in granular rocks?: Geophysical Research Letters, 25(3), 381–384.

Makse, H.A., N. Gland, D.L. Johnson, and L. Schwartz, 1999, Why effective medium theory fails in granular materials: Physical Review Letters, 13, 1–4.

Makse, H.A., N. Gland, D.L. Johnson, and L. Schwartz, 2004, Granular packings: Nonlinear elasticity, sound propagation and collective relaxation dynamics: Physical Review E, 70, 061302 1–19.

Mavko, G., T. Mukerji, and J. Dvorkin, 1998, The Rock Physics Handbook: Cambridge University Press.

Mindlin, R.D., 1949, Compliance of bodies in contact, Journal of Applied Mechanics, 16, 259–268.

Mueth, M.D., H.M. Jaeger, and S.R. Nagel, 1998, Force distribution in granular medium: Physical Review E, 57, 3164–3169.

Muhlhaus, H.B., and F. Oka, 1996, Dispersion and wave propagation in discrete and continuous models for granular materials: International Journal of Solids and Structures, 33, 2841–2858.

Murphy, W.F., 1982, Effect of microstructure and pore fluids on the acoustic properties of granular sedimentary materials: Ph.D. dissertation, Stanford University.

Norris, A.N., and D.L. Johnson, 1997, Non-linear elasticity of granular media: Journal of Applied Mechanics, 64, 39–49.

Pasternak, E., H.B. Muhlhaus, and A.V. Dyskin, 2006, Finite deformation model of simple shear of fault with microrotations: apparent strain localisation and en-echelon fracture pattern: Philosophical Magazine, 86, 3339–3371.

Ødegaard, E., and P. Avseth, 2004, Well log and seismic data analysis using rock physics templates: First Break, 22, 37–43.

Sava, D., and B. Hardage, 2006, Rock physics models of gas hydrates from deepwater, unconsolidated sediments: 76th Annual International Meeting, SEG, Expanded Abstracts, 1913–1917.

Spencer, J.W. Jr., M.E. Cates, and D.D. Thompson, 1996, Frame moduli of unconsolidated sands and sandstones: Geophysics, 59, 1352–1361.

Vega, S., 2003, Intrinsic and stress-induced velocity anisotropy in unconsolidated sands: Ph.D. thesis, Stanford University.

Walton, K., 1987, The effective moduli of a random packing of spheres: Journal of the Mechanics and Physics of Solids, 33, 213–226.

White, J.E., 1983, Underground Sound: Application of Seismic Waves, Elsvier, New York.

Winkler, K.W., 1983, Contact stiffness in granular and porous materials: Comparison between theory and experiment: Geophysical Research Letters, 10, 1073–1076.

Winkler, K.W., 1979, The effect of pore fluids and frictional sliding on seismic attenuation: Ph.D. dissertation, Stanford University.

Xu, S., 2002, Stress-induced anisotropy in unconsolidated sands and its effect on AVO analysis: 72th Annual International Meeting, SEG, Expanded Abstracts, 105–108.

Zimmer, M., M. Prasad, G. Mavko, and A. Nur, 2007, Seismic velocities in unconsolidated sands. Part 1—Pressure trends from 0.1 to 20 Mpa: Geophysics, 72, E1–E13.

Zimmer, 2003, Seismic velocities in unconsolidated sands, Ph.D. Thesis, Stanford University, Stanford, CA.

V. *Dynamics of landslides*

Thermo-poro-mechanical effects in landslide dynamics

Liran Goren
Department of Environmental Sciences, Weizmann institute of Science, Rehovot, Israel

Einat Aharonov
Institute of Earth Sciences, Hebrew University, Givat Ram, Jerusalem, Israel

Mark Anders
Lamont Doherty Earth Observatory of Columbia University, Palisades, NY, USA

ABSTRACT: Landslides are a significant worldwide natural hazard. Landslides also play a leading role in the morphological evolution of the Earth and other planets. Yet many aspects of the dynamics of slide initiation and motion remain unclear. We present analysis pertaining to the role of thermo-poro-mechanical (TPM) mechanisms in these dynamics. TPM affects arise when a porous, fluid-filled, shear zone is heated, either via frictional heating during sliding or from external sources such as dikes. If the shear zone is confined, and fluid diffusion is slow, elevated temperature will cause pore pressure to rise, reducing the layer strength and its resistance to sliding. This mechanism affects both the initiation and dynamics of slide motion. We present four separate studies, all linked by the common thread of TPM: (1) The enigmatic triggering of the largest known subaerial landslide, the Heart Mountain slide, is suggested to occur when magmatic dikes intruded in close sequence into a confined fluid-filled layer. The dikes thermal and mechanical affects elevated pore fluid pressure until the fluid pressure exceeded the lithostatic stress, allowing a whole mountain range to detach and slide. (2) The stability of the initial stages of the sliding process, whether slides will accelerate catastrophically or arrest on the same slope, is shown to be affected by the TPM mechanism. Here, an inherent bifurcation in the TPM mechanism controlled by minute change of system parameters results in regime shifts. (3) The volume effect of large catastrophic landslides, where larger slides exhibit longer runout distances, is explained by a combination of the TPM mechanism with depth decreasing permeability. Larger slides, with deeper shear zones residing in a lower permeability region, will reach larger sliding distances due to poorer pore pressure relaxation from the shear zone, with good agreement to field data. (4) The travel course of the Heart Mountain landslide is predicted by the TPM model with time dependent permeability, but results in unrealistic high temperatures.

1 INTRODUCTION

Landslides occur in a variety of geological and geomorphological settings including mountain ranges, stand-alone mountains, volcanic islands, submarine continental slopes and even gentle subaerial slopes. Despite their great importance both as a geohazard threatening lives and infrastructure and as a process that shapes the surface of Earth and possibly other planets, there are still many unanswered questions regarding their physics. For example: What are the processes involved in triggering of landslides? What determines the stability of the sliding process, i.e. whether a slide will accelerate catastrophically or quickly halt? And what determines the sliding distance (runout) of landslides in general, and why do larger slide exhibit longer runouts (volume effect)?

Various mechanisms have been invoked to address the physics of the different stages of the sliding process, from landslide triggering with specific application to the emplacement of the Heart Mountain landslide in northwestern Wyoming (e.g. Bucher 1947, Pierce 1957, Hughes 1970, Straw & Schmidt 1981, Melosh 1983, Beutner & Craven 1996), via landslides initial stability (e.g. Davis et al. 1990, Iverson 2000, Iverson et al. 2000, Mangeney-Castelnau et al. 2003, Savage &

Iverson 2003, Helmstetter et al. 2004, Iverson 2005, Mangeney et al. 2007, Pouya et al. 2007), to the long runouts of large slides (e.g. Voight & Faust 1982, Vardoulakis 2000, 2002, Campbell et al. 1995, Shaller & Smith-Shaller 1996, Dade & Huppert 1998, Kelfoun & Druitt 2005). Out of the many processes considered in the study of landslides physics, this work focuses on the Thermo-Poro-Mechanical (TPM) mechanism that in some cases acts in parallel to the other processes that control landslide motion. In the TPM process, temperature rise due to frictional heating at the base of the shearing block and emplacement of hot intrusions causes pore fluid pressure (PP) along the shear zone to increase via thermal pressurization. The PP increase in turn, lowers the effective normal stress and the shear resistance, thus allowing larger sliding velocities and distances to develop. Large velocity then increases the rate of heat generation and causes the temperature to increase.

In this work, the TPM mechanism is invoked to address three major questions in the field of landslide dynamics and kinematics: (1) The triggering of the Heart Mountain landslides, i.e. the process that detached the block from the slope, (2) The initial stability of the sliding process that dictates sliding kinematics, and (3) The sliding distance of catastrophic landslides, which in some cases is far greater than expected. Despite decades of investigation and large amount of published work, the above questions are still in the heart of discussion. For that reason, they are revisited here from the perspective of a TPM model which was recently shown to play a major role in controlling both landslide motion (e.g. Voight & Faust 1982, Vardoulakis 2000, 2002, Goren & Aharonov 2007, 2009) and motion along faults (e.g. Garagash & Rudnicki 2003, Rempel & Rice 2006, Rice 2006).

To explain the triggering mechanism of the Heart Mountain landslide (HML), it is argued, that magmatic dike injections played a critical role by mechanically and thermally causing PP elevation and destabilizing the upper block of the slide. The model quantitatively assesses the evolution of PP as a function of heat production and the changes in the state of stresses caused by the volcanic intrusions. The TPM mechanism is also shown to play an important role in controlling slides stability: whether slides are arrested or accelerated. It is demonstrated that stability arises from the competition between gravitational driving forces and frictional resisting forces controlled by pore fluid pressure generation and dissipation. The long travel distances of landslides and their relation to slide volumes are also studied with the same mechanism, showing that depth dependent hydraulic diffusivity results in longer runout for larger slides. Then, the unique travel course of the HML is investigated and the exact travel distance is predicted by the same mechanism.

It is satisfying that a single model can be applied to various stages of the sliding process that were previously treated with different models.

2   THERMO-PORO-MECHANICAL MECHANISM

In this section, a general model that couples thermal, poroelastic and mechanical considerations of a landslide is developed. The model assumes a simplified 1D block mass of thickness D residing on a frictional slope tilted γ (Fig. 1). When sliding is initiated, the mass slides as an intact body, and deformation is concentrated along a thin shear zone with thickness d at the base of the mass, where $D \gg d$. The shear zone is assumed fully fluid saturated as the water table is assumed to be at height w above the shear zone, so that prior to any process of pore fluid pressurization, $p_0 = \rho_f g(w - z) \cos \gamma$ for $z \leq w$, and $p_0 = 0$ for $z > w$, where the coordinate z is taken perpendicular to the slope, $z = 0$ is the slide base and $z = D$ is the surface. The model is developed as a set of coupled partial differential equations describing the temporal and the 1D spatial evolution of the excess PP, p, (with respect to the hydrostatic PP, p_0), and temperature, θ, and the temporal evolution of the sliding velocity, v. The boundary conditions are always impermeable base of the shear zone, $\partial p / \partial z (z = 0, t) = 0$, and upper drained surface $p(z = D, t) = 0$. The model is investigated both analytically and numerically. When numerical simulations are performed, the TPM model is solved using an explicit finite differences scheme.

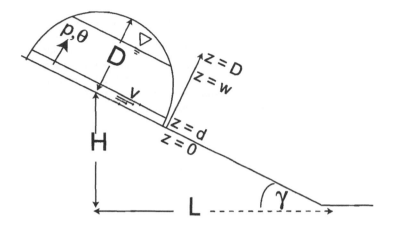

Figure 1.   Schematic representation of a simulated sliding rock column.

2.1   *Pressure equation*

A description of the PP is derived from pore fluid mass conservation consideration (Garagash & Rudnicki 2003):

$$\dot{\zeta} + \frac{\partial q}{\partial z} = 0, \tag{1}$$

where $\dot{\zeta} \equiv \partial \zeta / \partial t$ is the variation rate of increment of fluid content per area in s^{-1} and q is the fluid flow rate in m/s. $\dot{\zeta}$ may be expressed as:

$$\dot{\zeta} = S_\sigma \dot{p} - \alpha \dot{\theta} - \frac{1}{3H} \dot{\sigma}. \tag{2}$$

where S_σ and $1/H$ both in Pa^{-1} are the unconstrained specific storage, and the poroelastic expansion coefficient, respectively (Wang 2000), σ is an excess stress in Pa, α is the effective thermal expansion coefficient taken as $\alpha = \Phi \alpha_f + (1 - \Phi) \alpha_s - \alpha_{sk}$, where α_f, α_s and α_{sk} are the fluid, solid, and skeleton expansion coefficients (Vardoulakis 2002) and Φ is the porosity. A simpler reduced form of equation (2) is derived in Wang (2000, equation 1.12), where the thermal effect, $\alpha \dot{\theta}$, is not considered and the quantities, ζ, p, and σ are not differentiated with respect to time. Furthermore, the factor $1/3$ appears here in the σ term to indicate uniaxial stress, while in Wang (2000, equation 1.12) the principal stresses are implicitly assumed equal so that the factor of $1/3$ disappears. In equation (2) the effect of pore space compaction and dilation which may occur along the shear zone during initial stages of sliding (Iverson 2000, Garagash & Rudnicki 2003, Iverson 2005) is neglected, as if the shear zone is already in its critical porosity. Fluid flow within the slide is assumed to occur via porous flow, following Darcy's law:

$$q = -\frac{k}{\eta} \frac{\partial p}{\partial z}, \tag{3}$$

where k is the permeability in m^2, η is the fluid viscosity in Pa s, and $\partial p / \partial z$ is the excess PP gradient. Combining equations (1)–(3), the pore fluid continuity equation may be written as

$$\dot{p} = \frac{1}{S_\sigma} \frac{\partial}{\partial z} \left(\frac{k}{\eta} \frac{\partial p}{\partial z} \right) + \Lambda \dot{\theta} + \frac{B}{3} \dot{\sigma}. \tag{4}$$

where $k/\eta S_\sigma$ is the hydraulic diffusivity (Wang 2000), $\Lambda = \alpha/S_\sigma$ is the thermal pressurization coefficient (Voight & Faust 1982, Vardoulakis 2000, Rempel & Rice 2006, Rice 2006), and
$B = (S_\sigma H)^{-1}$ is Skempton's coefficient (Wang 2000). Equation (4) describes the different processes that affect PP: The first term on the right-hand side is PP diffusion, the second term describes
thermal pressurization that is controlled by the value of Λ that determines the response of the
PP within a confined porous volume to temperature variations, and the third term is pore fluid
mechanical pressurization controlled by B, Skempton's effect.

2.2   *Temperature equation*

The temperature, θ, change within the shear zone is derived from energy conservation considerations:

$$\rho c \dot\theta = -\rho_f c_f q \frac{\partial \theta}{\partial z} + \frac{\partial}{\partial z} k_\theta \frac{\partial \theta}{\partial z} + \tau(p) \frac{\partial v}{\partial z} + S \quad \text{for } 0 < z \le d, \tag{5}$$

where $\rho = (1 - \Phi)\rho_s + \Phi\rho_f$ and $c = (1 - \Phi)c_s + \Phi c_f$ are the density in kg m^{-3} and specific
heat in J kg^{-1}°C^{-1} of the solid-fluid mixture (Vardoulakis 2000), and subscripts s and f stand
for the solid and fluid respectively. The first term on the right-hand side of equation (5) expresses
heat advection with pore fluid. The second term expresses heat conduction where k_θ is the thermal
conductivity of the saturated rock in Wm^{-1}°C^{-1}. The third term expresses heat production at the
shear zone by frictional heating, where τ is the shear stress and $\partial v/\partial z$ is the velocity gradient along
the shear zone, assumed here linear (Aharonov & Sparks 2002). Thus, $\partial v/\partial z = (v - v_0)/d = v/d$,
where $v_0 = 0$ is the velocity at the base of the shear zone. The forth term on the right-hand side,
S, expresses heat source such as dikes intrusion that rise the temperature of their environment
by cooling and solidifying. The temperature above the shear zone evolves similarly (but without
frictional heating term):

$$\rho c \dot\theta = -\rho_f c_f q \frac{\partial \theta}{\partial z} + \frac{\partial}{\partial z} k_\theta \frac{\partial \theta}{\partial z} + S \quad \text{for } d < z \le D. \tag{6}$$

2.3   *Mechanical equations*

A block becomes gravitationally unstable on a slope and will start sliding when the shear stress
along its shear zone, τ, equals the gravitational driving force:

$$\tau(p) = C + \mu(\rho g D \cos\gamma - p_0 - p) = \rho g D \sin\gamma, \tag{7}$$

where C is the tensile strength of the block, and μ is the friction coefficient. After triggering,
sliding kinematics is controlled by:

$$\dot v = g \sin\gamma - \mu \left[g \cos\gamma - \frac{p_0}{\rho D} - \frac{p}{\rho D} \right]. \tag{8}$$

2.4   *Overview*

Equations (4)–(6) and (8) form a complete set for the PP, temperature and velocity of a sliding body.
This set encapsulates several feedback mechanisms that control the mechanics and kinematics of
the sliding process. One such feedback involves temperature rise by frictional heating, thermal
pressurization, reduction of shear stress and frictional resistance and increase of sliding velocity.
Higher velocity in turn leads to faster generation of heat, a process that enhances the feedback.
On the other hand, decrease of shear stress reduces the rate of heat production which will tend to
oppose the feedback. Another example is the connection between PP gradients and PP relaxation by
diffusion: Localized generation of PP along the shear zone has the consequent of reducing frictional

resistance, but also leads to fast PP relaxation by diffusion with the opposite effect. Accumulation of PP above the shear zone prevents effective relaxation and may also lead to frictional resistance reduction.

The TPM model developed here is next applied to various stages of the sliding process: triggering, initial stability and runout. In each stage, different terms in equations (4)–(8) might become either irrelevant or negligible and thus at each stage we neglect different terms.

3 TRIGGERING

This section focuses on the process of landslides triggering by PP elevation along a zone that will develop to be the sliding shear zone, upon which motion is initiated. The processes of PP elevation discussed here are thermal and mechanical. Hydraulic processes, such as rainfall and snow melt infiltration and accumulation along the shear zone (Sidle & Ochiai 2006) are not accounted despite their documented importance (e.g. Iverson 2000).

Sliding will initiate when the excess PP reaches a critical value, p_c, dictated by the criterion established in equation (7):

$$p_c = \frac{C}{\mu} + \rho g D \left(\cos \gamma - \frac{\sin \gamma}{\mu} \right) - p_0.$$

(9)

Equations (4)–(6) are used in the determination of the evolution of p until it meets p_c. Here, frictional heating is irrelevant as it comes into action only after the sliding process is triggered, when $v \neq 0$, thus the third term on the right-hand side of equation (5) is ignored.

3.1 *Triggering of the Heart Mountain landslide*

The TPM model is used here to explain the enigmatic triggering of the Eocene-age Heart Mountain landslide (HML), the largest subaerial landslide known, that slid on an extremely shallow slope dipping $\gamma = 2°$. HML, named after one of its upper sliding block fragments, covers an area of roughly 3,400 km^2 with a toe that thrust out over 45 km. The upper block deposits of the HML are composed of a collection of Paleozoic rocks sitting on top of Tertiary sediments in northwestern Wyoming (Aharonov & Anders 2006), varying in size between hundred of meters to 8 km (Pierce 1973). The shear zone of the HML is located along the Big Horn Dolomite formation with thickness of 175 m, that is bounded above and below by low-porosity shale horizons.

The emplacement mechanism of the HML is argued to be catastrophic in nature (Aharonov & Anders 2006, and references therein). However, a basic question remains as to what caused the initial movement on a near-horizontal surface. Previously proposed mechanisms include lifting of the slide mass on a cushion of air resulting from pressurized volcanic gases, vibration of the lower plate during eruption-related earthquakes, or by earthquake induced 'acoustic fluidization', recurrent eruptions at volcanic centers resulting in 'explosive pressure' buildup of pore fluids that exceeded lithostatic loading, and a vertical fracture filled with fluid (magma and water/steam) (Bucher 1947, Pierce 1957, Hughes 1970, Voight 1973, Straw & Schmidt 1981, Melosh 1983, Beutner & Craven 1996). Here we focus on the effect of dike intrusion in elevating PP thermally and mechanically.

Along the HML shear plane, PP elevation is evident by the character of the basal conglomerate layer that exhibits characteristics of extensive fluidization during the movement phase with a network of clastic dikes penetrating into the upper plate. Fluidization is most easily explained in the presence of pressurized water or gases. The source for pore water pressurization may lie in a phase of extensive dikes intrusion into the upper plate that is dated to be close in time to the emplacement of the HML (Aharonov & Anders 2006).

The effect of dikes injection is dual: they increase horizontal stresses and heat the surrounding layers. Here, we investigate the cumulative long-term effect of many dikes intruding sequentially into the fluid-saturated layer: (1) For the mechanical effect, consider n vertical dikes of thickness w_d, intruding from a single volcanic stock of radius r_0, into a layer of horizontal diameter $2r$. The inserted dikes wedges increase the layer perimeter by nw_d. It can be shown that injection of n radial dikes into the intruded layer causes an excess average azimuthal stress of:

$$\sigma = E\frac{V_i}{2}\ln\left(\frac{r}{r_0}\right), \tag{10}$$

where E is the layer's Young's modulus, and $V_i = nw_d/\pi r$ is the volume fraction of the intrusions with respect to the host rock, that depends on the number of injected dikes, n. (2) A volcanic intrusion has a temperature much larger than the temperature of the surrounding rock. The cooling of the intrusion by conduction and advection into the hosting rock and the latent heat released by solidification form together a heat source that raises the temperature of the hosting rock. The heat released from the intrusions per unit height is evaluated as:

$$Q_i = rnw_d\rho_i\left[c_i\left(\theta_i^0 - \theta^0 - \delta\theta\right) + L_i\right], \tag{11}$$

where subscript i refers to the intrusive body. θ^0 and θ_i^0 are the initial temperature of the host rock and intrusion, respectively, and L_i is the latent heat of solidification in J kg^{-1} (Aharonov & Anders 2006). If little heat is lost to the upper and lower layers confining the dolomite intruded layer, as will be established in the following, equation (11) may be equated to the heat absorbed by the hosting dolomite layer per unit height: $\pi r^2 \rho c\delta\theta$, to give (under the assumption of $V_i \ll 1$):

$$\theta = V_i\frac{\rho_i}{\rho c}\left[c_i\left(\theta_i^0 - \theta^0\right) + L_i\right]. \tag{12}$$

In terms of equations (5) and (6), the heat source term, S, may be evaluated as $S = \dot{V}_i\rho_i[c_i(\theta_i^0 - \theta^0) + L_i]$.

Next, to evaluate the importance of heat advection and conduction and PP diffusion away from the hosting dolomite layer, the characteristic time scales of these processes is derived. Heat conduction time scale is evaluated as: $t_{cond} = T^2\rho^c c^c/k_\theta^c$, where T is the thickness of the dolomite layer and superscript c refers to the bounding shales. For $\rho^c = 2675$ kg m^{-3}, $c^c = 1$ kJ kg^{-1}°C^{-1}, and $k_\theta^c = 1$ W m^{-1}°C^{-1}, $t_{cond} \sim 2600$ years. On the other hand from Turcotte & Schubert (1982, figure 9–21 and equation 9–152) fluids within the dolomite layer are expected to convect, with a plume traversing the layer in $\sim$100 days. Thus, it is predicted that although heat is quickly and evenly distributed within the dolomite layer by convection of fluid, it is retained in the dolomite because conduction through the bounding shales will be sluggish. For that reason, when considering the heat along the dolomite layer, heat conduction, second term in equations (5) and (6) is negligible. Next, the time scale of hydraulic diffusion of PP past the shales is derived: $t_{hyd} = T^2 S_\sigma \eta/k \sim 300$ years, accounting for shale permeability of 10^{-17} m^2. Therefore, by the same argument adopted for heat conduction, also PP diffusion away from the dolomite layer, first term on the right-hand side of equation (4) is negligible. By similar reasoning, heat advection, the first term on the right-hand side of equations (5) and (6) is also negligible as it is carried by pore fluid flow.

In this triggering problem, the focus of interest is the evolution of PP along the developing shear zone. The above time scales analysis indicating that PP and heat relaxation processes are negligible, leads to p and θ being constant along the dolomite layer and thus depth independent. Integrating with time the remaining terms of equation (4), and combining with equations (10) and (12), the excess PP along the dolomite layer of the HML is evaluated as:

$$p = V_i\left\{\Lambda\frac{\rho_i}{\rho c}\left[c_i\left(\theta_i^0 - \theta^0\right) + L_i\right] - \frac{BE}{6}\ln\left(\frac{r}{r_0}\right)\right\}, \tag{13}$$

Table 1.   Parameters used in section 3.1.

Symbol		Value
D	Upper block thickness	3000 m
Φ	Porosity	0.15
ρ_s	Solid density	2840 kg m^{-3}
ρ_f	Fluid density	1000 kg m^{-3}
ρ_i	Intrusion density	2900 kg m^{-3}
c_s	Solid specific heat	1000 J kg^{-1} °C^{-1}
c_f	Fluid specific heat	4187 J kg^{-1} °C^{-1}
c_i	Intrusion specific heat	1000 J kg^{-1} °C^{-1}
L_i	Latent heat of solidification	320 kJ kg^{-1}
θ^0	Initial temperature of dolomite	87.85°C
θ_i^0	Initial temperature of intrusion	1027.85°C
B	Skempton's coefficient	0.7
E	Young's modulus	40 GPa
r_0	Radius of volcanic stock	1 km
r	Radius of intruded layer	10 km
w_d	Intrusion thickness	2 m
Λ	Thermal pressurization coefficient	1.5×10^6 Pa°C^{-1}
γ	Slope angle	2°

and p depends linearly on V_i, i.e. on the number of injected dikes. Aharonov & Anders (2006) calculate that for the HML, equation (9) is satisfied when $p_c \sim 49$ MPa, after neglecting the tensile strength, C, for the many fractures in the block. Using the parameters of Table 1 it is found that equation (13) meets the critical pressure, p_c, when $n \sim 63$ (where n in the number of injected dikes). I.e. for any $n > 63$ sliding is expected to initiate. Indeed, field count of the number of dikes greatly exceed 63.

In the next sections, the sliding process after triggering is investigated with the TPM model.

4   STABILITY

Commonly, large-scale landslides (with failure volumes larger than 10^5 m^3 (Sidle & Ochiai 2006)) are characterized by very high sliding velocities of 10–100 m/s. Two examples are the notorious 1963 Vaiont landslide and the 1987 Val Pola landslide, in the Italian Alps. Both landslides accelerated catastrophically (Vaiont after a long episode of creeping), and halted only after sliding down the whole slope over which the sliding mass initially resided, when hitting a topographical barrier (valley floor for the Vaiont and the opposing slope for the Val Pola) (Crosta et al. 2004, Helmstetter et al. 2004, Genevois & Ghirotti 2005).

In contrast to such catastrophic events, there are instances in which large slope failures are arrested after they have moved a very short distance. These can also be regarded as landslides as they are downward movement of slope forming material composed of natural rocks and soils (Sidle & Ochiai 2006), but their potential to develop high velocities and large travel distances does not come into play for reasons discussed below. Such an arrested sliding motion is suggested to have occurred as a result of the 1949 eruption of Cumbre Vieja volcano at the Island of La Palma (Day et al. 1999, Ward & Day 2001). Gabet & Mudd (2006) describe similar arrested sliding motion as slumps, whose center of mass traveled 1–2 m and their deposits did not evacuate the meters long scars. Similarly, Bartelt et al. (2007) refer to such events as 'starving avalanches' and describe them as small debris avalanches that did not run out onto the valley bottom, but stopped close to where they began.

These observations suggest that gravitational mass movements down slopes may either develop into catastrophic landslides (like Vaiont and Val Pola) or into a self-arresting motion that halts on the slope over which it was initiated. Therefore, the question that needs to be addressed is: what determines the stability of the sliding process of landslides? Various explanations were proposed by previous studies referring to this question: Variation of mechanical parameters in a poro-elastic perspective (Davis et al. 1990, Iverson et al. 2000, Helmstetter et al. 2004, Iverson 2005), and competing infiltration and seepage of fluid (Iverson 2000, Pouya et al. 2007) did not predict full arrest of initially accelerating slides; instead, slides either accelerate or attain a constant velocity. However, models that accounts for thickness variations of the sliding mass along the down-slope axis do observe a regime in which full arrest of slides occurs (Mangeney-Castelnau et al. 2003, Savage & Iverson 2003, Mangeney et al. 2007). Here, we propose that a TPM mechanism may also play an important role in controlling whether slides are arrested or accelerated.

In this section the sliding mass is assumed homogeneous, with constant physical properties with depth. A shear zone is already formed and the mass is detached from its surrounding. For this reason, the last Skempton term in equation (4) is ignored because external stress cannot affect the sliding mass, and the last heating source term in equations (5) and (6) is ignored for the same reason. Heat conduction time scale is evaluated with Table 2 parameters, and the shear zone thickness, d, taken as the length scale, $t_{cond} = d^2 \rho c / k_\theta \approx 10^5$ s, and is found to be very long compared to the onset of sliding process that is modeled in this section (few seconds). In contrast to heat, the characteristic hydraulic diffusivity time scale is much shorter, $t_{hyd} = d^2 \eta S_\sigma / k \approx 0.04$ s. The short time scale for PP diffusion, t_{hyd}, suggests that hydraulic diffusivity plays a crucial role in this TPM application. Heat advection and heat generation characteristic time scales are $t_{adv} = (d\eta/\chi k)(\partial p/\partial z)^{-1}$, where $\chi = \rho_f c_f / \rho c$, and $t_{gen} = d/v$, respectively. These quantities are harder to estimate a-priori as they depend on dynamic variables. However, numerical testing reveals that addition or emission of heat advection changes the overall travel distance of the landslides considered here at most by 10%, and hence heat advection is neglected in the following. Thus, the temperature is controlled solely by frictional heating and the set of equations for the problem of stability is thus reduced to:

$$\dot{p} = \frac{k}{S_\sigma \eta} \frac{\partial^2 p}{\partial z^2} + \frac{v \Lambda \mu}{\rho c d} (\rho g D \cos \gamma - p_0 - p) \quad \text{for } 0 < z \leq d$$

$$\dot{p} = \frac{k}{S_\sigma \eta} \frac{\partial^2 p}{\partial z^2} \quad \text{for } d < z \leq D$$

$$\dot{v} = g \sin \gamma - \mu \left[g \cos \gamma \left(1 - \frac{\rho_f}{\rho} \right) - \frac{p}{\rho D} \right] \tag{14}$$

4.1 *Numerical bifurcation*

First, the system of equations (14) is solved numerically with water table at the top of the sliding mass, $w = D$. Table 2 presents parameter values used in simulations. The initial conditions are $v_{init} = 0$ and $p_{init} > 0$ corresponding to initial small fluid pressurization within the shear zone by either fluid addition due to rain infiltration, or by mechanical or thermal pressurization as considered for the case of the HML is section 3.1.

Figure 2 shows PP evolution within the shear zone, $\hat{p} = p/\rho g D$, and the sliding velocity, $\hat{v} = v/\sqrt{gD}$, where the ˆ notation indicates non-dimensional variables. Three sets of simulations are performed. In each set all the parameters are maintained constant except for one. In Figure 2a the permeability, $\hat{k}$, is varied, in Figure 2b, the thermal pressurization coefficient, $\hat{\Lambda}$, is varied and in Figure 2c the friction coefficient, μ, in varied. All simulations exhibit similar behavior during sliding onset, with initially velocity increase and PP decrease. However, after some time the behavior bifurcates between two regimes depending on the exact value of the parameter that is varied. When changing the permeability, for $\hat{k} < \hat{k}_c$, where $\hat{k}_c$ is the critical permeability, $\hat{v}$ and $\hat{p}$ rise and sliding accelerates. A distinctively different sliding regime is observed when $\hat{k} > \hat{k}_c$: here, after ∼2.5 s from the sliding onset, sliding decelerates. For $\hat{k} \gtrsim \hat{k}_c$, the slide halts after about 14 seconds having moved 0.4 m and attaining a maximum velocity of ∼4.2 cm/s. When $\hat{\Lambda}$ is varied,

Table 2. Parameters used in sections 4.

Symbol		Value
d	Shear zone thickness	0.1 m
D	Block thickness	100 m
Φ	Porosity	0.3
ρ_s	Solid density	2700 kg m^{-3}
ρ_f	Fluid density	1000 kg m^{-3}
c_s	Solid specific heat	1000 J kg^{-1} °C^{-1}
c_f	Fluid specific heat	4187 J kg^{-1} °C^{-1}
Λ	Thermal pressurization coefficient	0.3 × 10^6 Pa °C^{-1}
k	Permeability	7 × 10^{-13} m^2
μ	Friction coefficient	0.5
η	Fluid viscosity	10^{-3} Pa s
S_σ	Unconstrained specific storage	3 × 10^{-9} Pa^{-1}
γ	Slope angle	15°
l	Diffusion length scale	0.28 m
k_θ	Thermal conductivity	0.38 Wm^{-1}°C^{-1}
p_{init}	Initial conditions	0.15MPa

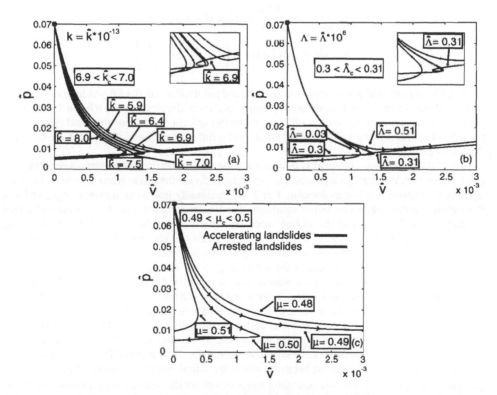

Figure 2. Numerical simulation results of the evolution of non-dimensional PP along the shear zone, $\hat{p}$, and velocity, $\hat{v}$, where temporal evolution is depicted by the direction of the arrows. Parameters from Table 2 are kept constant, except for one parameter that is varied in each set of simulations: Permeability is varied in (a), thermal pressurization coefficient in (b), and friction coefficient in (c). Simulations show a bifurcation between two distinct kinematics regimes: arrested slides where PP and velocity decrease until full arrest (dark gray), and catastrophic slides with PP rise and acceleration (light gray). The regime in which a slide reside depends on the exact value of the parameter that is varied with a small change of this parameter leading to different kinematics regime.

arrested slides occur when a value smaller than a critical $\hat{\Lambda}_c$ is used and accelerating for larger values. Similarly, increasing μ above μ_c, sliding is arrested, and decreasing it below μ_c, sliding is catastrophic.

The loops along the $\hat{k} = 6.9$, and $\hat{\Lambda} = 0.31$ curves in the inset of Figures 2a and 2b, respectively, result from the evolution of PP outside the shear zone by diffusion. Initially, the curve follows the arrested regime behavior, but then, PP outside of the shear zone accumulates to high enough values so that the diffusion rates from the shear zone slow down and the curve loops to continue along a catastrophic regime. This kind of interesting behavior is expected for parameters near the bifurcation values $\hat{k}_c$, $\hat{\Lambda}_c$, and μ_c.

To understand how minute changes in a single parameter may lead to such diverging dynamical behavior, a simplified form of equation set (14) is studied analytically.

4.2   *Analytical source of bifurcation*

For the purpose of studying the dynamics of equation set (14) within the shear zone, the equation that describes PP evolution outside the shear zone is ignored and the PP diffusion term is linearized. The second derivative of the PP, $\partial^2 p / \partial z^2$, is approximated as p/l^2, where l is a diffusion length scale in meters dictating the distance over which PP is relaxed. The equations to solve are then a function of time only:

$$\dot{p} = \frac{k}{S_\sigma \eta l^2} p + \frac{\nu \Lambda \mu}{\rho c d} \left(\rho g D \cos \gamma - p_0 - p \right)$$

$$\dot{v} = g \sin \gamma - \mu \left[g \cos \gamma \left(1 - \frac{\rho_f}{\rho} \right) - \frac{p}{\rho D} \right] \tag{15}$$

This non-linear system of equations has only one fixed point (p^*, v^*) in the PP—velocity space, that by physical considerations is located in the positive quarter of the phase space (Goren & Aharonov 2009). Next, the nature of the fixed point is examined by exploring the eigenvalues and eigenvectors of system (15) and is found to be a saddle point. (For a full exploration of the system see Goren & Aharonov (2009)).

Figure 3 shows the phase plane of system of equations (15), with the parameters given in Table 2. Figure 3 demonstrates that the eigenvectors (solid lines) divide the phase plane into four domains. Any initial condition that lies in domains 1 or 2 will eventually lead to an increase of $\hat{p}$ and $\hat{v}$ and catastrophic sliding, while any initial condition that lies in domains 3 or 4 will eventually lead to a decrease of $\hat{p}$ and $\hat{v}$ until the sliding block will reach zero velocity and halt. In short, the eigenvector, u_2, acts as a separatrix distinguishing between two stability regimes, to its right and top—catastrophic sliding, and to its left and bottom—arrested sliding.

Here, we focus on initial conditions for which $\hat{p}_{init} > p^*$ and $\hat{v}_{init} = 0 < v^*$, following the numerical simulations. Such initial conditions may lie either in domain 1 or 4. Initial conditions that corresponds to domains 2 and 3 are not discussed here, but are reviewed in Goren & Aharonov (2009). It is found that changes in some of the system parameters affect both the location of the fixed point and the slope of the eigenvectors and may thus affect sliding stability. A permeability increase, shifts the fixed point to the right and rotates u_2 clockwise, so that an initial condition that originally resides in domain 1 can move to domain 4, i.e. sliding is stabilized by permeability increase. Figure 3 demonstrates this behavior, where the initial condition is marked by 'x'. Taking a permeability of $\hat{k} = 5.9$, the corresponding eigenvectors are the solid light gray lines. The initial condition lie in this case in domain 1 and will evolve as a catastrophic slide that follows the dashed light gray curve. In contrast, when taking permeability $\hat{k} = 6.9$, the corresponding eigenvectors are the solid red lines. The initial conditions lie in this case in domain 4 and will evolve following the dashed dark gray curve. Sliding starts by acceleration, but then the mass starts decelerating until sliding is arrested. A similar situation occurs for other parameters: decreasing the thermal pressurization coefficient, Λ, or increasing the friction coefficient, μ, have the same affect of

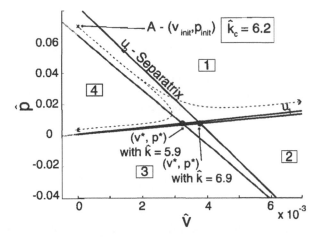

Figure 3.   Phase plane of equation set (15) with two sets of solutions for two choices of permeability. Solid lines are the eigenvectors, dashed curves show time evolution trajectories, circles are the fixed points and the '*x*' represents possible initial condition. Physical solutions exist only to the right of the shaded area, where $\hat{v} > 0$. Light and dark gray corresponds to solutions with permeabilities of $\hat{k} = 5.9$ and $\hat{k} = 6.9$, leading to catastrophic and arrested sliding, respectively. It is found analytically that for the parameters of Table 2 $\hat{k}_c = 6.2$. The small deviation from the value of $\hat{k}_c$ found in the numerical analysis (Fig. 2a) results from the simplification involved in the linearization of the diffusion term.

shifting the fixed point to the right and rotating u_2 clockwise, thus stabilizing landslides, as was observed numerically in section 4.1. Also, increasing the shear zone thickness, d, or decreasing the slope angle γ, have as well the same affect of stabilizing landslides.

From this section it is concluded that landslide triggering is not sufficient to ensure a catastrophic sliding process, as there are some set of mechanical, hydraulic and geometrical parameters that will act to stabilize an accelerating landslides. The understanding that a minute change in one of these parameters is sufficient to stabilize or catastrophically accelerate a slide is a new insight from this model.

After investigating the initial stability of landslides, the next section studies the runout distance of catastrophic slides.

5   RUNOUT

One of the features that has drawn much scientific interest is that large slides often exhibit a perplexing 'long runout' behavior—a travel distance so large that it seemingly violates mechanical considerations of a mass sliding down a frictional slope. Field data also shows that the apparent friction coefficient H/L (where H is the drop height and L the travel distance, Figure 1) decreases with increasing slide volume V (Legros 2002, Dade & Huppert 1998). The observed V–H/L relation cannot be explained by a simple model of a block sliding on a slope, as such a model yields volume independent travel distances (Shaller & Smith-Shaller 1996).

Previous studies addressing the long runout distances of slides used TPM models (e.g. Voight & Faust 1982, Vardoulakis 2000, 2002), discrete granular models (e.g. Campbell et al. 1995), and other mechanisms (for review see Shaller & Smith-Shaller 1996). The volume effect, that larger slides travel further, was offered to be the result of a constant stress resisting law (Dade & Huppert 1998, Kelfoun & Druitt 2005), but the origin of such a law is still missing.

Here again the TPM model is used to study both the long travel distances of landslides and their relation to slide volumes. As the TPM model discussed here has only one spatial dimension— landslide thickness, D, the observed (but not well understood) relation $V = D^3/\epsilon^2$ where $\epsilon = 0.05$

(Malamud et al. 2004, and references therein) is used. This relation enables comparing model results to field data.

Similarly to the stability investigation, (section 4), Skempton's effect on the PP in equation (4) and the heat source term in equations (5) and (6) are ignored, and heat conduction and advection are neglected, because here as well the sliding duration that is studied is small (an order of a minute). It is important to note however, that when considering natural long runouts with sliding duration of the order of 400 s, the processes of heat conduction and advection may become important.

5.1   *Simulations*

Numerical simulations are performed of a block mass sliding down a slope. When the slope terminates, the mass continues to slide along a horizontal strata, until it halts when $v = 0$. Table 3 lists the parameters used in the simulations.

First, two simple scenarios are studied: (1) Undrained shear zone—in this case PP cannot diffuse away from the shear zone, and it accumulates by thermal pressurization but is not relaxed. Our analysis of this situation shows that PP rises quickly. If by the time the mass has reached the horizontal starta, the PP becomes equal to the confining stress, $\tau = 0$, the mass will continue to slide indefinitely, as there is no force resisting the continuation of motion (Goren & Aharonov 2007). (2) A shear zone with constant permeability with depth. In this case our simulation results show that long runout distances are achieved with longer distances for smaller permeabilities. However, all slides regardless of slide thickness travel the same distance. Indeed, Goren & Aharonov (2007) show that with depth independent permeability the TPM model is D independent and thus the same travel distance is expected for slides with different volumes.

Next, a more natural scenario is studied where the TPM model is investigated with depth dependent permeability, $k = k(z)$. Permeability reduction with depth is a well documented phenomena (Manning & Ingebritsen 1999). It is expected in clays due to plastic compaction with overburden (Gutierrez & Wangen 2005, and references therein). For porous rocks, permeability is reduced with depth due to brittle pore space collapse and pressure solution. Manning & Ingebritsen (1999) argue for a power-law relation between permeability and depth for the continental crust. Saar & Manga (2004) modify this relation by introducing a shallow exponential law so that the depth-permeability relation coincides with the more easily measurable surface permeability.

In the simulations, the effect of Saar & Manga (2004) permeability-depth law is studied together with four other permeability-depth relations that are depicted in Figure 4. A set of simulations is performed for each of the five permeability laws with slide thicknesses ranging from 5 m to 1000 m. The evolution of the sliding velocity, PP and temperature at the base of slides is presented in Figure 5a, 5b and 5d, respectively. While sliding down the slope, v, p and θ increase with sliding

Table 3.   Parameters used in section 5.1.

Symbol		Value
d	Shear zone thickness	0.02 m
D	Block thickness	5-1000 m
Φ	Porosity	0.2
ρ_s	Solid density	2700 kg m^{-3}
ρ_f	Fluid density	1000 kg m^{-3}
c_s	Solid specific heat	1000 J kg^{-1} °C^{-1}
c_f	Fluid specific heat	4187 J kg^{-1} °C^{-1}
Λ	Thermal pressurization coefficient	0.5×10^6 Pa °C^{-1}
μ	Friction coefficient	0.5
η	Fluid viscosity	10^{-3} Pa s
S_σ	Unconstrained specific storage	3×10^{-9} Pa^{-1}
γ	Slope angle	$\tan^{-1} \mu$
H	Drop height	5 m

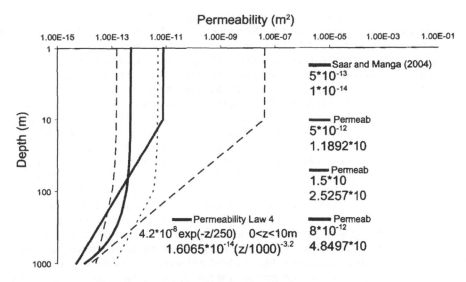

Figure 4. The five depth-permeability laws used in the simulations.

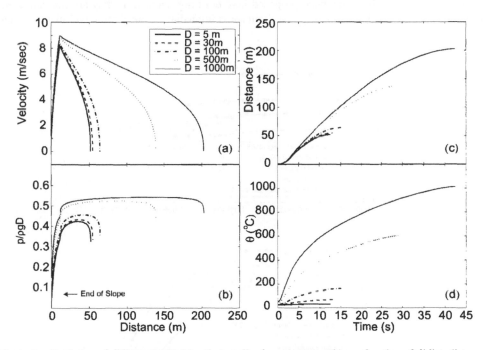

Figure 5. Evolution of sliding velocity (a) and normalized pore pressure (b) as a function of sliding distance, and sliding distance (c) and temperature (d) as a function of sliding duration, for landslides with different thicknesses using permeability law 2.

distance, at a rate that increases with slide thickness, D. Upon reaching the end of the slope, as the slide continues sliding on the plane, the shear stress increases immediately, leading to a jump in PP (Fig. 5b). However, because gravity does not drive continued sliding on the plane, the mass starts to decelerate (Fig. 5a). As expected, the lower velocities result in lower rates of heat production, and after some delayed time PP decreases. Simulations end when the mass velocity reaches zero.

As the basal temperature continues to rise during sliding (Fig. 5d), the larger slides may reach temperature high enough to induce melting (Rempel & Rice 2006). Also, for the larger slides water may transition into their critical state. These processes are beyond the scope of this study.

Figure 5 suggests that frictional heating, and by that lubrication of the shear zone by high PP, may induce long runout landslides, with p, θ and v increasing with increasing slide sizes. Figure 6a and 6b demonstrate this behavior explicitly, comparing the apparent friction coefficient H/L as a function of landslide volume, V, between numerical slides and field data of subaerial volcanic (6a), and non-volcanic (6b) landslides (Legros 2002). The field data is depicted by grey diamond symbols. Results of numerical simulations are presented as connected symbols. Points produced using the same permeability law are connected by a curve. The trend of the curved lines in Figures 6a and 6b indicate that the simulations successfully produce the volume effect observed in the field. The line generated with Saar & Manga (2004) permeability law (Fig. 6a), represents a qualitative average for the volcanic slides field data, while the lines produced by permeability laws 1 and 2 (in 6a) and permeability laws 3 and 4 (in 6b) envelope the volcanic and non-volcanic field data, respectively. Thus, these permeability laws represent upper and lower bounds for permeability profiles at the locations of the slides, assuming they were generated by the mechanism discussed here. A small group of volcanic slides lies outside the model envelop. These slides are characterized by small volume and hence shallow shear zones (up to 15 m) and show high values of apparent friction coefficient, $H/L \geq 0.5$. A possible interpretation is that this group represents dry slides. Due to their small volume, it is reasonable to speculate that the depth of their shear zone lies above the water table and thus the mechanism proposed here does not operate, as they are not saturated. Alternatively they break up so easily that drainage is immediate. Indeed the travel distance they

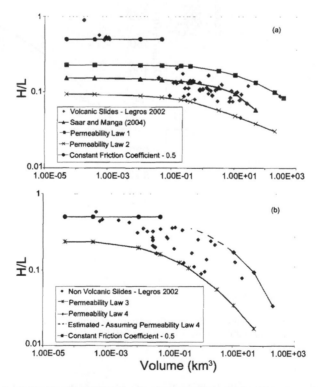

Figure 6.   Comparisons between field data of volcanic (a) and non-volcanic (b) subaerial landslides (Legros 2002), and simulation results with depth-dependent permeability. Simulation results successfully produce the volume effect. Each curved line represents a series of simulations with different thicknesses and the same permeability law taken from Figure 4. Dashed light grey areas represent the field of possible $V - H/L$ values enveloped by the simulation curves.

exhibit is normal, accordingly they do not represent long runout landslides but normal landslides. Such a group is also identified among the non-volcanic slides, but here it exhibits smaller values of friction coefficient and thus may be interpreted as representing both dry and partly saturated slides. It is also observed that the field data for the non-volcanic slides is more scattered. This scattering may be attributed to the wide range of lithologies and degrees of saturation they represent.

The study of long runout reveals that the TPM mechanism can explain the long runout distances of landslides. When combined with depth-dependent permeability, the volume effect is predicted by the model, and comparison to field data shows good fit.

6 THE LONG RUNOUT OF THE HEART MOUNTAIN LANDSLIDE

Here, the tools developed in the previous sections are used to investigate the runout process of the Heart Mountain landslide (HML) after triggering. In section 4, the stability of slides was analyzed assuming the controlling parameters remain constant with time. However, for any given natural slide, there are parameters such as the permeability that are expected to vary with time or sliding distance. Permeability evolution with sliding motion is expected by fractures formation which will lead to its increase. Therefore, a sliding block that starts with a relatively low permeability in the catastrophic regime, might eventually self-stabilize by a dynamic permeability increase. Such a scenario is hypothesized to apply to the HML.

HML is considered a long runout landslide, as it exhibits the main perplexing features of long runouts: very long travel distance along a shallow dipping slope. Fragments of the HML upper block are found as far as 105 km from the breakaway fault. This landslide accelerated along a shallow slope of 2°, and attained a travel distance of tens of kilometers, stopping on a slope with the same 2° dip (Pierce 1973). Following the triggering mechanism discussed here in section 3.1 and in Aharonov & Anders (2006), it is concluded that the PP along the evolving shear zone exceeded the value of the confining stress allowing the initial motion. As the runout distance is so great, it is concluded that the sliding block slid catastrophically with increasing PP and velocity. By the notion adopted in section 4 it is speculated that initially HML was placed in domain 1 of the phase space (Fig. 3), with elevated PP, zero velocity and catastrophic travel course. The Heart Mountain fragment is located 82 km from the breakaway fault and is deposited on a 2° slope. Thus, the regime must have changed from catastrophic to arrested to allow halting on the same slope. It is speculated that the process that allowed this change of regime from catastrophic to arrested is the extensive fragmentation the upper block suffered during the sliding process which must have led to a permeability increase.

After presenting the general framework for the sliding process of the HML, we next investigate it in details using the TPM model.

6.1 *Geometry and parameters*

Pierce (1973) supplies a thorough review of the geometry and field relation of the HML. We briefly repeat it here. Prior to sliding, the upper plate block occupied an area of 1300 km^2, bounded in its northwest end by a high-angle breakaway fault dipping to the southeast and to its southeast by a transgressive fault dipping 10° to the northwest. The distance between the two ends is 56 km (Fig. 7). The sliding block became separated from the surface over which it slid along a bedding plane dipping 2° to the southeast. This is also the sliding direction. To the southeast of the transgressive fault the upper plate blocks continued sliding along an Eocene land surface dipping as well 2° to the southeast. Prior to movement, 2–4 km of younger rocks overlay the detachment plane (Aharonov & Anders 2006). The upper plate blocks thus slid along the detachment fault past the transgressive fault and then along the Eocene land surface. Not all the sliding blocks climbed past the transgressive fault as the landslide deposits span the vast area between the breakaway fault in the northwest to McColloch Picks about 50 km southeast of the transgressive fault marking the remotest location of the landslide deposits.

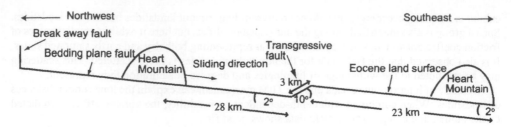

Figure 7.   The travel course of the Heart Mountain block starting from the center of mass of the upper plate, along the bedding plane, past the reversely dipped transgressive fault and continued along the Eocene land surface.

Here, we model the sliding process of the block composing Heart Mountain itself (Fig. 7). The process of fragmentation of the upper sliding block is not accounted for, but only the travel distance. Heart Mountain was deposited about 23 km to the southeast of the transgressive fault. Its travel distance along the bedding plane detachment fault up to the transgressive fault is not known and hence it is taken here to be 28 km, as if initially it was the center of mass of the upper plate. The block composing the Heart Mountain thus climbed the 3 km across the transgressive fault separating the bedding plane detachment fault from the Eocene surface plane.

6.2   *Simulations*

For simulating the travel course of the Heart Mountain fragment, initial and boundary conditions for the excess PP and temperature are first defined:

$$p(z, t = 0) = \rho g(D - z) \left[\cos \gamma - \frac{\sin \gamma}{\mu} \right] - p_0(z) \quad \text{for } 0 < z \leq 175\,\text{m}$$

$$p(z, t = 0) = 0 \quad \text{for } 175\,\text{m} < z \leq D. \tag{16}$$

Equation set (16) expresses elevated excess PP along the bottom 175 m, within the dolomite layer of the upper block. PP is elevated to the critical value according to equation (9). Pore fluid pressurization resulted from mechanical and thermal stresses, as discussed in section 3.1. Above the dolomite layer, excess PP is zero and the conditions are hydrostatic with water table at $w = D$. The initial temperature distribution is taken to follow a geothermal gradient of 25°C km^{-1}, and the bottom 175 m experience an additional temperature rise of 5.5°C due to dikes heating (Aharonov & Anders 2006). The boundary conditions for the temperature are constant temperature at the top of the foot block (over which the mass is sliding) as the slide velocity is high enough to prevent its heating, and constant temperature at the top of the slide.

As in sections 4 and 5, Skempton's effect on PP in equation (4) and the heat source in equations (5) and (6) are ignored, as we study the sliding process after the mass is detached from its surrounding. Then, equation (4)–(6) and (8) are studied numerically accounting for the temperature conduction, advection, and generation by frictional heating, and PP diffusion and generation by thermal pressurization. For simplicity, we assume that the permeability increases linearly with sliding distance, s, i.e. $\Delta k / \Delta s$ is a constant with units of m. The choice of linear permeability increase is compatible with Scholz (2002, figure 2.15) that shows a linear increase in the generation of wear material with displacement along faults. It is possible to use more complex laws for the evolution of the permeability, but it is of interest to observe that a simple linear permeability increase law may result in a long runout landslide mimicking the travel course of Heart Mountain block.

Simulation results performed with parameters from Table 4 show that as the upper block starts sliding along the bedding plane detachment fault it accelerates (Fig. 8a), heating occurs (Fig. 8d) and PP increases (Fig. 8b). Acceleration continues until the reversely dipped transgressive fault is reached. The maximal velocity of the upper block is attained at the base of the bedding plane

Table 4. Parameters used in section 6.2.

Parameter		Value
d	Shear zone thickness	1 m
D	Upper block thickness	3000 m
Φ	porosity	0.15
ρ_s	Solid density	2840 kg m^{-3}
ρ_f	Fluid density	1000 kg m^{-3}
c_s	Solid specific heat	1000 J kg^{-1} °C^{-1}
c_f	Fluid specific heat	4187 J kg^{-1} °C^{-1}
k_i	Initial permeability	2.29 × 10^{-13} m^2
$\Delta k / \Delta s$	Permeability increment	4.52168 × 10^{-14} m
Λ	Thermal pressurization coefficient	1.5 MPa °C^{-1}
μ	Friction coefficient	0.5
η	Fluid viscosity	1 × 10^{-3} Pa s
S_σ	Unconstrained specific storage	3 × 10^{-9} Pa^{-1}
k_θ	Thermal conductivity	0.38 Wm^{-1} °C^{-1}

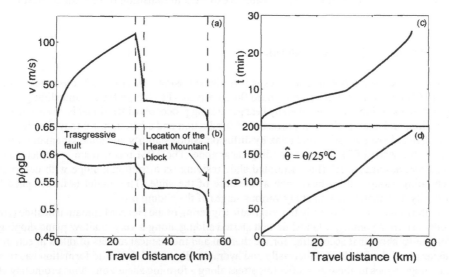

Figure 8. Heart Mountain Landslide simulation results showing the evolution of the sliding velocity (a) dimensionless PP at the shear zone (b) the sliding duration (c) and the dimensionless temperature at the shear zone (d). The distance at which the simulated slide halted corresponds to the nowadays location of the Heart Mountain block. Temperature predictions surpass the dissociation temperature of dolomite, indicating that an additional heat sink is missing.

(Fig. 8a), and is equal to 109 m/s (about a third of the speed of sound in air). PP starts relaxing after peaking at about 1.5 km from the slide onset. An additional moderate rise of PP occurs from about 14 km from the sliding onset till the transgressive fault is reached. When climbing the transgressive fault both PP and velocity decrease dramatically while temperature production is slowed. Upon reaching the Eocene land surface, PP and velocity first decrease slowly and then abruptly, until the mass stops. The travel duration from motion onset to halting after 54 km along course is 26 minutes (Fig. 8c). Sliding along the bedding plane detachment fault lasted about 9 minutes, sliding past the transgressive fault takes another minute, and the slide along the Eocene land surface lasted 16 minutes.

The TPM model successfully produces the travel distance of the Heart Mountain fragment of the upper plate, using linear increase of the permeability with sliding distance. However, the model predicts that the temperature within the shear zone rises to more than 4000°C (Fig. 8d) due to the slowness of the temperature relaxation processes of heat conduction and advection. This temperature, being unrealistic, lies above the temperature of the critical point of water (~374 °C), where pore fluid water will transform to supercritical water. Upon phase transition, the basic physical properties of the pore filling material such as: density, viscosity, thermal conductivity, compressibility and thermal expansivity are expected to vary significantly (in an order or two orders of magnitude).

Moreover, at temperature of about ~745°C, dolomite under pressure of 70 MPa (expected PP within the shear zone) will start to dissociate into calcite, periclase and carbon dioxide, $CaMg(CO_3)_2 \rightleftarrows CaCO_3 + MgO + CO_2$ (Harker & Tuttle 1955). Indeed, Beutner & Gerbi (2005) describe field evidence from the detachment plane of the HML pointing to end products of dolomite dissociation. This reaction being endothermic is potentially an additional heat dissipation mechanism that may help in relaxing the high temperatures.

To summarize this section, it is found that the TPM mechanism together with time dependent permeability evolution may reproduce the exact runout distance of the HML given the topography. However, the unrealistic temperatures predicted by the model indicate that an additional heat dissipation mechanisms (such as dolomite dissociation) should be considered, together with temperature dependent variations of the physical properties of the pore filling fluid (in particular the changes expected due to the fact that the heated water is expected to transition to its critical phase).

7   SUMMARY AND CONCLUSIONS

In this work the mechanics of landslides is investigated using a thermo-poro-mechanical mechanism (TPM) that couples temperature, pore fluid pressure (PP), slope failure and sliding velocity. Temperature variations are controlled by heat generating processes of frictional heating and heating by intrusion, and heat dissipation processes of conduction and advection. PP may rise by thermal and mechanical pressurization and relax by diffusion. Slope stability and slide kinematics are controlled by the value of PP along the sliding shear zone. The TPM mechanism is implemented in a model that assumes a 1D fluid saturated slide residing on a frictional slope with deformation concentrating along a thin shear zone at the base of the slide. The model is investigated both numerically and analytically to study various stages of the sliding process.

First, the model is used to study the enigmatic triggering of the Heart Mountain landslide (HML) in northwestern Wyoming, a landslide that started sliding along a very shallow plane dipping 2°. Model results show that accounting for the thermal and mechanical affects of dikes injection into the upper plate, the PP along a thermally and hydraulically sealed dolomite formation has risen to large enough values to allow zero effective stress along a forming shear zone, thus promoting slope failure along the dolomite formation.

Next, the TPM model is applied to the problem of sliding stability to explain the observation that some slides are catastrophic in nature, (gaining high velocities and potentially very large travel distances), while others halt spontaneously on the slope after sliding short distances. Here, the model demonstrates numerically the emergence of a bifurcation between two sliding kinematic regimes—catastrophic and arrested. An analytical study allows deep understanding of the source of bifurcation by unraveling the dependence between sliding kinematics and the mechanical and geometrical parameters of the sliding mass, such as permeability, thermal pressurization coefficient, and friction coefficients. For example: low permeability is shown to lead to catastrophic sliding while large permeability stabilizes a slide and leads to its fast arrest.

After investigating the initial stability of the sliding process, the TPM model is used to study the long runout distances of catastrophic slides and the perplexing observed volume effect that larger slide have longer runouts. The model reveals that positive feedback between frictional heating, thermal pressurization, reduction of shear stress, and velocity increase lead to long runouts. When combining the model with permeability that decreases with depth, results show that large slides with deep shear zones that lie within a lower permeability domain travel further, in good agreement with field data.

Finally, the conclusions drown from the triggering, stability and the long runout studies are applied to explain the travel course of the HML: its catastrophic acceleration, long runout, and final arrest along the same low dipping slope. Here, the TPM model is combined with a time dependent permeability that increases linearly with sliding distance, as if mass fracturing and breakage during sliding increases permeability. It is found that the exact travel distance of the Heart Mountain block may be predicted by the model, but unrealistic high temperatures. It is offered that temperature dependent physical properties should be considered for the pore fluid, and that an additional heat sink resulting from the endothermic reaction of dolomite dissociation may have operated, lowering the temperatures considerably.

The ability of this relatively simple TPM mechanism to explain diverse observations related to sliding kinematics and mechanics is unique, and may be indicative of its strength and importance.

REFERENCE

Aharonov, E. & Anders, M.H. 2006. Hot water: A solution to the Heart Mountain detachment problem? *Geology*, 34(3), 165–168. 10.1130/G22027.1.

Aharonov, E. & Sparks, D. 2002. Shear profiles and localization in simulations of granular materials. *Phys. Rev. E*, 65(5). 10.1103/PhysRevE.65.051302.

Bartelt, P., Buser, O. & Platzer, K. 2007. Starving avalanches: Frictional mechanisms at the tail of finite-sized mass movements. *Geophys. Res. Lett.*, 34. 10.1029/2007GL031352.

Beutner, E.C. & Craven, A.E. 1996. Volcanic fluidization and the Heart Mountain detachment, Wyoming. *Geology*, 24, 595–598.

Beutner, E.C. & Gerbi, G.P. 2005. Catastrophic emplacement of the Heart Mountain block slide, Wyoming and Montana, USA. *Geol. Soc. America Bull.*, 117, 724–735.

Bucher, W.H. 1947. Heart Mountain problem. In D.L. Jr. Blackstone & C.W. Sternberg (Eds.), *Field Conference in the Bighorn Basin* (pp. 189–197): Wyoming Geological Association Field Conference Guidebook.

Campbell, C.S., Cleary, P.W. & Hopkins, M. 1995. Large-scale landslide simulations: Global deformation velocities and basal friction. *J. Geophys. Res.*, 100(B5), 8267–8283.

Crosta, G., Chen, H. & Lee, C. 2004. Replay of the 1987 Val Pola landslide, Italian Alps. *Geomorphology*, 60, 127–146. 10.1016/j.geomorph.2003.07.015.

Dade, W.B. & Huppert, H.E. 1998. Long-runout rockfalls. *Geology*, 26(9), 803–806.

Davis, R.O., Smith, N.R., & Salt, G. 1990. Pore fluid frictional heating and stability of creeping landslides. *Int. J. Num. Anal. Meth. Geomechanics*, 14, 427–443.

Day, S.J., Carracedo, J.C., Guillou, H. & Gravestock, P. 1999. Recent structural evolution of the Cumbre Vieja volcano, La Palma, Canary Islands: volcanic rift zone reconfiguration as a precursor to volcano flank instability? *J. Volc. Geotherm. Res.*, 94, 135–167.

Gabet, E.J. & Mudd, S.M. 2006. The mobilization of debris flows from shallow landslides. *Geomorphology*, 74, 207–218. 10.1016/j.geomorph.2005.08.013.

Garagash, D.I. & Rudnicki, J.W. 2003. Shear heating of a fluid-saturated slip-weakening dilatant fault zone 1. limiting regimes. *J. Geophys. Res.*, 108(B2 2121). 10.1029/2001JB001653.

Genevois, R. & Ghirotti, M. 2005. The 1963 Vaiont landslide. *Giornale di Geologia Applicata*, 1, 41–52.

Goren, L. & Aharonov, E. 2007. Long runout landslides: The role of frictional heating and hydraulic diffusivity. *Geophys. Res. Let.*, 34(L07301). 10.1029/2006GL028895.

Goren, L. & Aharonov, E. 2009. On the stability of landslides: A thermo-poro-elastic approach. *Earth and Planet. Sci. Lett.*, 277(3–4). 10.1016/j.epsl.2008.11.002.

Gutierrez, M. & Wangen, M. 2005. Modeling of compaction and overpressuring in sedimentary basins. *Marine and Petroleum Geology*, 22. 10.1016/j.marpetgeo.2005.01.003.

Harker, R.I. & Tuttle, O.F. 1955. Studies in the system $CaO-MgO-CO_2$; Part 1, The thermal dissociation of calcite, dolomite and magnesite. *American Journal of Science*, 253, 209–224.

Helmstetter, A., Sornette, D., Grasso, J.R., Andersen, J.V., Gluzman, S. & Pisarenko, V. 2004. Slider block friction model for landslides: Application to Vaiont and La Clapière landslides. *J. Geophys. Res.*, 109. 10.1029/2002JB002160.

Hughes, C.J. 1970. The Heart Mountain detachment fault—A volcanic phenomenon? *The Journal of Geology*, 78, 107–116.

Iverson, R.M. 2000. Landslide triggering by rain infiltration. *Water Resour. Res.*, 36(7), 1897–1910.

Iverson, R.M. 2005. Regulation of landslide motion by dilatancy and pore pressure feedback. *J. Geophys. Res.*, 110(F02051). 10.1029/2004JF000268.

Iverson, R.M., Reid, M.E., Iverson, N.R., LaHusen, R.G., Logan, M., Mann, J.E., & Brien, D.L. 2000. Acute sensitivity of landslide rates to initial soil porosity. *Science*, 290(5491), 513–516. 10.1126/science.290.5491.513.

Kelfoun, K. & Druitt, T.H. 2005. Numerical modeling of the emplacement of Socompa rock avalanche, Chile. *J. Geophys. Res.*, 110(B12202). 10.1029/2005JB003758.

Legros, F. 2002. The mobility of long-runout landslides. *Eng. Geol.*, 63, 301–331.

Malamud, B.D., Turcotte, D.L., Guzzetti, F. & Reichenbach, P. 2004. Landslide inventories and their statistical properties. *Earth Surf. Process. Landforms*, 29(6). 10.1002/esp.1064.

Mangeney, A., Tsimring, L.S., Volfson, D., Aranson, I.S., & Bouchut, F. 2007. Avalanche mobility induced by the presence of an erodible bed and associated entrainment. *Geophys. Res. Lett.*, 34(L22401). 10.1029/2007GL031348.

Mangeney-Castelnau, A., Vilotte, J.-P., Bristeau, M.O., Perthame, B., Bouchut, F., Simeoni, C., & Yerneni, S. 2003. Numerical modeling of avalanches based on Saint Venant equations using a kinetic scheme. *J. Geophys. Res.*, 108(B11). 10.1029/2002JB002024, 2003.

Manning, C.E. & Ingebritsen, S.E. 1999. Permeability of the continental crust: Implications of geothermal data and metamorphic systems. *Rev. Geophys.*, 37(1), 127–150.

Melosh, H.J. 1983. Acoustic fluidization. *American Scientist*, 71, 158–165.

Pierce, W.G. 1957. Heart Mountain and South Fork detachment thrusts of Wyoming. *American Association of Petroleum Geologists Bulletin*, 41, 591–626.

Pierce, W.G. 1973. Principle features of the Heart Mountain fault and the mechanism problem. In K.A. De Jong & R. Scholten (Eds.), *Gravity and Tectonics* (pp. 457–471). New York: John Wiley and Sons.

Pouya, A., Léonard, C., & Alfonsi, P. 2007. Modelling a viscous rock joint activated by rainfall: Application to the La Clapière landslide. *Int. J. Rock Mech. Min. Sci.*, 44, 120–129. 10.1061/j.ijrmms.2006.05.004.

Rempel, A.W. & Rice, J.R. 2006. Thermal pressurization and onset of melting in fault zone. *J. Geophys. Res.*, 111(B09314). 10.1029/2006JB004314.

Rice, J.R. 2006. Heating and weakening of faults during earthquake slip. *J. Geophys. Res.*, 111(B05311). 10.1029/2005JB004006.

Saar, M.O. & Manga, M. 2004. Depth dependence of permeability in the Oregon Cascades inferred from hydrologic, thermal, seismic and magnetic modeling constraints. *J. Geophys. Res.*, 109(B04204). 10.1029/2003JB002855.

Savage, S.B. & Iverson, R.M. 2003. Surge dynamics coupled to pore-pressure evolution in debris flows. In D. Rickenmann & C.-L. Chen (Eds.), *In Debris-Flow Hazards Mitigation: Mechanics, Prediction and Assessment* (pp. 503–514). Rotterdam: Millepress.

Scholz, C.H. 2002. *The Mechanics of Earthquakes and Faulting*. Cambridge, UK: Cambridge University Press.

Shaller, P.J. & Smith-Shaller, A. 1996. Review of proposed mechanisms for sturzstroms (long-runout landslides). In P.L. Abbott & D.C. Seymour (Eds.), *Sturzstroms and Detachment Faults* (pp. 285–202). Anza-Borrego Desert State Park, California: South Coast Geological Society, Inc., Santa Ana, CA.

Sidle, R.C. & Ochiai, H. 2006. *Landslides: Processes, Prediction and Land Use*. Washington, DC: American Geophysical Union.

Straw, W.T. & Schmidt, C.J. 1981. Heart Mountain detachment fault: A phreatomagmatic-hydraulic hypothesis. *Geological Society of America Abstracts with Programs*, 13(7), 562.

Turcotte, D.L. & Schubert, G. 1982. *Geodynamics: applications of continuum physics to geological problems*. London: John Wiley and Sons.

Vardoulakis, I. 2000. Catastrophic landslides due to frictional heating of the failure plane. *Mech. Coh. Frict. Mat.*, 5, 443–467.

Vardoulakis, I. 2002. Dynamic thermo-poro-mechanical analysis of catastrophic landslides. *Géotechnique*, 52(3), 157–171.

Voight, B. 1973. The mechanics of retrogressive block-gliding, with emphasis on the evolution of the Turnagain Heights landslide, Anchorage, Alaska. In K.A. De Jong & R. Scholten (Eds.), *Gravity and tectonics* (pp. 97–121). New York: John Wiley and Sons.

Voight, B. & Faust, C. 1982. Frictional heat and strength loss in some rapid landslides. *Géotechnique*, 32(1), 43–54.

Wang, H.F. 2000. *Theory of Linear Poroelasticity with Applications to Geomechanics and Hydrogeology*. Princeton, NJ: Princeton University Press.

Ward, S.N. & Day, S. 2001. Cumbre Vieja Volcano—potential collapse and tsunami at La Palma, Canary Island. *Geophys. Res. Lett.*, 28(17), 3397–3400.

Mechanisms of fluid overpressurization related to instability of slopes on active volcanoes

Derek Elsworth
Energy and Mineral Engineering, Penn State University, University Park, PA, USA
G³ Center and EMS Energy Institute, Penn State University, University Park, PA, USA

B. Voight
Geosciences, Penn State University, University Park, PA, USA

J. Taron
Energy and Mineral Engineering, Penn State University, University Park, PA, USA
G³ Center and EMS Energy Institute, Penn State University, University Park, PA, USA

ABSTRACT: Excess fluid pressures exert important controls on the stability of lava domes and of the flanks of volcanoes. Migrating overpressures reduce the shear strength of the edifice and may control the timing, morphology, and energetics of failure. Excess pressures may be developed both directly from magma degassing, and indirectly from the interaction of magma with infiltrating rainwater or groundwater. Interior gases influence the strength of the volcanic pile, and hence its stability, in at least two ways. In the fractured and solidified outer carapace high gas contents reduce effective stresses and concomitantly lower shear strength. In the dome interior, magmas which avoid the off-gassing of volatiles exhibit a low and primarily cohesive strength. Signatures of these various processes are evident in the extensive record of collapses which chart the episodic growth and destruction of the lava dome at Soufriere Hills volcano, Montserrat. Mechanisms include (1) interior pressurization by magma degassing, and (2) the interaction of rainwater with the hot dome rind. The influence of gas overpressures applied interior to a brittle carapace is typified by the response to episodes of cyclic inflation, where collapse may be delayed and may be triggered at inferred pressures below the peak reached in the prior cycle. Similar influences on timing, and in collapse style are present for rainfall-triggered events where deluges beyond a given intensity and duration are required to promote failure, and the style of collapse is influenced by the antecedent conditions of gas pressurization within the lava dome. In all instances, interior gas overpressures or the presence of a segregated plastic core are both viable mechanisms to promote a switch between shallow instability of the dome carapace to deep transection of the dome core. Such switching to a more hazardous and mobile failure mode may occur absent the usual seismic, geodetic, or chemical signatures which herald a collapse event, and poses special challenges in monitoring for hazard assessment.

1 INTRODUCTION

Lava dome collapse represents an important and potentially hazardous feature in the life cycle of silicic volcanoes (Miller 1994). Collapse of the highly gas-charged dome materials may spawn hazardous and highly mobile pyroclastic flows (Nakada & Fujii 1993; Abdurachman et al. 2000), which can potentially remove greater than 90% of the dome structure, and involve tens of millions of cubic meters of hot tephra.

Recent dome-building activity observed at Soufrière Hills volcano (SHV), Montserrat has significantly illuminated the mechanisms of dome growth and collapse (Sparks et al. 1998; Watts et al. 2002; Calder et al. 2002; Norton et al. 2002; Carn et al. 2004). Many collapses have resulted from the complementary and potentially additive effects of dome oversteepening (Fink & Griffiths 1998; Sparks et al. 2000) and interior gas-pressurization (Voight & Elsworth 2000; Elsworth & Voight 2001). Such collapse modes adequately match observed near-dome tilt (Watson et al. 2000; Widiwijayanti et al. 2005), RSAM (Miller et al. 1998), and gas discharge histories that span major dome collapse events. However, collapses that occurred during periods of "residual volcanic

275

activity", 3 July 1998, and those that lacked short-term precursory seismic signals, 20 March 2000 and 29–30 July 2001 (Carn et al. 2004), cannot be explained by slope oversteepening or traditional mechanisms of gas overpressurization. While largely characterized by a deficiency in traditional pre-collapse indicators, these events have directly coincided with periods of intense precipitation (Matthews et al. 2002; Norton et al. 2002; Herd et al. 2003; Matthews & Barclay 2004; Elsworth et al. 2004; Carn et al. 2004).

In the following we examine the role of (1) interior pressurization by magma degassing, and (2) the interaction of rainwater with the hot dome rind in contributing to styles and the timing of dome collapse.

2   SUPPORT FOR RAINFALL MECHANISMS: SHV

In common with other volcanoes, including Mount St. Helens (Mastin 1994), Merapi (Voight et al. 2000), and Unzen (Yamasato et al. 1998), a number of large dome collapses at SHV have been associated with heavy rainfall (Matthews et al. 2002; Matthews & Barclay 2004; Elsworth et al. 2004). To examine rainfall driven collapse modes, this study focuses on the 20 March 2000 and 29–30 July 2001 events, each of which interrupted a period of active dome growth, exhibited an absence of elevated precursory seismic activity, and proceeded concurrently with an intense rainfall event, with retrogressive collapses initiated in the latter stages of heavy rain-fall, likely contributing to a deepening failure surface. Cyclic evolution of the dome, segregated by these rainfall triggered collapse events, has been extensively charted by the Montserrat Volcano Observatory (MVO).

By mid March 2000, andesitic growth within the July 1998 collapse scar had reached $\sim$29 $\times$ 106 m^3, of which $\sim$28 $\times$ 106 m^3 ($\sim$95%) was removed through numerous pyroclastic flows over a period of $\sim$5 h on 20 March 2000 (Carn et al. 2004). The July 2001 collapse was similar in form, but accumulated a total collapse volume of $\sim$45 $\times$ 10^6 m^3 (Carn et al. 2004), or approximately 50% of active dome volume (utilizing unpublished data from MVO to interpolate extrusion rates from measured active dome volumes on 8 December 2000 and 23 September 2001). Real-time rainfall and seismic data for the July collapse indicate a corresponding and drastic increase in rockfall-type seismic activity (used as a proxy for the evolving instability of the dome) as rainfall approached its maximum intensity of 50 mm in 2 h (e.g., Figure 2 of Elsworth et al. 2004).

3   GAS OVERPRESSURES

Explosive eruptions of the Soufrière Hills volcano, Montserrat, B.W.I., directly followed major collapses of the lava dome on September 17 1996, and June 25, August 3, and September 21 1997 (Robertson et al. 1998; Cole et al. 1998; Young et al. 1998), indicating the presence of volatile-rich magma high in the conduit and suggesting a possible role of gas pressurization in dome instability. Tilt deformation of the edifice prior to several events confirmed shallow pressurization and indicated a coincidence of collapse with the timing of peak pressurization (Voight et al. 1998a). These instances, and a number of analogous cases elsewhere (Newhall & Melson 1987; Sato et al. 1992; Miller 1994; Voight et al. 1998b; Elsworth & Voight 1995; Voight & Elsworth 1997), imply two types of gravitational lava dome failure—one with little or no gas over-pressure, and the other with significant gas overpressure that influences the failure process (Newhall & Voight 1997). A discussion of proposed mechanisms follows.

3.1   *Failure geometry*

The potential failure geometry is shown in Figure 1 (Voight & Elsworth 2000), and is described by a hemispherical dome with interior gas pressures, seated upon an inclined failure surface. Interior gas pressures reduce effective stresses, and drive the dome towards collapse.

Factor of safety (*Fs*) as a function of basal failure plane inclination is noted on Figure 2 for a variety of conditions, for material of cohesion $c = 0.5$ MPa and friction angle $\phi = 25°$, unless

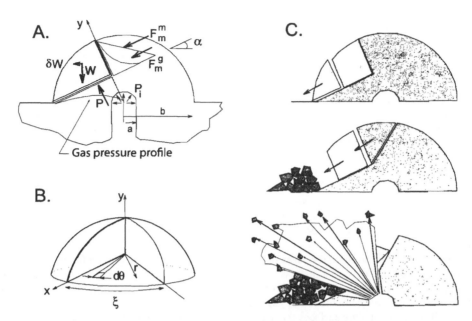

Figure 1. Schematic views of dome collapse. (A). Section through a dome of external radius, b, subject to gas pressure, P_i, in a cavity of radius, a. The diffusive gas pressure acts on the boundaries of a failing block of weight, W. This block rests on a plane, inclined at angle, α, and is acted upon by uplift force, P, and downslope forces (F^* in text) representing gas pressures, F_m^g, or magmastatic pressures, F_m^m, acting on the block rear. (B). The three-dimensional geometry is defined by the sector angle, ξ. (C). Failure initiates with release of a toe-block, with failure retrogressing to unload the pressurized core, resulting in the potential for spontaneous disintegration or a directed explosion.

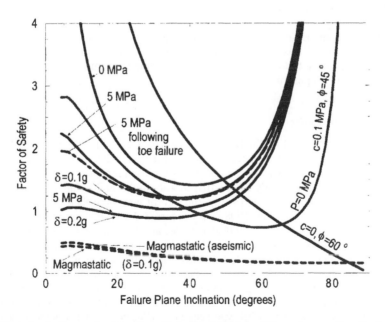

Figure 2. Variation in factor of safety, Fs, with inclination of the failure plane, α. Results for a dome of external radius $b = 200$ m and cavity radius $a = 15$ m within material of $c = 0.5$ MPa and $\alpha = 25°$, unless otherwise noted. All results are for gas pressure loading of the block rear, except magmastatic (shown dashed). Idealized failure plane pivots upwards from dome base. Seismic accelerations are $\delta = 0.1$ g and $\delta = 0.2$ g.

otherwise noted, and holding fluid diffusivity (κ) constant at 10^3 m^2/day. If the external cara-pace of the dome is a purely frictional, interlocked, fractured crystalline solid ($c = 0$, $\phi = 60°$), exfoliation of blocks can occur on a steeply inclined surface that does not transect the dome core (Figure 2). In the absence of internal pressure and for interior strength parameters of $c = 0.1$ MPa and $\phi = 45°$, failure on a less-steeply-inclined basal surface is possible, but the fail-ure surface still does not penetrate deeply into the dome interior. However, although the actual strength properties of hot dome lava ($\sim$800°C) with some residual melt have not yet been fully measured, its frictional property may be much reduced from these supposed values. Assuming, then, a more cohesive and less frictional core, say $c = 0.5$ MPa and $\phi = 25°$, an unpressurized dome is more stable than for the previous case. However, as internal gas pressure builds, stability is reduced on a critical failure plane inclined at about 35° to the horizontal (Figure 2). Spalling at the block toe has a minor influence on this result, but strong shaking by volcanic earthquakes or tremor (Voight et al. 1998a; Newhall & Voight 1997), back-scarp "magmastatic" pressuriza-tion by viscous extrusion from the dome core (Elsworth & Voight 1995), or raising the cavity gas pressure ($P_i > 5$ MPa), can all reduce block stability on a more-shallow basal surface to $Fs \approx 1$ (Figure 2).

Furthermore, in these last three cases the sensitivity of factor of safety with failure plane incli-nation is reduced (i.e. the curve of Fs versus plane inclination flattens in Figure 2), so that minor amounts of strength heterogeneity could result in critical ($Fs \approx 1$) failure surface inclinations near-ing 10°–20°. Thus, the combination of gas pressurization and augmentation by seismic loads can result in deep-seated failure surfaces that can approach the highly pressurized core.

The magmastatic case values in Figure 2 (combined with a gas-pressurized basal surface) imply failure at all geometries. These cases are extreme examples in that full development of magmatic pressures along sector fractures may be unlikely in general. However the example illustrates that localized injections of magma can play an important role on stability.

4   RAINFALL TRIGGERING

Anecdotal evidence implicates deluges accompanying storms events as a trigger in the collapse of metastable lava domes. Despite significant supporting evidence, it is difficult to reconcile the role of rainfall in inducing fluid overpressures in any traditional manner, since the dome rocks are typically hot and will support little infiltration.

4.1   *Observed failure modes*

Two failure styles are typified by the Montserrat collapses of 3 July 1998 and 20 March 2000. The earlier event removed a large volume, but limited fraction ($\sim$20%), of an immense metastable dome erupted 4 months previously. In contrast, the latter event removed $\sim$90% of a much smaller but newly grown dome. The 29 July 2001 collapse was similar to 20 March 2000, but double its size (Matthews et al. 2002), and the 13 July 2003 event was more than double that of 29 July 2001 (Herd et al. 2003). The 3 July 1998 event occurred during the period of no dome growth between March 1998 and November 1999 (Norton et al. 2002) and removed $\sim$20% of an over-sized lava dome ($\sim$110 $\times$ 10^6 m^3) that had been erupted (and partly eroded) from November 1995 to February 1998. The collapse left a canyon-like slot scar in which a new dome grew. This collapsed on 20 March 2000; $\sim$90% of the dome was removed. In each case, retrogres-sive collapses were initiated in the latter-open to the east. In contrast, a smaller dome ($\sim$27 $\times$ 10^6 m^3), which had grown at $\sim$2.5 m^3/s since November 1999 failed during heavy rainfall, generating a sharp increase in rockfall-type seismicity (Figure 3), and in some cases were fol-lowed by elevated gas flux measurements immediately after the collapse (Norton et al. 2002). The collapses occurred as semi-continuous to sequential failures over periods of several hours. A model is proposed here to explain these observed collapses, for which seismic precursors were largely absent.

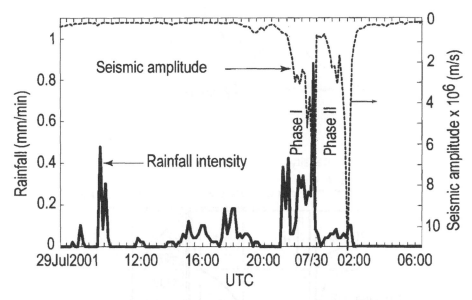

Figure 3.   Correlation of rainfall intensity and seismic amplitude (inverted) for collapse of 29 July 2001. Rainfall intensity recorded by University of East Anglia rain gauges at St. Georges Hill (4 km west of dome) (solid line, read on left axis). Seismic amplitude (short dashed line, read on right axis) recorded at Windy Hill digital seismic station. Antecedent rainfall began at 0800, in multiple bursts reaching 2 hour duration, and may have been important in priming system for failure. Heaviest rainfall was from 2100 UTC (07/29) to midnight (07/30) and ceased at 0230. Dome collapse began at ~2200 and peaked at 1150; second phase of collapse resumed at 0030 and peaked at 0200.

4.2   *Rainfall-induced failure mechanisms*

We consider multiple potential mechanisms that may have contributed to collapse (Elsworth et al. 2004; Taron et al. 2007). Conventional mechanisms not involving rain include slope oversteepening (Sparks et al. 2000), gas overpressurization of the dome interior (Voight & Elsworth 2000; Elsworth & Voight 2001), and hydrothermal weakening of the dome or its substrate. Storm-triggered destabilization of the steep apron of dome talus has been observed on Montserrat on a number of occasions, e.g., 14 October 2001 (unpublished data from the Montserrat Volcano Observatory), and it is possible that larger failures could then result if unstable, oversteepened lava is thereby exposed. A traditional mechanism for storm-triggered rockslides is rain in-filling of joints that elevates destabilizing pore pressures, although such a mechanism is unlikely to work in hot lava because of rapid vaporization of the infiltrating fluid. The extension of surface cracks in lava by rainfall quenching is likely to contribute to failure by the degradation of the mass (fractured rock) strength of the dome materials, although elevation of interior fluid (gas) pressures appears necessary to generate the scale of failures observed. Consequently, alternative mechanisms are desirable for some of the observed rain-triggered dome-removing failures on Montserrat, with collapse scars that cut deeply into the dome interior (Sparks et al. 2000).

4.3   *Failure model*

We consider the limit equilibrium stability of a dome where the trigger for failure is the augmentation of interior gas pressures, as infiltrating rainwater stanches the escape of magmatic gases through the fractured hot dome carapace (Figure 4A). The dome becomes less stable as interior gas pressures build and will ultimately fail if a critical, but undefined, overpressure is reached. Gas overpressure is limited to the static pressure present at the infiltration front within the fractured carapace, defined as the product of penetration depth (d) and unit weight ($\rho_w g$) of the infiltrating fluid (Figure 4B).

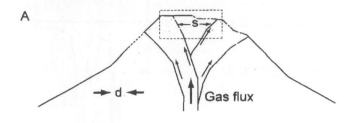

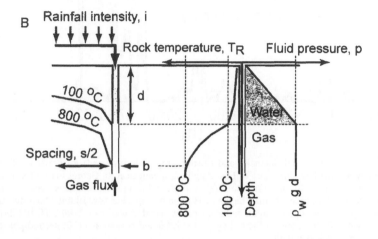

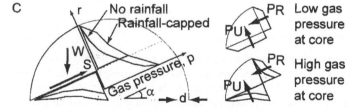

Figure 4. Schematics of dome geometry, infiltration into carapace, and stability analysis. (A) Gas flow in dome is localized on relict shear surfaces (d = depth; s = fracture spacing). Insets show locations of B and C. (B) Infiltrating water penetrates fractures to a depth d, enabled by locally depressed 100°C isotherm, and builds water pressure to $p = \rho_w g d$ at infiltration front. (C) Existing gas pressures (dark shading) are augmented (unshaded) by stanched gas flow, increasing weakening (P_U) and disturbing (P^R) fluid forces acting on detached failing dome sector of weight W, held by shear resistance, S.

Consequently, instability may be indexed to the anticipated depth of liquid infiltration, and this depth in turn is limited by the vaporization of the infiltrating fluid.

4.3.1 *Mechanical instability*

We simplify the dome geometry to accommodate the approximate spherical symmetry of gas flow, discharged from a central conduit (Figure 4A & C). The stability of an isolated block on the dome flank is indexed through the ratio of forces resisting downslope movement to those promoting it, as a factor of safety (Fs; Voight & Elsworth 2000). For a degassing vent, gas pressures diminish radially outward from the conduit (Figure 4C) and apply net uplift (P_U) and downslope (P_R) forces to the block isolated on a detachment plane inclined at angle α (discussed extensively in Voight & Elsworth 2000). It is important here to note that capping gas pressures at a peak magnitude of $p_b + \rho_w g d$, where p_b is fluid displacement or "bubbling" pressure, at a depth d beneath the saturated carapace or occluded fracture augments the uplift (P_U) and downslope (P_R) forces that act additively to destabilize the block (Figure 4B). This augmentation in pressure (and hence destabilizing effect) is greatest when gas discharge from the dome core is high (Figure 4C, lower right inset), but is also present for low fluxes or where vaporization around the liquid infiltration front self-generates overpressure (Figure 4C, upper right inset). In either case, the limiting pressure at depth d is $p_b + \rho_w g d$.

4.3.2 *Rainfall infiltration*
For 75 mm of rainfall over 3 h, the resulting hydraulic penetration is smallest for the very narrow spacing of fractures ($s < 0.2$ m) or equivalent porous medium, where the dome carapace is quenched to a maximum depth of about one-third the storm total rainfall. Above this spacing ($s > 0.2$ m), hydraulic penetration depth (d) grows linearly with spacing to reach 8 m for fractures spaced 80 m apart in rocks of 800°C and 20 m penetration for rocks at 400°C. For fractures spaced only 5m apart, penetration depths decrease to 0.5 m (800°C) and 1.4 m (400°C). The water plugging of the most widely spaced, and most highly gas-conductive, fractures (Figure 3A) will cause the greatest reduction in gas flow and the largest corresponding increase in trapped overpressures. These highly conductive fractures are the focus of this work.

4.3.3 *Anticipated magnitudes of interior gas overpressures*
Gas overpressures are evaluated for rain infiltration into a representative large, near-dormant dome (July 1998) with non-negligible (but unmeasured) effusive gas activity (Edmonds et al. 2003) and with a surface that may have been multiply quenched and chilled by previous storms. For an average carapace temperature of 400°C and lower bound permeability of $k = 10^{-12}$ m^2 (Melnik & Sparks 2002), fractures spaced between 5 and 80 m apart may be penetrated, in a given intense storm, to depths of 1.5 to 21.6 m. These depths represent limits on passive interior gas pressurization to ~20 m of static head (0.2 MPa). For $k = 10^{-12}$ m^2, the role of fluid-displacement pressure is negligible (Taron et al. 2007), and the capping pressure magnitude is adequately (and conservatively) represented by the static pressure head at the infiltration front as $\rho_w g d$.

4.4 *Evaluated modes of collapse*

Idealized collapse modes are examined for simplified dome geometries of large (350 m) and small (200 m) relative heights, for varied inclinations (α) of an assumed detachment plane, and under varied conditions of interior gas pressurization and rainfall capping of the carapace (Figure 5). Consistent cohesive strengths of 0.5 MPa and friction angles of 25° are derived from inverse analyses of spine expulsions (Sparks et al. 2000; Voight & Elsworth 2000). For simplicity, a uniform material having these rock-mass parameters is assumed in the analysis, but we recognize that such complexly extruded domes are not actually uniform. Results are similar to those obtained using other reasonable parameter choices.

Absent gas pressurization, a saturated carapace as thick as 50 m exerts a negligible impact on instability (not illustrated). However, the large dome is metastable when unpressurized, and uniform interior pressures corresponding to an infiltration depth of 10 m are adequate to induce failure ($Fs < 1$) at an inclination of 35°–55° (Figure 5). Such a collapse is roughly comparable to the July 1998 failure that produced, following retrogression, a canyon-like slot in the dome. For strength parameters consistent with the previously stated values, the unpressurized small dome is stable, but may be brought close to instability by steady core pressures of 5 MPa (Figure 4). Pressure augmentation by liquid infiltration to only 20 m (trapped uniform pressures of 0.2 MPa) is sufficient to promote low-angle failure as shallow as 35°–40° and to remove ~20% of the edifice. If interior gas pressures are further augmented at the dome core, e.g., to 10 MPa (Figure 5), then a failure surface could drive preferentially on a low-angle (~10°–20°), potentially capable of piercing the dome core and unroofing the conduit. Although such failure geometry could roughly simulate the geometry of the March 2000 collapse, 10 MPa overpressure seems excessive for a small dome. We emphasize that scar geometry is ultimately conditioned by the characteristics of retrogressive failure, and not by the geometry of the initial dome structure.

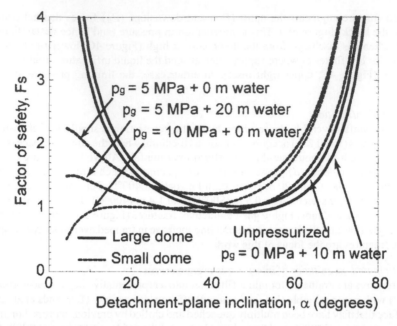

Figure 5. Variation in factor of safety for small and large domes, subject to applied radially diminishing core pressures (maximum core values of 0, 5, and 10 MPa) and supplemented by uniform gas pressures (0, 10, and 20 m of water head).

4.5 *Correlations with observed collapse*

These proposed mechanistic models broadly reproduce observed timing and geometries of recent rainfall-preceded collapses of the lava dome at Montserrat (Elsworth et al. 2004; Taron et al. 2007). The collapse of an oversized and metastable dome (July 1998)—during a period of volcanic repose and absent obvious precursors other than a rainstorm—is consistent with triggering by high-level gas pressurization within the dome. Elevated gas flux measurements immediately after the collapse (Norton et al. 2002) suggest that pressurized gas existed within the dome prior to the collapse, capped by rainfall percolating into the carapace. The near-complete removals of lava domes in March 2000 and July 2001, both in a period of reinitiated effusive activity and absent other precursors, are consistent with gas pressurization of the dome core, critically augmented by the partial sealing of the dome carapace. In each case, collapse geometry and timing are reasonably consistent with available geometric, transport, and strength parameters used in modeling, although it should be appreciated that the collapses are also influenced strongly by retrogressive failure processes that ensue once the key blocks have failed.

Understanding the complex mechanisms of rainfall-triggered instability is important because such failures can occur without warning from standard solid-earth precursory signatures and yet may generate extremely hazardous, large-volume, gas-charged dome-collapse pyroclastic flows and surges. Such correlations emphasize the need to include rainfall monitoring with traditional volcano monitoring methods in order to aid the anticipation of hazardous collapses.

5 CONCLUSIONS

Observations of dome collapse at Montserrat illustrate a variety of interesting features, some of which are described here. Close-in measurements of dome tilt have noted that collapse events do not occur synchronously with peak dome tilt—rather collapses are observed on the down-swing from peak pressurization, and processes of transient inflation are implicated. The penetration of the interior pressure pulse into the outer carapace of the dome is one mechanism whereby this asynchronous failure may occur, and is consistent with predictions.

The role of rainfall in triggering slope instability also deviates from expected norms in these unusual environments. The hot surface of the dome cannot be infiltrated under normal circumstances—the high temperature repels the influent pulse. Where the deluge is sufficiently intense, quenching by conduction and by latent heat is capable of temporarily cooling the surface, and of allowing infiltration to sufficient depths to generate excess pressures in the dome interior, and to push the dome towards failure.

ACKNOWLEDGEMENTS

This work is a result of partial support from National Science Foundation grants CMS-9908590, EAR-9909673, and EAR-0116826. The generous support of colleagues at the Montserrat Volcano Observatory (MVO) is acknowledged, and we thank A. Matthews and J. Barclay, University of East Anglia, for permission to use rain-gauge data.

REFERENCES

Abdurachman, E.K., Bourdier, J.L. & Voight, B., 2000. Nuees ardentes of 22 Novermber 1994 at Merapi Volcano, Java, Indonesia. *Journal of Volcanology and Geothermal Research* 100(1–4): 345–361.

Calder, E.S., Luckett, R., Sparks, R.S.J. & Voight, B., 2002. Mechanisms of lava dome instability and generation of rockfalls and py-roclastic flows at Soufrière Hills volcano, Montserrat. In: Druitt, T.H., Kokelaar, B.P. (Eds.), *The Eruption of Soufrière Hills Volcano, Montserrat, from 1995 to 1999. Geological Society [London] Memoirs* 21: 173–190.

Carn, S.A., Watts, R.B., Thompson, G. & Norton, G.E., 2004. Anatomy of a lava dome collapse: the 20 March 2000 event at Soufrière Hills Volcano Montserrat. *Journal of Volcanology and Geothermal Research* 31: 241–264.

Cole, P. et al. 1998. Pyroclastic flows generated by gravitational instability of the 1996–1997 lava dome of Soufrière Hills Volcano, Montserrat. *Geophysical Research Letters* 25: 3425–3428.

Edmonds, M., Pyle, D.M., Oppenheimer, C.M. & Herd, R.A., 2003. SO_2 emissions 1995–2001 from Soufrière Hills volcano, Montserrat WI and their relationship to conduit permeability, hydrothermal interaction and degassing regime. *Journal of Volcanology and Geothermal Research* 124: 23–43.

Elsworth, D. & Voight, B., 1995. Dike intrusion as a trigger for large earthquakes and the failure of volcano flanks. *Journal of Geophysical Research* 100(B4): 6005–6024.

Elsworth, D. & Voight, B., 2001. The mechanics of harmonic gas-pressurization and failure of lava domes. *Geophysical Journal International* 145: 187–198.

Elsworth, D., Voight, B., Thompson, G. & Young, S.R., 2004. A thermal-hydrologic mechanism for rainfall-triggered collapse of lava domes. *Geology* 32(11): 969–972. Doi:10.1130/G20730.1.

Fink, J.H. & Griffiths, R.W., 1998. Morphology, eruption rates, and rheology of lava domes: Insights from laboratory models. *Journal of Geophysical Research* 103(B1): 527–545.

Herd, R., Edmonds, M., Strutt, M. & Ottermeiler, L., 2003. The collapse of the lava dome at Soufrière Hills volcano, 12–15 July 2003. *Eos (Transactions, American Geophysical Union)* 84(46): F1596.

Mastin, L., 1994. Explosive tephra emissions at Mount St. Helens, 1989–1991—The violent escape of magmatic gas following storms. *Geological Society of America Bulletin* 106: 175–185.

Matthews, A., Barclay, J., Carn, S., Thompson, G., Alexander, J., Herd, R., & Williams, C., 2002. Rainfall-induced volcanic activity on Montserrat. *Geophysical Research Letters* 29(13): Doi: 10.1029/2002GL014863.

Matthews, A.J. & Barclay, J., 2004. A thermodynamical model for rainfall-triggered volcanic dome collapse. *Geophysical Research Letters* 31(5), L05614. Doi:10.1029/2003GL019310.

Melnik, O. & Sparks, R.S.J., 2002. Dynamics of magma ascent and lava extrusion at Soufrière Hills volcano, Montserrat. In: Druitt, T.H., and Kokelaar, B.P. (Eds.), *The eruption of Soufrière Hills volcano, Montserrat, from 1995 to 1999: Geological Society [London] Memoir* 32: 153–172.

Miller, T., 1994. Dome growth and destruction during the 1989–1990 eruption of Redoubt volcano. *Journal of Volcanology and Geothermal Research* 62: 197–212.

Miller, A.D. et al. 1998. Seismicity associated with dome growth and collapse at the Soufrière Hills volcano, Montserrat. *Geophysical Research Letters* 25(18): 3401–3404.

Nakada, S., Fujii, T., 1993. Preliminary report on the activity at Unzen volcano (Japan). November 1990–November 1991: dacite lava domes and pyroclastic flows. *Journal of Volcanology and Geothermal Research* 54: 319–333.

Newhall, C.G. & Melson, W.G., 1987. Explosive activity associated with growth of volcanic domes. *Journal of Volcanology and Geothermal Research* 17: 111–131.

Newhall, C.G. & Voight, B., 1997. A survey of precursors to dome collapse. *Merapi Decade Int. Workshop II:* 48–49.

Norton, G.E. et al., 2002. Pyroclastic flow and explosive activity of the lava dome of Soufrière Hills volcano, Montserrat, during a period of virtually no magma extrusion. In Druitt, T.H., and Kokelaar, B.P., (eds.), *The eruption of Soufrière Hills volcano, Montserrat, from 1995 to 1999: Geological Society [London] Memoir* 21: 467–482.

Robertson, R. et al. 1998. The explosive eruption of Soufrière Hills Volcano, Montserrat, West Indies, September 17, 1996. *Geophysical Research Letters* 25: 3429–3433.

Sato, H., Fujii, T. & Nakada, S., 1992. Crumbling of dacite dome lava and generation of pyroclastic flows at Unzen volcano. *Nature* 360: 664–666.

Sparks, R.S.J. et al. 1998. Magma production and growth of the lava dome of the Soufrière Hills volcano, Montserrat, West Indies: November 1995 to December 1997. *Geophysical Research Letters* 25(18): 3421–3424.

Sparks, R.S.J., Murphy, M.D., Lejeune, A.M., Watts, R.B., Barclay, J. & Young, S.R., 2000. Control on the emplacement of the andesite lava dome of the Soufrière Hills volcano, Montserrat by degassing-induced crystallization. *Terra Nova* 12: 14–20.

Taron, J., Elsworth, D., Thompson, G. & Voight, B., 2007. Mechanisms for rainfall-concurrent lava dome collapse at Soufrière Hills volcano, 2000–2002. *Journal of Volcanology and Geothermal Research* 160: 195–209.

Voight, B. & Elsworth, D., 1997. Failure of volcano slopes. *Geotechnique* 47(1): 1–31.

Voight, B. et al. 1998a. Remarkable cyclic ground deformation monitored in real time on Montserrat and its use in eruption fore-casting. *Geophysical Research Letters* 25: 3405–3408.

Voight, B. et al. 1998b. Deformation and seismic precursors to dome collapse pyroclastic flows at Merapi volcano, Java, 1994–1998. *Eos (Transactions, American Geophysical Union)* 79(45): F1001.

Voight, B. & Elsworth, D., 2000. Instability and collapse of lava domes. *Geophysical Research Letters* 27: 1–4.

Voight, B., Constantine, E., Siswowidjoyo, S. & Torley, R., 2000. Historical eruptions of Merapi volcano, central Java, Indonesia, 1768–1998. *Journal of Volcanology and Geothermal Research* 100: 69–138.

Watson, I.M. et al. 2000. The relationship between degassing and ground deformation at Soufrière Hills volcano, Montserrat. *Journal of Volcanology and Geothermal Research* 98(1–4): 117–126.

Watts, R.B., Herd, R.A., Sparks, R.S.J. & Young, S.R., 2002. Growth patterns and emplacement of the andesite dome at Soufrière Hills volcano, Montserrat. In: Druitt, T.H., Kokelaar, B.P. (Eds.), *The Eruption of Soufrière Hills Volcano, Montserrat, from 1995 to 1999. Geological Society [London] Memoirs* 21: 115–152.

Widiwijayanti, C., Clarke, A., Elsworth, D. & Voight, B., 2005. Geodetic constraints on the shallow magma system at Soufrière Hills volcano, Montserrat. *Geophysical Research Letters* 32, L11309. Doi:10.1029/2005GL022846.

Yamasato, H., Kitagawa, S. & Komiya, M., 1998. Effect of rainfall on dacitic lava dome collapse at Unzen volcano, Japan. *Papers in Meteorology and Geophysics* 48(3): 73–78.

Young, S.R. et al. 1998. Overview of the eruption of Soufrière Hills Volcano, Montserrat, July 18, 1995, to December 1997. *Geophysical Research Letters* 25: 3389–3392.

Author index